New Frontiers in Developmental Biology

New Frontiers in Developmental Biology

Editor: Leonardo Reeves

Callisto Reference
www.callistoreference.com

Callisto Reference,
118-35 Queens Blvd., Suite 400,
Forest Hills, NY 11375, USA

Visit us on the World Wide Web at:
www.callistoreference.com

ISBN: 978-1-64116-560-0 (Hardback)

Cataloging-in-Publication Data

New frontiers in developmental biology / edited by Leonardo Reeves.
p. cm.
Includes bibliographical references and index.
ISBN 978-1-64116-560-0
1. Developmental biology. 2. Biology. 3. Growth. I. Reeves, Leonardo.
QH491 .N49 2022
571.8--dc23

Table of Contents

Preface

I am honored to present to you this unique book which encompasses the most up-to-date data in the field. I was extremely pleased to get this opportunity of editing the work of experts from across the globe. I have also written papers in this field and researched the various aspects revolving around the progress of the discipline. I have tried to unify my knowledge along with that of stalwarts from every corner of the world, to produce a text which not only benefits the readers but also facilitates the growth of the field.

The study of the process through which animals and plants develop and grow is known as developmental biology. There are various aspects which are covered under developmental biology such as asexual reproduction, metamorphosis and the biology of regeneration. Some of the processes which are studied under this discipline are morphogenesis, cell differentiation and regional specification. Research in developmental biology makes use of a small number of model organisms. These non-human organisms are chosen with the aim that discoveries which are made in the study of these models will provide insights into the workings of other organisms. This book unfolds the innovative aspects of developmental biology which will be crucial for the progress of this field in the future. It presents researches and studies performed by experts across the globe. Those in search of information to further their knowledge will be greatly assisted by this book.

Finally, I would like to thank all the contributing authors for their valuable time and contributions. This book would not have been possible without their efforts. I would also like to thank my friends and family for their constant support.

Editor

1

Hermes (Rbpms) is a Critical Component of RNP Complexes that Sequester Germline RNAs during Oogenesis

Tristan Aguero [1], Yi Zhou [1], Malgorzata Kloc [2], Patrick Chang [3], Evelyn Houliston [3] and Mary Lou King [1,*]

Academic Editors: Robert W. Dettman and Andy Wessels

[1] Department of Cell Biology, University of Miami School of Medicine, Miami, FL 33136, USA; t.aguero@med.miami.edu (T.A.); yzhou1999@126.com (Y.Z.)

[2] Houston Methodist Research Institute, The Houston Methodist Hospital, Bertner Avenue, Houston, TX 77030, USA; mkloc@houstonmethodist.org (M.K.)

[3] Laboratoire de Biologie du Développement de Villefranche-sur-mer (LBDV), Sorbonne Universités, UPMC Univ Paris 06, CNRS, Villefranche-sur-mer 06230, France; chang@obs-vlfr.fr (P.C.); houliston@obs-vlfr.fr (E.H.)

* Correspondence: mking@med.miami.edu (M.L.K.)

Abstract: The germ cell lineage in *Xenopus* is specified by the inheritance of germ plasm that assembles within the mitochondrial cloud or Balbiani body in stage I oocytes. Specific RNAs, such as *nanos1*, localize to the germ plasm. *nanos1* has the essential germline function of blocking somatic gene expression and thus preventing Primordial Germ Cell (PGC) loss and sterility. Hermes/Rbpms protein and *nanos* RNA co-localize within germinal granules, diagnostic electron dense particles found within the germ plasm. Previous work indicates that *nanos* accumulates within the germ plasm through a diffusion/entrapment mechanism. Here we show that Hermes/Rbpms interacts with *nanos* through sequence specific RNA localization signals found in the *nanos*-3′UTR. Importantly, Hermes/Rbpms specifically binds *nanos*, but not *Vg1* RNA in the nucleus of stage I oocytes. *In vitro* binding data show that Hermes/Rbpms requires additional factors that are present in stage I oocytes in order to bind *nanos1*. One such factor may be hnRNP I, identified in a yeast-2-hybrid screen as directly interacting with Hermes/Rbpms. We suggest that Hermes/Rbpms functions as part of a RNP complex in the nucleus that facilitates selection of germline RNAs for germ plasm localization. We propose that Hermes/Rbpms is required for *nanos* RNA to form within the germinal granules and in this way, participates in the germline specific translational repression and sequestration of *nanos RNA*.

Keywords: *Xenopus* oogenesis; nanos1 RNA localization; germline; germinal granules; RNP particle; *nanos* ribonucleoprotein complex

---

1. Introduction

Localization of specific RNAs to subcellular domains is one mechanism by which cells restrict protein synthesis in time and space. During *Xenopus* oogenesis, selected RNAs are localized and retained within the vegetal cortex at two distinct time periods. Most RNAs essential to forming the germline localize very early in oogenesis within a macroscopic structure called the mitochondrial cloud (MC) or Balbiani body [1–8]. There, the germ plasm assembles and contains all the components, including germinal granules, required and sufficient to determine germ cell identity [9–12]. One known component of germinal granules is *nanos* RNA whose product is essential to the preservation of the germline in many diverse species including *Drosophila*, *Xenopus*, and mouse [3,13–18].

Another set of RNAs, previously distributed throughout stage I oocytes, is actively localized via a microtubule kinesin-dependent mechanism at later oocyte stages [19–21]. RNAs following this so-called late pathway include the transcription factor *VegT* and a *TGFβ* family member *Vg1*. The products of these genes are required for specifying and patterning the endoderm and mesoderm [22–25]. Why *Vg1* remains uniformly distributed in the cytoplasm while *nanos* RNA accumulates in the MC remains an unanswered question. The selection process for the different localization pathways is not well understood but is essential for the creation of the future germline and primary germ layers.

Time-lapse Confocal microscopy and FRAP (Fluorescence Recovery After Photobleaching) analysis show that injected fluorescently labeled *nanos* RNA form particles that disperse evenly throughout the ooplasm in stage I oocytes. Over time, these particles became progressively immobilized, but only within the MC where they form larger aggregates reminiscent of germinal granule formation [6]. Identification of the cis- and trans-acting factors involved in the selection process for either the early or late localization pathway is certainly an important step towards a full mechanistic understanding of RNA localization. Although there are exceptions [26], virtually all localization signals (LS) reside in the 3′UTR, consist of multiple elements, and display considerable functional redundancy [27,28]. Clustering of these repeated elements may be critical to facilitating interactions between different proteins in the localization machinery [29–31].

Vg1 and *VegT* are directed to the vegetal pole by a 340-nt localization signal (LS) in their 3′UTR. *In vitro* and *in vivo* UV crosslinking analyses reveal six proteins that interact directly with both the *Vg1* and *VegT*-LS [30–32]. Within this signal are certain small repeated elements called E2 and VM1 that are required for localization [33–38] and have been found to function as binding sites for the specific RNA-binding proteins Vg1RBP/Vera and hnRNP I, respectively. *Vg1*-LS mutants in E2 and VM1 sites fail to bind their respective proteins *in vitro* and fail to localize *in vivo* [33,35,38–40]. Two other proteins Prrp and Xstau also bind *Vg1* RNA and co-localize with it at the vegetal cortex [21,41,42]. RNA localization begins as a recognition event, most likely in the nucleus, and has been linked to splicing events [42–44]. More recently, Vg1RBP/Vera and hnRNP I were found to bind to each other and to *Vg1* in the nucleus while Prrp and Xstau were recruited to the RNP complex only in the cytoplasm [42]. These findings strongly suggest that RNA binding to distinct proteins in the nucleus segregates the early and late pathways.

The *nanos*-3′UTR has two different localization signals. The 240 nt mitochondrial cloud localization signal (MCLS) is both required and sufficient to direct *nanos* into the MC [4,6]. In addition, the 160 nt germinal granule localization element (GGLE) is required to direct *nanos* into germinal granules, an event that requires the prior functioning of the MCLS [45]. The *nanos* MCLS was shown to bind directly to Vg1RBP/Vera and hnRNP I *in vitro*, consistent with its ability to use the late pathway after injections into late staged oocytes [29,46]. However, endogenous Vg1RBP/Vera appears excluded from the MC as shown by immunolocalization [6]. Furthermore, the ER-*nanos* association/entrapment event does not involve Vg1RBP/Vera, a protein implicated in linking *Vg1* RNA to the ER [33]. How then can the early and late pathways be distinguished and sorted into different cellular domains? Clearly, proteins that bind early pathway RNAs like *nanos*, but not *Vg1* RNA must exist in the stage I oocyte.

The nature of the RNA-protein interactions operating in the early pathway that mediate the steps of RNA selection, entrapment, and translational regulation remain unknown. Complicating our understanding of these processes are RNAs such as *hermes/rbpms* that localize using both early and late pathways [25]. Here we describe work on the RNA binding protein Hermes/Rbpms. Hermes/Rbpms is an RNA Recognition Motif (RRM) family member originally found to play a role in embryonic heart development [47] and later re-discovered in a screen for vegetally localized maternal RNAs [25]. Functional studies have linked it with myocardial differentiation [48] and cell division within the vegetal hemisphere [25], but its mode of operation in these events remains unclear although a negative role in translation has been proposed [49].

We find that Hermes/Rbpms protein is present in the MC, throughout the cytoplasm and in the nucleus of stage I oocytes. However, unlike other RNA-binding proteins, we confirm that Hermes/Rbpms co-localizes with *nanos* in the MC, concentrating in the germ plasm forming region and within the germinal granules [49]. Importantly, we show that Hermes/Rbpms protein forms distinct particles and associates with *nanos* but not *Vg1* or *VegT* RNA within the nucleus. Hermes/Rbpms binding to nanos 3′UTR requires an unknown component present in stage I but not stage VI oocyte extracts. The UGCAC repeats essential for *nanos* RNA MC localization are also required for Hermes/Rbpms binding to the MCLS. In addition, Hermes/Rbpms binds the GGLE domain, a region that does not contain UGCAC elements but does contain two VM1 hnRNP I binding sites. Further, we find that the terminal 34 amino acids in Hermes/Rbpms, a region conserved with human RBPMS, is required to form homodimers as well as to bind the *nanos* 3′UTR.

Taken together, our observations show that Hermes/Rbpms is an important component of the *nanos* RNP particle. Hermes/Rbpms functions as a homodimer in concert with other proteins to facilitate granule formation. We propose that Hermes/Rbpms binds *nanos* RNA in the nucleus in association with hnRNP I, initiating a series of events that results in *nanos* but not *Vg1* entering the early pathway. We suggest that, by incorporating *nanos* RNA into a Hermes/Rbpms nuclear particle, *nanos* RNA is effectively prevented from being translated after it exits the nucleus. To our knowledge, this is the first evidence for a nuclear RNA-binding protein specific for the early pathway.

2. Experimental Section

2.1. Oocytes and Microinjection

Stage I/II oocytes were obtained after collagenase treatment of surgically removed ovarian tissue as previously described [6]. Myc-Hermes/Rbpms mRNA was transcribed with mMessage mMachine (Ambion) and 300 pg injected into stage I/II oocytes. Injected oocytes were cultured for two days in oocyte culture medium (OCM: 60% Liebovitz L-15 (Gibco), 0.04% bovine serum albumin, 1 mM L-glutamine, and 5 μg/mL gentamycin).

2.2. Myc-Hermes/Rbpms Immunoprecipitation

After the two-day culture period, the nuclear and cytoplasmic fractions were manually isolated in OR2 buffer. After injection of 1nl of water, the nucleus emerged from the needle-puncture made in the plasma membrane. Whole oocytes, ooplasms, and nuclear samples were homogenized in YSS buffer (50 mM Tris pH 8.0, 50 mM NaCl, 0.1% NP40, 0.1U/μL superRNAsin, 1X protease inhibitors from Roche, 0.5 mM DTT, 100 mM sucrose) and centrifuged at 16,000 g for 3 min to remove the insoluble materials. Anti-myc antibody (9E10, Affymetrix) and protein G beads were added to the supernatant. The mixture was incubated at 4 °C overnight. The beads were then washed with YSS buffer four times. The bound RNAs were recovered with proteinase K digestion and analyzed by RT-PCR using the primers: Nanos-F, gaggctacacttgccctttg and Nanos-R, gcccattagtggtgcagaat; Vg1-F, atgcctattgcttctatttgc and Vg1-R, ggtttacgatggtttcactca and VegT-F, caagtaaatgtgagaaaccg and VegT-R, caaatacacacacatttccc). Possible cross contamination between ooplasm and nuclei during isolation procedures was monitored by western blotting with anti-alpha tubulin (12G10; 0.5 μg/mL) and anti-nucleoplasmin (B7-1A9; 1.5 μg/mL) antibodies respectively (Developmental Studies Hybridoma Bank).

2.3. Mutagenesis, Cloning, and RNA Synthesis

Hermes/Rbpms deletion mutants were constructed as follows:

Hermes/Rbpms deletion A (a.a. 101–121 were deleted): the N terminal fragment was amplified with Hermes/Rbpms ATTTAGGTGACACTATAGAgtacaccatgagcggcatcaagtcagaca (forward) and Hermes/Rbpms-DA1 (CTTTGTGTTGGCCTTTGC). The C terminal fragment was amplified with Hermes/Rbpms-DA2 (GCAAAGGCCAACACAAAGCACTTCATTGCACGAGAT) and Hermes/Rbpms TTTTTTTTTTTTTTTTTAACAAAACTGCCGAGACT (reverse 1). The two fragments were then fused by PCR. Hermes/Rbpms deletion B (a.a. 122–161 were deleted): the N terminal fragment was amplified with Hermes/Rbpms ATTTAGGTGACACTATAGAgtacaccatgagcggcatcaagtcagaca (forward) and Hermes/Rbpms-DB1 (TGCGCCAAGTGCTGGGTG). The C terminal fragment was amplified with Hermes/Rbpms-DB2 (CACCCAGCACTTGGCGCAGCTTTCACATACCCTGCT) and Hermes/Rbpms TTTTTTTTTTTTTTTTTAACAAAACTGCCGAGACT (reverse 1). The two fragments were then fused by PCR. Hermes/Rbpms deletion C (a.a. 164–198 were deleted) was amplified with Hermes/Rbpms ATTTAGGTGACACTATAGAgtacaccatgagcggcatcaagtcagaca (forward) and Hermes/Rbpms TTTTTTTTTTTTTTTTTAAGCAGCATGTGGAATGGC (reverse 2). The resulting PCR products from Hermes/Rbpms deletion mutants were used as templates for *in vitro* transcription (mMessage mMachine, Ambion, Austin, TX, USA). The RNAs were translated in reticulocyte lysates following the manufacturer's instructions (Promega).

Myc-Hermes/Rbpms: plasmid pCS2+6Xmyc Hermes/Rbpms was designed as detailed in Song *et al.*, 2007 [49]. hnRNP I: plasmid PSP64TSN-RLMCS-p60-Flag was a gift from Dr. K. Mowry (Brown University). Myc-Nanos: Nanos coding region was amplified with MT-nanos cagcttgaattca ATGGATGGCGGTCTCTGC [49] (forward) and MT-nanos actagtctcgagTCAGTGTCTCAGCTTTGG (reverse). The PCR product was digested with EcoRI and XhoI and cloned into the pCS2+MT vector (from D. Turner).

GST-Hermes/Rbpms-C: The Hermes/Rbpms C terminal region was amplified with GST-Hermes/Rbpms-C ACTGACGAATTCATGGCC AAGAACAAACTA (forward) and GST-Hermes/Rbpms (AGCTATCTCGAGTTAACAAAACTGCCGAGA (reverse). The PCR product was digested with EcoRI and XhoI and cloned into pGEX-5x-1.

Sense RNAs were transcribed from NotI-linearized plasmids with SP6 RNA polymerase and the mMessage mMachine kit (Ambion, Austin, TX). For fluorescent probes, 600 μM Alexa 488-UTP or tetramethylrhodamine-UTP (Molecular Probes, Eugene, OR, USA) was used exactly as described in Chang *et al.*, 2004 [6]. Alexa 488-labeled *nanos*-3UTR was injected 20 h before fixation and antibody staining.

2.4. *In Vitro Translation*

^{35}S-Methionine labeled Hermes proteins were synthesized in rabbit reticulocyte lysates (Promega) using capped *in vitro* transcribed RNA following the manufacturer's protocol. For radiolabeling, 20 μCi of ^{35}S-methionine was included (New England Nuclear, Boston, MA, USA, NEG009A, 1200 Ci/mmol). Samples were immunoprecipitated under denaturing conditions with anti-myc antibody and analyzed by SDS-PAGE. The bound protein was quantitated using a PhosphorImager™ (Storm 840, Molecular Dynamics/GE Healthcare Life Sciences, Pittsburgh, PA, USA) with the ImageQuant software package (8.1, GE Healthcare Life Sciences, Pittsburgh, PA, USA).

2.5. *RNA-AADA Pull-Down*

RNA was transcribed with mMessage Machine (Ambion) and purified by LiCl precipitation. 0.2–0.5 μg of RNA was oxidized with 1 mg/mL of sodium periodate in 0.1 M NaAc (pH 5.0) at room temperature for one hour. The oxidized RNA was then immobilized on adipic acid dehydrazide agarose (AADA) beads (Sigma, St. Louis, MO, USA) in 0.1 M NaAc (pH 5.0) at 4 °C over night. The RNA-AADA beads were washed with 1x TPB buffer (60 mM Tris pH7.4, 10 mM MgCl2, 80 mM NaCl, 0.1% Triton X-100, 10% glycerol).

Stage I/II oocytes were homogenized in the presence of 1x protease inhibitor (Roche), DTT (0.3 mM) and heparin (5 mg/mL). Insoluble materials were removed by centrifugation at 16,000 g for 3 min. The supernatant was mixed with *in vitro* translated Myc-Hermes/Rbpms labeled with ^{35}S-methionine for 5 min at room temperature. The mixture was then precleared for 30 min with AADA agarose beads prewashed with 1x TPB buffer at room temperature. After centrifugation at 16,000 g for 3 min, the supernatant was mixed with RNA-AADA beads at room temperature for 5 min.

The beads were then washed three times with 1x TPB buffer. The radio-labeled bound proteins were analyzed by SDS-PAGE and visualized using a PhosphorImager™.

2.6. Yeast-Two-Hybrid

We identified proteins interacting with Hermes/Rbpms using the Two-Hybrid System 3 from Clontech following their detailed protocol (Clontech Yeast Protocol Handbook, Matchmaker 3 manual, 2 Hybrid System TRAFO Protocol). System 3 uses yeast strain AH109, which includes four reporter genes (ADE2, HIS3, lacZ, and MEL1) to help reduce false positives. High stringency conditions were used plating yeast on SD/-Ade/-His/-Leu/-Trp/X-a-gal plus 10mM 3-AT following the procedures presented by Gietz and Schiestl [50]. A *Xenopus* oocyte two-hybrid cDNA library cloned in pGAD10 was obtained from Clontech. Two hundred micrograms of plasmid DNA from each library was transformed into competent yeast L40-coat/WT cells as described by Clontech. The Hermes/Rbpms coding region was used as bait and amplified with pAS1-Hermes/Rbpms AGCTAGGCCATGG AGagcggcatcaagtcagac (forward) and pAS1-Hermes/Rbpms AGTCAGGTCG ACTTAACAAAACTGCCGAGA (reverse). The PCR product was cloned into the pAS1 vector (from Dr. S. Elledge) using NcoI and SalI. Hermes/Rbpms was fused to the C terminus of Gal4 DNA binding domain by PCR.

2.7. Immunofluorescence Microscopy and Antibodies

Oocytes were fixed and processed exactly as described in Chang *et al.*, 2004 [6] for immunofluorescence microscopy. Anti-Hermes/Rbpms antibody was generated against a nonconserved peptide region (AHFIARDPYDLTGAA) and purified as described in Zearfoss *et al.*, 2004 [25]. Anti- Vg1RBP/Vera antibodies from Nancy Standart (University of Cambridge, Oxford, UK), anti-alpha-tubulin mouse mAb DMIA (Sigma), used at 1:1000, and preabsorbed rhodamine- or fluorescein isothiocyanate-labeled anti-Ig antibodies (Jackson Immunoresearch Laboratories, West Grove, PA, USA), used at 1:75 to 1:150. Anti-hnRNP I antibodies were a gift from K. Mowry (Brown University, RI, USA) used at 1:100 and secondary antibody was either FITC-Goat anti-rabbit IgG 1/100 or Alexa® 568-conjugated goat anti–rabbit (ThermoFisher Scientific , NY, USA) at 1:100 [42]. Rat monoclonal anti-GRP94 1/250 and Rhodamine-anti-rat 1/75 was used as an ER marker as described in Beckhelling *et al.*, 2003 [51] (StressGen Biotechnologies, Victoria, BC, Canada). Imaging on the Leica SP2 confocal.

2.8. Ultrastructural Analysis

All hybridization and post-hybridization steps, embedding, and sectioning were done as described in Kloc *et al.* 2001 [10]. *Nanos* (Xcat2) RNA and Hermes/Rbpms protein were identified by *in situ* hybridization as described in Song *et al.*, 2007 [49] and Kloc *et al.*, 2002 [3].

2.9. Ethical Statements

The animal protocols used in this work were evaluated and approved by the Institutional Animal Care and Use Committee (IACUC) of the University of Miami. All activities are in compliance with federal, state, and institutional regulations. The University was granted full accreditation by the Association for Assessment and Accreditation of Laboratory Animal Care, International (AAALAC) in February 2005 and received its current re-accreditation in 22 October 2013. In addition, University of Miami is licensed by the U.S. Department of Agriculture (USDA) and has filed a Letter of Assurance with the Office of Laboratory Animal Welfare (OLAW), U.S. Department of Health and Human Services (DHHS).

3. Results

3.1. Hermes/Rbpms Co-Localizes with Nanos RNA within Germinal Granules

In agreement with Zearfoss *et al.*, 2004 [25], we found that Hermes/Rbpms is present within the mitochondrial cloud (MC) and throughout the cytoplasm of stage I oocytes. However, we also found Hermes/Rbpms protein in the nucleus (Figure 1). Hermes/Rbpms RNA follows the same cellular expression pattern as its protein and suggests that Hermes/Rbpms may have several roles during oogenesis depending on which RNA it may bind [49,52]. Previous work showed that Hermes/Rbpms protein over-expressed at the vegetal, but not the animal pole, formed large aggregates that contain RNA [47,48,53]. We also confirmed by ultrastructural immune-localization that within the MC, Hermes/Rbpms and *nanos* RNA are predominantly localized within the germinal granules (Figure 1; [49]). These observations led us to ask if Hermes/Rbpms protein and *nanos* RNA directly interact with each other.

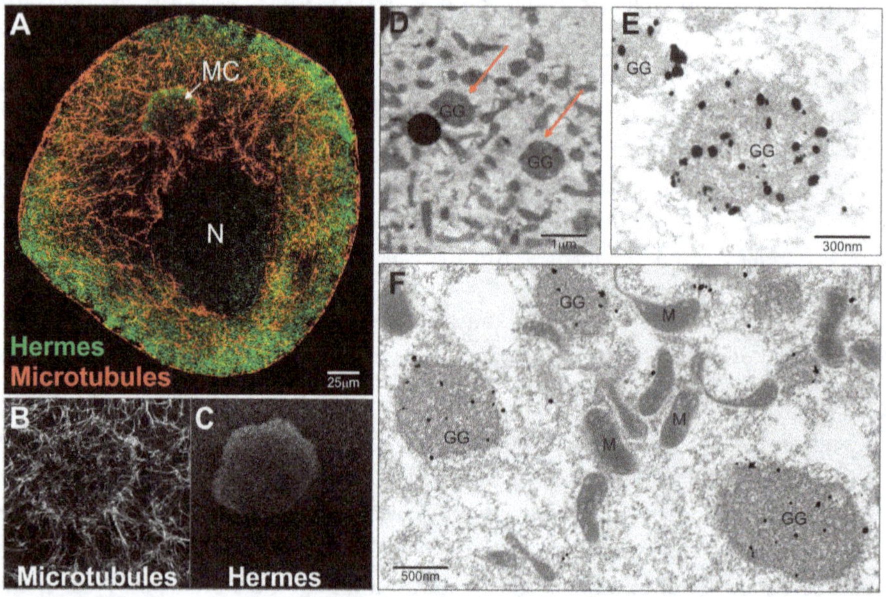

Figure 1. Hermes/Rbpms protein and *nanos* RNA localize within germinal granules unique to germ plasm. Stage I oocyte. (**A**) Immunofluorescence (IF) and confocal microscopy showing Hermes/Rbpms protein (**green**) is ubiquitous in stage I oocyte including the mitochondrial cloud (MC), nucleus (N) and ooplasm; microtubules (**red**). Enlarged image from (**A**) showing microtubule organization around MC (**B**) and Hermes/Rbpms (**C**); Note how Hermes/Rbpms is enriched within the germ plasm region of the MC; (**D**) Electron micrograph showing a region of germ plasm within the MC of stage I oocyte. Red arrows indicate two germinal granules; (**E**) *Nanos* RNA is localized within germinal granules in stage I oocytes as is Hermes/Rbpms protein (**F**). Electron microscopy *in situ* hybridization with antisense Digoxigenin labeled *nanos* RNA was visualized with nanogold conjugated anti-Dig antibody and silver enhancement. Electron microscopy immunostaining with Hermes/Rbpms polyclonal antibody [49], anti-rabbit nanogold conjugated secondary antibody and silver enhancement. Scale bars are as indicated. GG (germinal granules); M (mitochondria); N (nucleus).

3.2. Hermes/Rbpms Protein Specifically Binds the Nanos Mitochondrial Cloud Localization Signal

To address whether Hermes/Rbpms specifically binds *nanos* RNA, the *nanos* 3′UTR was chemically cross-linked to AADA beads and the beads incubated with reticulocyte lysate containing *in vitro* translated radio-labeled Myc-Hermes/Rbpms protein [54,55]. Bound protein was analyzed by SDS-PAGE and visualized by autoradiography. Myc-Hermes/Rbpms was pulled down with the *nanos* 3′UTR, but only when the lysate was supplemented 1:1 with stage I oocyte extract (+). These findings strongly suggested another protein(s) is required for Hermes/Rbpms to efficiently associate with the *nanos* 3′UTR [48]. Alternatively, Hermes/Rbpms may need to be modified in some way before being competent to bind *nanos*. Importantly, Hermes/Rbpms did not bind the *Vg1* localization signal (LS), suggesting a level of discrimination between these two RNAs at the level of Hermes/Rbpms protein binding (Figure 2). Galactosidase (Gal) served as a negative control and showed little binding capability.

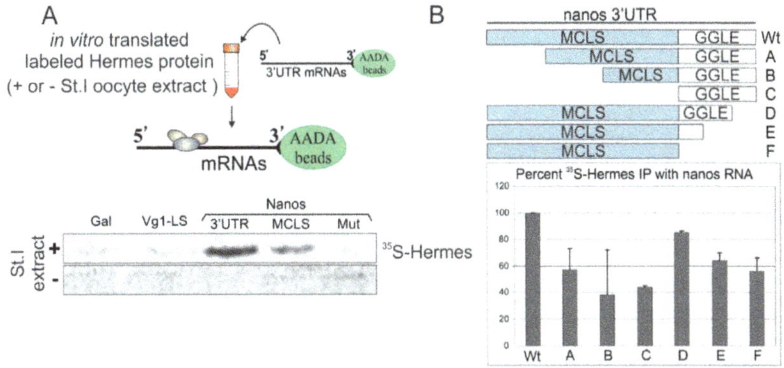

Figure 2. Hermes/Rbpms specifically associates with *nanos* RNA in a UGCAC dependent manner. (**A**) Hermes/Rbpms interacts with *nanos* but not *Vg1* RNA. *Vg1* 3′UTR Localization Experimental design is shown at top. *Vg1* Localization Signal (LS), *nanos1* 3′UTR, *nanos* Mitochondrial Cloud Localization Signal (MCLS), and a *nanos* mutant MCLS (mut) were tested for Hermes/Rbpms binding. Individual RNAs were immobilized on AADA agarose beads and mixed with *in vitro* translated Hermes/Rbpms protein labeled with ^{35}S-methionine. Bound proteins were analyzed by SDS-PAGE and autoradiography. *Galactosidase* (Gal) RNA served as a negative control. Hermes/Rbpms only bound the *nanos1-3′UTR* and only when reactions included stage I/II oocyte extracts and not reticulocyte lysates alone. Note: the 3′UTR binds more Hermes/Rbpms than the MCLS alone suggesting the GGLE may also bind Hermes/Rbpms; (**B**) Both the MCLS and GGLE are required for wild-type levels of Hermes/Rbpms binding and require stage I oocyte extract. The deletion mutants diagramed were immobilized on AADA beads and analyzed as described in Methods. Error bars indicate standard deviation of three experiments. Hermes/Rbpms binding to the full-length *nanos* 3′UTR is set at 100%. Note: both regions contribute to Hermes binding.

Within the 240 nt MCLS are six UGCAC motifs previously shown to be essential for *nanos* RNA to localize within the MC. The UGCAC sequences likely mediate redundant protein binding to the signal [6,28,29,56]. In the pull-down assay, Hermes/Rbpms consistently bound the *nanos* 240 nt-MCLS, but with reduced affinity compared to the full length 3′UTR. To test the importance of the MCLS UGCAC repeats for Hermes/Rbpms binding, we introduced substitution mutations in all six, altering them to non-cognate sequences UUGGG or AGGCC. Previous studies with Par-Clip indicate CAC sites are preferred for Hermes/Rbpms binding [52]. But we found that simply changing the UGCAC repeats to UUCAC prevented Hermes/Rbpms binding to the 240 nt-MCLS and significantly impaired MC entrapment [6]. Hermes/Rbpms failed to bind to the *nanos* 240 nt-MCLS lacking the UGCAC repeats in the pull-down assay (Mut in Figure 2). Thus, the UGCAC sites required for localization are also required for Hermes/Rbpms to bind the *nanos* MCLS. These results show that Hermes/Rbpms binds *nanos* RNA specifically in a UGCAC dependent manner and loss of Hermes/Rbpms binding is correlated with loss of *nanos* RNA localization into the MC. However, the results do not address whether Hermes/Rbpms directly interacts with the UGCAC repeats.

The 160 nt germinal granule localization element (GGLE) just downstream of the MCLS directs *nanos* into germinal granules, an event that requires the prior functioning of the MCLS [45]. The reduced binding of Hermes/Rbpms to the 240 nt-MCLS compared to the entire 3′UTR raised the possibility that Hermes/Rbpms also associated with the GGLE. To determine if Hermes/Rbpms could bind the GGLE alone, deletion mutants were introduced into the 3′UTR by PCR. All deletions showed a decreased affinity to Hermes/Rbpms, indicating both the MCLS and GGLE are required for optimal binding (Figure 2B) The GGLE alone retained 40% binding activity compared to 58% by the MCLS. The GGLE region does not contain any UGCAC sites but several VM1 sites. Thus both regions of the *nanos* 3′UTR contribute to Hermes/Rbpms binding (Figure 2B; compare lanes C and F with WT). These results indicate that Hermes/Rbpms can associate with the *nanos* GGLC in the absence of UGCAC sites or the MCLS, suggesting a role in germinal granule formation.

3.3. Mapping Hermes/Rbpms for Regions Required for Nanos-Hermes/Rbpms Interaction

Proteins involved in RNA localization often function as dimers or oligomers, operating as partners with other proteins to confer specificity to the RNA recognition event [15,57]. Previous work has shown that Hermes/Rbpms forms homodimers and can oligomerize [48]. Therefore, we asked what region of Hermes/Rbpms was required for its interaction with *nanos* RNA and whether that region was required for Hermes/Rbpms to form homodimers. Hermes/Rbpms protein has an N terminal RRM domain required but not sufficient for binding polyA+ RNAs [48]. Keeping the RRM domain intact, we introduced three deletion mutations within the C-terminus and determined what region was required for *nanos* binding. Each deletion mutant as well as wild-type Hermes/Rbpms was first translated *in vitro* and shown to translate equally well by western blotting (Figure 3A). Each myc-tagged Hermes/Rbpms protein was then tested as previously described for the ability to bind *nanos* RNA immobilized on AADA beads in the presence of stage I oocyte extract. The results showed that only the terminal 34 amino acids were absolutely required for *nanos* binding in the presence of the RRM (Figure 3A, mutant C). Therefore, the Hermes/Rbpms RRM domain was not sufficient for nanos RNA interaction, but binding also required the very hydrophilic C-terminus of Hermes/Rbpms (Figure 3C).

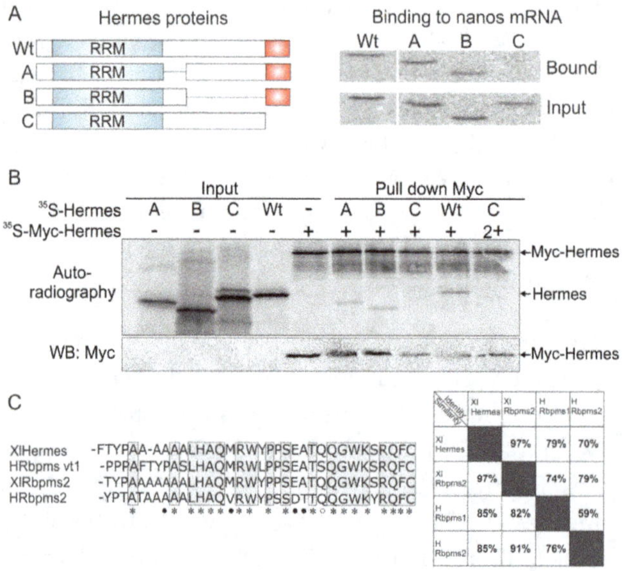

Figure 3. The carboxyl terminal 34 amino acids of Hermes/Rbpms are required for *nanos1* binding and for Hermes/Rbpms dimerization. (**A**) Three deletion mutants of Hermes/Rbpms are diagramed showing the RRM domain and the hydrophilic carboxyl terminal 34 amino acids (**red** box). ^{35}S-methionine labeled Hermes/Rbpms and its deletion mutants were analyzed for their ability to be pulled-down by *nanos1* 3′UTR immobilized on AADA beads. Wild-type and mutant protein (input) used in reaction and bound are shown by SDS-PAGE. Note that only the last 34 amino acids were required for *nanos* binding in the presence of the Hermes/Rbpms RRM; (**B**) The presence of Hermes/Rbpms homodimers was detected by autoradiography as two bands. Note that Hermes/Rbpms missing the terminal 34 amino acids failed to form a dimer. As *nanos* RNA was not present, Hermes/Rbpms formed homodimers in the absence of RNA; (**C**) C-terminal region involved in Hermes/Rbpms homodimerization and binding to the *nanos* 3'UTR. Alignment of conserved terminal 34 amino acids (AA) of Human and *Xenopus* Hermes/Rbpms proteins. An * (asterisk) indicates a fully conserved AA. Black dots indicate conservation between groups of strongly similar properties (scoring > 0.5 in the Gonnet PAM 250 matrix). A white dot indicates weakly similar properties (scoring ⩽ 0.5 in the Gonnet PAM 250 matrix), Clustal Omega, EMBL-EBI, UK. Table shows the high level of identity and similarity between frog and human Hermes/Rbpms proteins.

To determine what region of Hermes/Rbpms is required for homodimerization, Hermes/Rbpms wild-type and mutant proteins, both with or without myc tags, were translated *in vitro* in the presence of ^{35}S-methionine. Each synthesized radiolabelled Hermes/Rbpms protein was mixed with its myc-tagged counterpart and incubated prior to immunoprecipitation (Figure 3B). After immunoprecipitation with anti-Myc antibody, the co-precipitates were analyzed by western blotting and autoradiography (PhosphorImager). Myc-Hermes/Rbpms migrates at 38 kDa while wild-type Hermes/Rbpms at 28 kDa. Homodimers between myc-Hermes/Rbpms and Hermes/Rbpms are easily detected as two bands. Only mutant C, missing the last 34 amino acids, failed to form a homodimer (Figure 3B). Taken together, the results indicate that the same 34 amino acids are required for Hermes/Rbpms to form homodimers and to be competent to associate with *nanos* RNA. Interestingly, the last 34 amino acids are 85% conserved between *Xenopus* and human Hermes/Rbpms (Figure 3C).

3.4. Hermes/Rbpms Interacts with hnRNP I

Hermes/Rbpms binding to *nanos* required the addition of oocyte extract to the assay (Figure 2A) suggesting the presence of a co-factor that facilitates Hermes/Rbpms binding. Alternatively, or in addition, an oocyte factor may promote a post-translational modification required for *nanos* association. We favor the former explanation, as endogenous and translated Hermes/Rbpms appeared to have the same molecular mass after gel fractionation. To identify oocyte proteins capable of interacting with Hermes/Rbpms, we screened an expression library (Clontech) made from *Xenopus* oocyte mRNA containing 2.5×10^6 independent clones with an average insert size of 1.5 kb in a yeast-two-hybrid assay. The known RNA-binding protein hnRNP I was identified from approximately 700×10^3 clones (Figure 4). Previous work had shown that hnRNP I is present in early staged oocytes and has a role in *Vg1* RNA localization [39]. hnRNP I seemed a strong candidate and was selected for further analysis. We asked if the Hermes/Rbpms:hnRNP I interaction in the yeast-2-hybrid assay required the last 34 amino acids (ΔC162-196) of Hermes/Rbpms. A similar failure in binding of hnRNP I would suggest this protein is the missing factor found in oocyte extracts required for Hermes/Rbpms binding to *nanos*. However, mutC-Hermes/Rbpms still interacted efficiently with hnRNP I while SNF4, the negative control, did not (Figure 4). Therefore, Hermes/Rbpms-hnRNP I interaction does not require the terminal 34 residues.

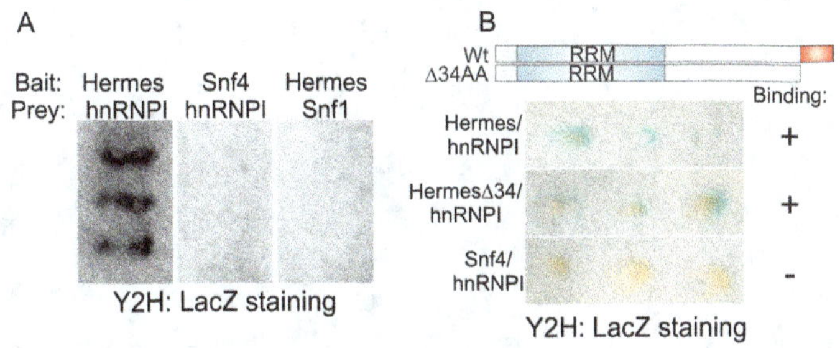

Figure 4. Hermes/Rbpms interacts with another Nanos1 binding protein: To identify proteins that might interact with Hermes/Rbpms during early oogenesis, a *Xenopus* cDNA library (Clontech) was screened using a yeast two-hybrid (Y2H) system. (**A**) Full length Hermes/Rbpms fused with the DNA binding domain (BD) of the transcription factor Gal4 served as bait. Candidate interacting proteins were fused to the Gal4-activation domain (AD) and served as prey. The reporter gene B-galactosidase, driven by the Gal4 binding site, was positively transcribed as the result of Hermes/Rbpms and hnRNP I (heterogeneous nuclear ribonucleoprotein I) interaction. Two proteins involved in yeast glucose metabolism, Sfn4 and Sfn1, were used as controls to discard false positive results. Snf4/Gal4-BD was used as a bait and Snf1/Gal4-AD as a prey to check false interactions with hnRNP I and Hermes/Rbpms protein respectively; (**B**) The C-terminal 34 amino acids (in red) required for Hermes/Rbpms to dimerize and bind *nanos* RNA are not required to interact with hnRNP I. Hermes/Rbpms protein lacking the terminal 34 AA was used as bait and hnRNP I as prey. Induction of B-galactosidase indicated a positive interaction between bait and prey.

Xpat protein is a major structural component of the germ plasm and is thought to provide a scaffold for other germ plasm components [58]. To characterize the relationship between endogenous Hermes/Rbpms and Xpat within the MC, stage I/II oocytes were fix and stained with the corresponding antibodies followed by confocal microscopy. Dissimilar patterns were observed for the two proteins. Hermes/Rbpms appeared as fine particles, smaller and less distinct than those for Xpat, but falling within the larger defined Xpat structure. The Confocal images are consistent with the idea that Hermes/Rbpms is directly associated with localized RNAs whereas Xpat has more of a global role in germ plasm structure within which RNP particles are embedded, including germinal granules (Figure 5A,A'). Our results are consistent with those of Nijjar and Woodland [53,59]. Next we examined the two endogenous late pathway *Vg1* RNA binding proteins Vg1RBP/Vera and hnRNP I in relation to labeled injected early pathway *nanos* RNA. hnRNP I was found in the nucleus and appeared to partially co-localize with RNA just outside the MC (Figure 5B,D). Both Vg1RBP/Vera and hnRNP I however, are excluded from the MC suggesting the RNP particles destined for the germ plasm are different from those following the late *Vg1* localization pathway (Figure 5C,D).

We attempted to visually capture Hermes/Rbpms-*nanos* RNA interaction within the ooplasm during the process of RNA localization into the MC. Fluorescently tagged *nanos* 3′UTR RNA was injected into stage I oocytes, and after 5 or 20 h, oocytes were fixed. The 3′UTR of nanos RNA was used in order to eliminate the possibility of *nanos* translation, an event that could obscure RNA-protein interactions. Endogenous Hermes/Rbpms protein was identified by immunofluorescence after staining with an anti-Hermes/Rbpms antibody. Regardless of conditions or time, we could not detect any co-localization between injected *nanos* 3′UTR RNA and endogenous Hermes/Rbpms (data not shown). Thus, we were unable to use this approach to extend our analysis *in vivo*.

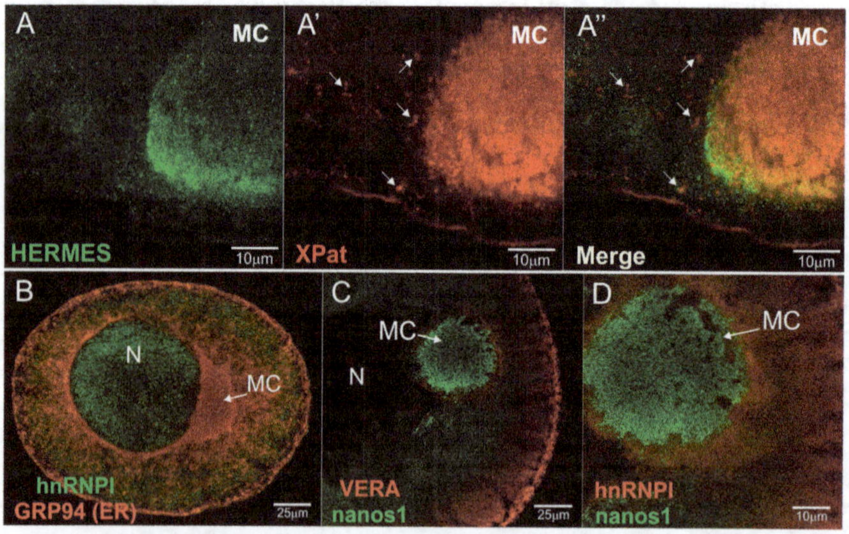

Figure 5. Cellular distribution of Hermes/Rbpms, hnRNP I, Vg1RBP/Vera, and Xpat proteins in the stage I/II oocytes. Confocal images showing immunofluorescence (IF) of stage I/II oocyte with (**A**) anti-Hermes/Rbpms (green) and (**A′**) anti-Xpat (red) antibodies; (**A″**) Superimposed image. White arrows indicate Xpat and Hermes/Rbpms particles are not identical; (**B**) Superimposed images of hnRNP I (**green**) and GRP94 (**red**) proteins. Co-staining with anti-GRP94 reveals the ER enriched within the MC, cortex, and perinuclear region. Note that hnRNP I is excluded from the MC but is present in the nucleus and cytoplasm; (**C**) Merged images of endogenous Vg1RBP/Vera (**red**); or (**D**) hnRNP I (**red**) with Alexa 488-labeled *nanos* 3′UTR injected 20 h before fixation (**green**). Note that while *nanos* RNA localizes to the MC, both Vg1RB/Vera and hnRNP I are excluded from it. N: nucleus; MC: mitochondrial cloud. Scale bars are as indicated.

3.5. Hermes/Rbpms Associates with Nanos RNA within the Nucleus of Stage I Oocytes

In our assay, *nanos* RNA associated with Hermes/Rbpms while the late pathway RNA *Vg1* did not. These results suggested that Hermes/Rbpms could discriminate between endogenous RNAs that employ different localization pathways. To determine if such discrimination occurs *in vivo*, myc-Hermes/Rbpms RNA was injected into stage I oocytes. The oocytes were cultured for two days to allow translation to occur (Figure 6A). The oocytes were then homogenized, and myc-antibody was used to immunoprecipitate (IP) the newly translated myc-Hermes/Rbpms from the supernatant. RNA was isolated from the IP pellet and the presence of *nanos* and *Vg1* RNAs determined by RT-PCR using specific primers. While *nanos* RNA was recovered in the precipitate, *Vg1* RNA was not detected (Figure 6B). From these results, we conclude that *nanos* RNA associates with Hermes/Rbpms *in vivo* in stage I oocytes.

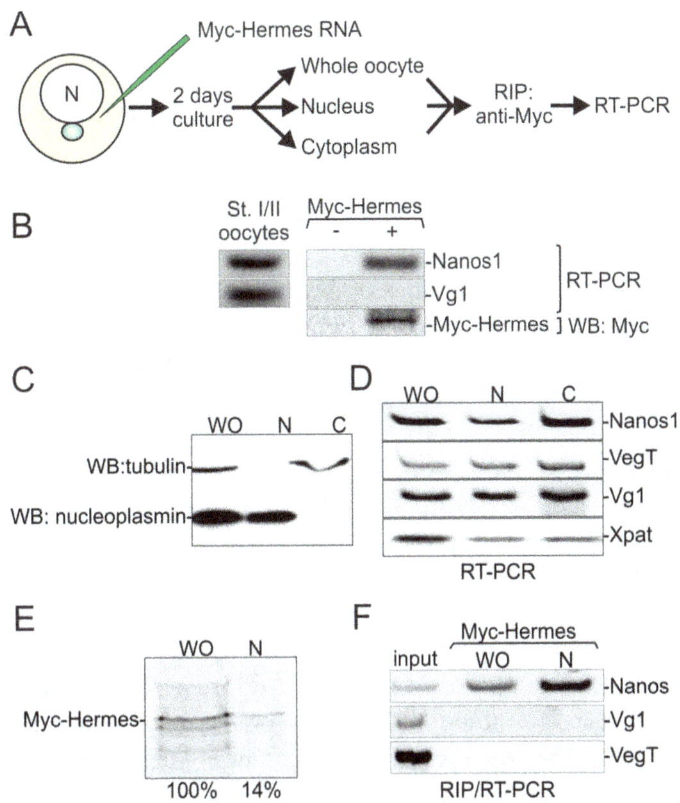

Figure 6. Hermes/Rbpms associates with *nanos* but not *Vg1* RNA in the nucleus. (**A**) Overall experimental design. Myc-Hermes/Rbpms mRNA was injected into stage I/II oocytes. After culturing for two days, either myc-Hermes/Rbpms protein was immunoprecipitated (IP) with anti-myc antibody from whole oocytes only or the nuclear (N) and cytoplasmic (C) fractions were manually isolated before IP with anti-myc antibody. In either design, RNA was extracted from the resulting pellet and analyzed with specific primers for *Vg1* or *nanos* RNA by RT-PCR; (**B**) Myc-Hermes/Rbpms interacts with *nanos* but not *Vg1* RNA *in vivo*. Control RT-PCR with whole oocytes shows that *nanos* and *Vg1* RNAs were present. Control western blot shows that myc-Hermes/Rbpms was translated after injection; (**C**) The nuclear and cytoplasmic fractions were manually isolated from stage I/II oocytes. Tubulin (cytoplasmic marker) and nucleoplasmin (nuclear marker) were detected by western blot analysis and show clean separation; (**D**) Late and Early Pathway RNAs accumulate in nucleus concurrently. RNAs were analyzed by RT-PCR using specific primers; (**E**) Western blot analysis showed newly translated Hermes/Rbpms in the nucleus; (**F**) Hermes/Rbpms interacts with *nanos* RNA in the nucleus, but not *Vg1* or *VegT*. WO: whole oocyte; Uninjected oocytes served as negative controls. Blue sphere (MC); N (nucleus).

As shown in Figure 1, some of the endogenous Hermes/Rbpms protein is nuclear, raising the possibility that Hermes/Rbpms associates with *nanos* RNA there during the earliest steps of RNA localization. We first determined if *nanos*, and two late pathway RNAs, *Vg1* and *VegT*, were present concurrently in stage I nuclei. Nuclei were manually isolated together with the enucleated cytoplasm. Western blotting with antibodies against tubulin and nucleoplasmin were used to assess the level of contamination between the isolated cytoplasm and nuclear samples respectively (Figure 6C). *VegT*, *Vg1* and *nanos* RNA were all shown to be present within nuclei of stage I oocytes and with similar nuclear to cytoplasmic ratios (Figure 6D). To determine if Hermes/Rbpms could discriminate between these three RNAs, Myc-*hermes/rbpms* RNA was injected into isolated stage I oocytes and oocytes cultured for two days. Cleared supernatants were prepared from whole and nuclear samples and Myc-Hermes/Rbpms immunoprecipitated from them. As expected, newly synthesized Myc-Hermes/Rbpms protein was found in the nucleus as well as the cytoplasm at the end of the two-day culture period (Figure 6E). RNA was isolated from the IP pellet and the presence of *nanos*, *Vg1* and *VegT* RNAs determined by RT-PCR using specific primers. While *nanos* RNA was recovered in the precipitate from the whole and nuclear fractions, *Vg1* and *VegT* RNA were not detected (Figure 6F). Taken together, our results show that interaction between Hermes/Rbpms and *nanos* RNA initiates within the nucleus. Further, this interaction is specific as *Vg1* and *VegT* RNAs are not recognized.

4. Discussion

Germline development requires proper assembly of specific RNAs and proteins within the germ plasm region of the MC. At the same time, RNAs encoding proteins with exclusive roles in somatic fates must be excluded. The consequence of mis-expression of somatic determinants within the germline is sterility. However, the mechanism(s) for sorting RNAs into their correct locations remains poorly understood. Early recognition by RNA-binding proteins within the nucleus is one attractive mechanism for initiating an essential sorting step in RNA localization. In this report, we show that Hermes/Rbpms recognizes *nanos*, an essential germline component, but not *Vg1* or *VegT* RNA, within the nucleus. Four lines of evidence support a role for Hermes/Rbpms as a key protein component of the *nanos* RNP particle: (1) Hermes/Rbpms co-localizes with *nanos* RNA within the MC and is a component of germinal granules ([25], Figure 1); (2) Hermes/Rbpms binds *nanos* but not *Vg1* RNA in both *in vitro* and *in vivo* assays (Figures 2 and 6); (3) The UGCAC repeats that are essential for *nanos* localization to the MC germ plasm are also required for Hermes/Rbpms binding to the MCLS (Figure 2B); (4) The C-terminal 34 residues of Hermes/Rbpms are required for Hermes/Rbpms to form homodimers. Deletion of these residues, even in the presence of the Hermes/Rbpms RRM, prevents Hermes/Rbpms from binding to *nanos*.

Studies on *Vg1* RNA have provided support that the late RNA localization pathway initiates in the nucleus [42,60]. Those studies show that Vg1RBP/Vera and hnRNP I directly contact each other and bind *Vg1* RNA in the nucleus. Interestingly, once the *Vg1* RNP exits the nucleus and enters the cytoplasm, additional proteins are now added and Vg1RBP/Vera and hnRNP I are no longer in direct contact. Staufen is one of the additional proteins found in the cytoplasmic Vg1RNP complex. Staufen likely initiates association with the motor protein kinesin and thus promotes active transport to the vegetal pole [21,42]. Therefore, a dynamic remodeling of the RNP complex takes place that could account for *Vg1* active transport to the vegetal cortex using the late pathway.

There are parallels to be drawn between the *Vg1* studies and those reported here. *Nanos* RNA passes through five cellular compartments before localizing into germinal granules: the nucleus, cytoplasm, MC, ER, and germ plasm. None of these steps require active transport. Each cellular location could involve remodeling, that is proteins added, subtracted, or modified. Hermes/Rbpms association with *nanos* RNA, but not with *Vg1* or *VegT* RNAs within the nucleus suggests that Hermes/Rbpms binding might initiate a sorting pathway that terminates with *nanos*/Hermes/Rbpms within the germinal granules.

In the assay shown in Figure 2, *nanos* pulled down Hermes/Rbpms only when the reticulocyte lysate was supplemented with a protein extract from stage I oocytes. These results strongly suggested other proteins were required for Hermes/Rbpms binding to the *nanos* 3′UTR [48]. Could hnRNP I be part of an early nuclear *nanos* RNP? *In vitro* UV-crosslinking experiments show a specific interaction between hnRNP I and *nanos* RNA, however an association *in vivo* remains unclear [6]. Interestingly, both Vg1RBP/Vera and hnRNP I are excluded from the MC, so any association with *nanos* would be temporary (Figure 5). hnRNP I does interact directly with Hermes/Rbpms protein absent any RNA as revealed in the yeast-2-hybrid assay (Figure 4). Hermes/Rbpms is also found throughout the oocyte and is known to interact with other RNAs [52]. Further studies are required to establish if indeed hnRNP I is part of the endogenous Hermes/Rbpms/*nanos* RNP particle.

Nijjar and Woodland [53] have recently identified four other proteins that directly interact with Hermes/Rbpms: Xvelo splice variant (SV), Xvelo-full length (FL) and RNA-binding proteins Rbm42b and Rbm24b. They used a bimolecular fluorescence complementation (BiFC) approach to show likely protein-protein interactions after injection of tagged candidates. However, their assays were done in fully-grown oocytes (stage VI) and not at stages when *nanos* normally localizes and a MC is present. The protein components of the *nanos* RNP will likely change depending on the oocyte stage examined and localization mechanism employed. Indeed, tagged-*nanos* RNA injected into late stage oocytes is capable of using active transport to the vegetal pole but also has been shown to enter germ plasm structures by diffusion and entrapment at these late stages [59]. Although these interactions and subcellular locations are based on over-expression of reporters in the ooplasm, they do indicate what is possible [59]. With these caveats in mind, their findings suggest that Hermes/Rbpms likely associates with Rbm42b in the nucleus. Rbm42b also co-localizes with *nanos* RNA in late stage oocytes. From our proteomic analysis of the MC, Rbm42b is present (unpublished data) and could be a good candidate for one of the factors required for Hermes/Rbpms *nanos* RNA binding. Their analysis further suggests that Rbm24b and XveloSV associate with each other in the cytoplasm, but not in the nucleus indicating dynamic remodeling likely does occur.

Hermes/Rbpms colocalization with Xvelo within the germ plasm particles is a significant finding [53]. Xvelo is the homologue to Bucky ball in zebrafish, a gene required for the formation of the MC (Balbiani body) and oocyte polarity. By an unknown mechanism, Bucky ball is also linked to the germline as its over-expression results in more PGCs being formed [61]. Endogenous VeloFL does not appear to be nuclear but is cytoplasmic and enriched within the MC [53]. Thus Xvelo may join the Hermes/Rbpms/*nanos* RNP after it exits the nucleus.

The question remains as to what function Hermes/Rbpms performs within the germ plasm. Hermes/Rbpms has been studied in a wide range of cell types and tissues [62–64]. A conserved aspect is Hermes/Rbpms ability to oligomerize with itself and other proteins to form RNP granules. Within these RNP granules, Hermes/Rbpms is associated with repression whether in retinal ganglion cells (RGCs) [62,63], *Xenopus* germinal granules [49,65] or mouse and human cell lines [64]. Hermes/Rbpms appears to negatively regulate RINGO/Spy and Mos, RNAs involved in meiotic maturation and early cleavage respectively [49]. Hermes/Rbpms was shown to repress activator protein-1 (AP-1) signaling through cFos, cJun, and Smad3, a pathway implicated in tumor growth and progression [64]. Hermes/Rbpms blocks the interaction of these complexes and their recruitment to the promoter regions of AP-1 target genes. Similarly, over-expression of Hermes/Rbpms causes significant reduction in NKx2-5 expression, a transcription factor required for heart development [48].

In RGCs, Hermes/Rbpms forms RNP granules by interacting with other RNA binding proteins NonO, PSF, and G3BP1. Proteonomic analysis of the *Xenopus* MC revealed the presence of NonO (our unpublished results). Hermes/Rbpms containing granules are transported down growing axons. Hermes/Rbpms gain-of-function and loss-of function experiments *in vivo* in RGCs reveal a complex phenotype consistent with a role in balancing synaptogenesis and axon arborization [62]. In that study, a dominant negative form of Hermes/Rbpms was constructed lacking the C-terminus. *In vivo*, this mutant failed to form granules and localize properly. Their results are consistent with our findings that the C-terminus is required for Hermes/Rbpms to oligomerize and to bind *nanos* RNA.

In our studies, Hermes/Rbpms was also shown to form RNP particles within the nucleus, cytoplasm and MC. Once in the germ plasm, Hermes/Rbpms forms germinal granules that contain *nanos* RNA (Figure 1) [49]. Although *nanos* is transcribed very early in oogenesis, it is not translated at any time during oogenesis, but only after fertilization [13,65]. Premature translation in oocytes results in abnormal development [66]. Therefore, as soon as *nanos* RNA is transcribed, it must be translationally silenced before its incorporation into germinal granules. Once within germinal granules, *nanos* is stable for many months.

5. Conclusions

Our data are consistent with a model where multiple copies of Hermes/Rbpms specifically bind *nanos* RNA in the nucleus of stage I oocytes. Hermes/Rbpms may require a co-factor to bind RNA efficiently as it contains only one RRM, perhaps not sufficient by itself to bind RNA (Figure 2; [48]). Nuclear Hermes/Rbpms recruits other proteins to the *nanos* RNP particle and these may be Rbm42b and XveloFL. The Hermes/Rbpms complex does not bind *Vg1* or *VegT* RNA. Upon exiting the nucleus, these proteins prevent *nanos* from being translated. The *nanos* RNP particle diffuses through the cytoplasm and is trapped on an ER component (Xpat?) within the MC [6,58]. Within the MC, other RNA binding proteins likely associate, but in the final germinal granule, Hermes/Rbpms is the only protein identified to date. Dynamic remodeling of the *nanos* RNP particle as it moves to the germ plasm and is incorporated within germinal granules is an important concept. Additional studies will be required to clarify the endogenous *nanos* RNP particle during its different localization steps that culminate in the germinal granule.

Acknowledgments: We would like to thank the Krieg lab for providing two deletion mutants and a HA tagged Hermes/Rbpms construct [48]. The anti-hnRNP I antibody was a generous gift from K. Mowry (Brown University, RI, USA). Nancy Standart (University of Cambridge, UK) kindly provided the anti-Vg1RBP/Vera antibody. We thank Karen Newman for her expert technical assistance and Kenneth Dunner Jr. at the MD Anderson Cancer Center for EM technical service (supported by Core Grant CA16672). This work was supported by NIH grant GM102397 to M.L.K. and an AHA postdoctoral fellowship to Y.Z. (0525528B).

Author Contributions: M.L.K. and E.H. conceived and designed the experiments; T.A., Y.Z., M.K., P.C., and E.H. performed the experiments; T.A. and M.L.K. analyzed the data; M.K., P.C., E.H. contributed reagents/materials/analysis tools; M.L.K. wrote the paper.

Conflicts of Interest: The authors declare no conflict of interest.

References

1. Heasman, J.; Quarmby, J.; Wylie, C.C. The mitochondrial cloud of *Xenopus* oocytes: The source of germinal granule material. *Dev. Biol.* **1984**, *105*, 458–469. [CrossRef]
2. Gard, D.L.; Cha, B.J.; King, E. The organization and animal-vegetal asymmetry of cytokeratin filaments in stage VI *Xenopus* oocytes is dependent upon F-actin and microtubules. *Dev. Biol.* **1997**, *184*, 95–114. [CrossRef] [PubMed]
3. Kloc, M.; Dougherty, M.T.; Bilinski, S.; Chan, A.P.; Brey, E.; King, M.L.; Patrick, C.W., Jr.; Etkin, L.D. Three-dimensional ultrastructural analysis of RNA distribution within germinal granules of *Xenopus*. *Dev. Biol.* **2002**, *241*, 79–93. [CrossRef] [PubMed]
4. Zhou, Y.; King, M.L. Localization of Xcat-2 RNA, a putative germ plasm component, to the mitochondrial cloud in *Xenopus* stage I oocytes. *Development* **1996**, *122*, 2947–2953. [PubMed]
5. Kloc, M.; Bilinski, S.; Dougherty, M.T.; Brey, E.M.; Etkin, L.D. Formation, architecture and polarity of female germline cyst in *Xenopus*. *Dev. Biol.* **2004**, *266*, 43–61. [CrossRef] [PubMed]
6. Chang, P.; Torres, J.; Lewis, R.A.; Mowry, K.L.; Houliston, E.; King, M.L. Localization of RNAs to the mitochondrial cloud in *Xenopus* oocytes through entrapment and association with endoplasmic reticulum. *Mol. Biol. Cell* **2004**, *15*, 4669–4681. [CrossRef] [PubMed]

7. Heinrich, B.; Deshler, J.O. RNA localization to the Balbiani body in *Xenopus* oocytes is regulated by the energy state of the cell and is facilitated by kinesin II. *RNA* **2009**, *15*, 524–536. [CrossRef] [PubMed]
8. Wilk, K.; Bilinski, S.; Dougherty, M.T.; Kloc, M. Delivery of germinal granules and localized RNAs via the messenger transport organizer pathway to the vegetal cortex of *Xenopus* oocytes occurs through directional expansion of the mitochondrial cloud. *Int. J. Dev. Biol.* **2005**, *49*, 17–21. [CrossRef] [PubMed]
9. Tada, H.; Mochii, M.; Orii, H.; Watanabe, K. Ectopic formation of primordial germ cells by transplantation of the germ plasm: Direct evidence for germ cell determinant in *Xenopus*. *Dev. Biol.* **2012**, *371*, 86–93. [CrossRef] [PubMed]
10. Kloc, M.; Bilinski, S.; Chan, A.P.; Allen, L.H.; Zearfoss, N.R.; Etkin, L.D. RNA localization and germ cell determination in *Xenopus*. *Int. Rev. Cytol.* **2001**, *203*, 63–91. [PubMed]
11. Houston, D.W.; King, M.L. Germ plasm and molecular determinants of germ cell fate. *Curr. Top. Dev. Biol.* **2000**, *50*, 155–181. [PubMed]
12. Houston, D.W.; King, M.L. A critical role for Xdazl, a germ plasm-localized RNA, in the differentiation of primordial germ cells in *Xenopus*. *Development* **2000**, *127*, 447–456. [PubMed]
13. Lai, F.; Zhou, Y.; Luo, X.; Fox, J.; King, M.L. Nanos1 functions as a translational repressor in the *Xenopus* germline. *Mech. Dev.* **2011**, *128*, 153–163. [CrossRef] [PubMed]
14. Lai, F.; Singh, A.; King, M.L. *Xenopus* Nanos1 is required to prevent endoderm gene expression and apoptosis in primordial germ cells. *Development* **2012**, *139*, 1476–1486. [CrossRef] [PubMed]
15. Forbes, A.; Lehmann, R. Nanos and Pumilio have critical roles in the development and function of Drosophila germline stem cells. *Development* **1998**, *125*, 679–690. [PubMed]
16. Kobayashi, S.; Yamada, M.; Asaoka, M.; Kitamura, T. Essential role of the posterior morphogen nanos for germline development in Drosophila. *Nature* **1996**, *380*, 708–711. [CrossRef] [PubMed]
17. Tsuda, M.; Sasaoka, Y.; Kiso, M.; Abe, K.; Haraguchi, S.; Kobayashi, S.; Saga, Y. Conserved role of nanos proteins in germ cell development. *Science* **2003**, *301*, 1239–1241. [CrossRef] [PubMed]
18. Wang, Z.; Lin, H. Nanos maintains germline stem cell self-renewal by preventing differentiation. *Science* **2004**, *303*, 2016–2019. [CrossRef] [PubMed]
19. Yisraeli, J.K.; Sokol, S.; Melton, D.A. A two-step model for the localization of maternal mRNA in *Xenopus* oocytes: Involvement of microtubules and microfilaments in the translocation and anchoring of Vg1 mRNA. *Development* **1990**, *108*, 289–298. [PubMed]
20. Betley, J.N.; Heinrich, B.; Vernos, I.; Sardet, C.; Prodon, F.; Deshler, J.O. Kinesin II mediates Vg1 mRNA transport in *Xenopus* oocytes. *Curr. Biol.* **2004**, *14*, 219–224. [CrossRef] [PubMed]
21. Yoon, Y.J.; Mowry, K.L. *Xenopus* Staufen is a component of a ribonucleoprotein complex containing Vg1 RNA and kinesin. *Development* **2004**, *131*, 3035–3045. [CrossRef] [PubMed]
22. Zhang, J.; Houston, D.W.; King, M.L.; Payne, C.; Wylie, C.; Heasman, J. The role of maternal VegT in establishing the primary germ layers in *Xenopus* embryos. *Cell* **1998**, *94*, 515–524. [CrossRef]
23. Joseph, E.M.; Melton, D.A. Mutant Vg1 ligands disrupt endoderm and mesoderm formation in *Xenopus* embryos. *Development* **1998**, *125*, 2677–2685. [PubMed]
24. Chan, A.P.; Kloc, M.; Etkin, L.D. fatvg encodes a new localized RNA that uses a 25-nucleotide element (FVLE1) to localize to the vegetal cortex of *Xenopus* oocytes. *Development* **1999**, *126*, 4943–4953. [PubMed]
25. Zearfoss, N.R.; Chan, A.P.; Wu, C.F.; Kloc, M.; Etkin, L.D. Hermes is a localized factor regulating cleavage of vegetal blastomeres in *Xenopus* laevis. *Dev. Biol.* **2004**, *267*, 60–71. [CrossRef] [PubMed]
26. Claussen, M.; Horvay, K.; Pieler, T. Evidence for overlapping, but not identical, protein machineries operating in vegetal RNA localization along early and late pathways in *Xenopus* oocytes. *Development* **2004**, *131*, 4263–4273. [CrossRef] [PubMed]
27. Kloc, M.; Zearfoss, N.R.; Etkin, L.D. Mechanisms of subcellular mRNA localization. *Cell* **2002**, *108*, 533–544. [CrossRef]
28. King, M.L.; Messitt, T.J.; Mowry, K.L. Putting RNAs in the right place at the right time: RNA localization in the frog oocyte. *Biol. Cell* **2005**, *97*, 19–33. [CrossRef] [PubMed]
29. Betley, J.N.; Frith, M.C.; Graber, J.H.; Choo, S.; Deshler, J.O. A ubiquitous and conserved signal for RNA localization in chordates. *Curr. Biol.* **2002**, *12*, 1756–1761. [CrossRef]
30. Bubunenko, M.; Kress, T.L.; Vempati, U.D.; Mowry, K.L.; King, M.L. A consensus RNA signal that directs germ layer determinants to the vegetal cortex of *Xenopus* oocytes. *Dev. Biol.* **2002**, *248*, 82–92. [CrossRef] [PubMed]

31. Lewis, R.A.; Kress, T.L.; Cote, C.A.; Gautreau, D.; Rokop, M.E.; Mowry, K.L. Conserved and clustered RNA recognition sequences are a critical feature of signals directing RNA localization in *Xenopus* oocytes. *Mech. Dev.* **2004**, *121*, 101–109. [CrossRef] [PubMed]
32. Mowry, K.L. Complex formation between stage-specific oocyte factors and a *Xenopus* mRNA localization element. *Proc. Natl. Acad. Sci. USA* **1996**, *93*, 14608–14613. [CrossRef] [PubMed]
33. Deshler, J.O.; Highett, M.I.; Schnapp, B.J. Localization of *Xenopus* Vg1 mRNA by Vera protein and the endoplasmic reticulum. *Science* **1997**, *276*, 1128–1131. [CrossRef] [PubMed]
34. Gautreau, D.; Cote, C.A.; Mowry, K.L. Two copies of a subelement from the Vg1 RNA localization sequence are sufficient to direct vegetal localization in *Xenopus* oocytes. *Development* **1997**, *124*, 5013–5020. [PubMed]
35. Deshler, J.O.; Highett, M.I.; Abramson, T.; Schnapp, B.J. A highly conserved RNA-binding protein for cytoplasmic mRNA localization in vertebrates. *Curr. Biol.* **1998**, *8*, 489–496. [CrossRef]
36. Schwartz, S.P.; Aisenthal, L.; Elisha, Z.; Oberman, F.; Yisraeli, J.K. A 69-kDa RNA-binding protein from *Xenopus* oocytes recognizes a common motif in two vegetally localized maternal mRNAs. *Proc. Natl. Acad. Sci. USA* **1992**, *89*, 11895–11899. [CrossRef] [PubMed]
37. Elisha, Z.; Havin, L.; Ringel, I.; Yisraeli, J.K. Vg1 RNA binding protein mediates the association of Vg1 RNA with microtubules in *Xenopus* oocytes. *EMBO J.* **1995**, *14*, 5109–5114. [PubMed]
38. Kwon, S.; Abramson, T.; Munro, T.P.; John, C.M.; Kohrmann, M.; Schnapp, B.J. UUCAC- and vera-dependent localization of VegT RNA in *Xenopus* oocytes. *Curr. Biol.* **2002**, *12*, 558–564. [CrossRef]
39. Cote, C.A.; Gautreau, D.; Denegre, J.M.; Kress, T.L.; Terry, N.A.; Mowry, K.L. A *Xenopus* protein related to hnRNP I has a role in cytoplasmic RNA localization. *Mol. Cell* **1999**, *4*, 431–437. [CrossRef]
40. Wilhelm, J.E.; Hilton, M.; Amos, Q.; Henzel, W.J. Cup is an eIF4E binding protein required for both the translational repression of oskar and the recruitment of Barentsz. *J. Cell Biol.* **2003**, *163*, 1197–1204. [CrossRef] [PubMed]
41. Zhao, W.M.; Jiang, C.; Kroll, T.T.; Huber, P.W. A proline-rich protein binds to the localization element of *Xenopus* Vg1 mRNA and to ligands involved in actin polymerization. *EMBO J.* **2001**, *20*, 2315–2325. [CrossRef] [PubMed]
42. Kress, T.L.; Yoon, Y.J.; Mowry, K.L. Nuclear RNP complex assembly initiates cytoplasmic RNA localization. *J. Cell Biol.* **2004**, *165*, 203–211. [CrossRef] [PubMed]
43. Nakamura, A.; Sato, K.; Hanyu-Nakamura, K. Drosophila cup is an eIF4E binding protein that associates with Bruno and regulates oskar mRNA translation in oogenesis. *Dev. Cell* **2004**, *6*, 69–78. [CrossRef]
44. Farina, K.L.; Singer, R.H. The nuclear connection in RNA transport and localization. *Trends Cell Biol.* **2002**, *12*, 466–472. [CrossRef]
45. Kloc, M.; Bilinski, S.; Pui-Yee Chan, A.; Etkin, L.D. The targeting of Xcat2 mRNA to the germinal granules depends on a cis-acting germinal granule localization element within the 3′UTR. *Dev. Biol.* **2000**, *217*, 221–229. [CrossRef] [PubMed]
46. Zhou, Y.; King, M.L. RNA transport to the vegetal cortex of *Xenopus* oocytes. *Dev. Biol.* **1996**, *179*, 173–183. [CrossRef] [PubMed]
47. Gerber, W.V.; Yatskievych, T.A.; Antin, P.B.; Correia, K.M.; Conlon, R.A.; Krieg, P.A. The RNA-binding protein gene, hermes, is expressed at high levels in the developing heart. *Mech. Dev.* **1999**, *80*, 77–86. [CrossRef]
48. Gerber, W.V.; Vokes, S.A.; Zearfoss, N.R.; Krieg, P.A. A role for the RNA-binding protein, hermes, in the regulation of heart development. *Dev. Biol.* **2002**, *247*, 116–126. [CrossRef] [PubMed]
49. Song, H.W.; Cauffman, K.; Chan, A.P.; Zhou, Y.; King, M.L.; Etkin, L.D.; Kloc, M. Hermes RNA-binding protein targets RNAs-encoding proteins involved in meiotic maturation, early cleavage, and germline development. *Differentiation* **2007**, *75*, 519–528. [CrossRef] [PubMed]
50. Gietz, R.D.; Schiestl, R.H. High-efficiency yeast transformation using the LiAc/SS carrier DNA/PEG method. *Nat. Protoc.* **2007**, *2*, 31–34. [CrossRef] [PubMed]
51. Beckhelling, C.; Chang, P.; Chevalier, S.; Ford, C.; Houliston, E. Pre-M phase-promoting factor associates with annulate lamellae in *Xenopus* oocytes and egg extracts. *Mol. Biol. Cell* **2003**, *14*, 1125–1137. [CrossRef] [PubMed]
52. Farazi, T.A.; Leonhardt, C.S.; Mukherjee, N.; Mihailovic, A.; Li, S.; Max, K.E.; Meyer, C.; Yamaji, M.; Cekan, P.; Jacobs, N.C.; *et al.* Identification of the RNA recognition element of the RBPMS family of RNA-binding proteins and their transcriptome-wide mRNA targets. *RNA* **2014**, *20*, 1090–1102. [CrossRef] [PubMed]

53. Nijjar, S.; Woodland, H.R. Protein interactions in *Xenopus* germ plasm RNP particles. *PLoS ONE* **2013**, *8*, e80077. [CrossRef] [PubMed]
54. Voeltz, G.K.; Ongkasuwan, J.; Standart, N.; Steitz, J.A. A novel embryonic poly (A) binding protein, ePAB, regulates mRNA deadenylation in *Xenopus* egg extracts. *Genes Dev.* **2001**, *15*, 774–788. [CrossRef] [PubMed]
55. Caputi, M.; Mayeda, A.; Krainer, A.R.; Zahler, A.M. hnRNP A/B proteins are required for inhibition of HIV-1 pre-mRNA splicing. *EMBO J.* **1999**, *18*, 4060–4067. [CrossRef] [PubMed]
56. Zhou, Y.; King, M.L. Sending RNAs into the future: RNA localization and germ cell fate. *IUBMB Life* **2004**, *56*, 19–27. [PubMed]
57. Sonoda, J.; Wharton, R.P. Recruitment of Nanos to hunchback mRNA by Pumilio. *Genes Dev.* **1999**, *13*, 2704–2712. [CrossRef] [PubMed]
58. Machado, R.J.; Moore, W.; Hames, R.; Houliston, E.; Chang, P.; King, M.L.; Woodland, H.R. *Xenopus* Xpat protein is a major component of germ plasm and may function in its organisation and positioning. *Dev. Biol.* **2005**, *287*, 289–300. [CrossRef] [PubMed]
59. Nijjar, S.; Woodland, H.R. Localisation of RNAs into the germ plasm of vitellogenic *Xenopus* oocytes. *PLoS ONE* **2013**, *8*, e61847. [CrossRef] [PubMed]
60. Lewis, R.A.; Gagnon, J.A.; Mowry, K.L. PTB/hnRNP I is required for RNP remodeling during RNA localization in *Xenopus* oocytes. *Mol. Cell. Biol.* **2008**, *28*, 678–686. [CrossRef] [PubMed]
61. Bontems, F.; Stein, A.; Marlow, F.; Lyautey, J.; Gupta, T.; Mullins, M.C.; Dosch, R. Bucky ball organizes germ plasm assembly in zebrafish. *Curr. Biol.* **2009**, *19*, 414–422. [CrossRef] [PubMed]
62. Hornberg, H.; van Wollerton Horck, F.; Maurus, D.; Zwart, M.; Svoboda, H.; Harris, W.A.; Holt, C.E. RNA-binding protein Hermes/RBPMS inversely affects synapse density and axon arbor formation in retinal ganglion cells *in vivo*. *J. Neurosci.* **2013**, *33*, 10384–10395. [CrossRef] [PubMed]
63. Furukawa, M.T.; Sakamoto, H.; Inoue, K. Interaction and colocalization of HERMES/RBPMS with NonO, PSF, and G3BP1 in neuronal cytoplasmic RNP granules in mouse retinal line cells. *Genes Cells* **2015**, *20*, 257–266. [CrossRef] [PubMed]
64. Fu, J.; Cheng, L.; Wang, Y.; Yuan, P.; Xu, X.; Ding, L.; Zhang, H.; Jiang, K.; Song, H.; Chen, Z.; *et al.* The RNA-binding protein RBPMS1 represses AP-1 signaling and regulates breast cancer cell proliferation and migration. *Biochim. Biophys. Acta* **2015**, *1853*, 1–13. [CrossRef] [PubMed]
65. MacArthur, H.; Bubunenko, M.; Houston, D.W.; King, M.L. Xcat2 RNA is a translationally sequestered germ plasm component in *Xenopus*. *Mech. Dev.* **1999**, *84*, 75–88. [CrossRef]
66. Luo, X.; Nerlick, S.; An, W.; King, M.L. *Xenopus* germline nanos1 is translationally repressed by a novel structure-based mechanism. *Development* **2011**, *138*, 589–598. [CrossRef] [PubMed]

2

Roles of the Hedgehog Signaling Pathway in Epidermal and Hair Follicle Development, Homeostasis, and Cancer

Yoshinori Abe and **Nobuyuki Tanaka** *

Department of Molecular Oncology, Institute for Advanced Medical Sciences, Nippon Medical School, 1-396 Kosugi-machi, Nakahara-ku, Kawasaki 211-8533, Japan; yoshiabe@nms.ac.jp
* Correspondence: nobuta@nms.ac.jp

Abstract: The epidermis is the outermost layer of the skin and provides a protective barrier against environmental insults. It is a rapidly-renewing tissue undergoing constant regeneration, maintained by several types of stem cells. The Hedgehog (HH) signaling pathway is one of the fundamental signaling pathways that contributes to epidermal development, homeostasis, and repair, as well as to hair follicle development and follicle bulge stem cell maintenance. The HH pathway interacts with other signal transduction pathways, including those activated by Wnt, bone morphogenetic protein, platelet-derived growth factor, Notch, and ectodysplasin. Furthermore, aberrant activation of HH signaling is associated with various tumors, including basal cell carcinoma. Therefore, an understanding of the regulatory mechanisms of the HH signaling pathway is important for elucidating fundamental mechanisms underlying both organogenesis and carcinogenesis. In this review, we discuss the role of the HH signaling pathway in the development and homeostasis epidermis and hair follicles, and in basal cell carcinoma formation, providing an update of current knowledge in this field.

Keywords: Hedgehog signaling; epidermis; hair follicle; development; homeostasis; basal cell carcinoma; signaling crosstalk

---

1. Introduction

Although the thickness of the skin is as little as 2 mm on average, it is a complex stratified organ consisting of an outer epidermis and appendages (for example, hair follicles, sebaceous glands, and sweat glands) separated from the underlying dermis by a basement membrane. It is the largest organ in the human body, by area and weight. Skin is the primary barrier that protects the body from major environmental stresses, such as dehydration, scratches, wounds, and microbial infection [1]. To repair damaged tissue and replace old cells, different stem cell pools resident in the skin contribute to the maintenance and repair of the various epidermal tissues, including interfollicular epidermis, hair follicles, sebaceous glands, and sensory touch domes [2,3]. The stem cells residing in the epidermis are maintained by various signaling pathways such as those activated by Wnt, transforming growth factor β (TGF-β), Notch, and Hedgehog (HH) [3]. These signaling pathways also have important roles in embryonic skin development [4]. However, the dysregulation of signaling pathways related to skin development and homeostasis induces epigenetic reprograming of skin cells, resulting in cancer development. In particular, aberrant activation of HH signaling is the major cause of basal cell carcinoma (BCC) [5]. Studies of tumorigenesis mediated by HH signaling activation have recently culminated in the USA Food and Drug Administration's (FDA) approval of GDC-0449 (vismodegib) [6], an oral inhibitor of the HH signaling mediator, Smoothened (SMO), as a therapeutic for treating advanced or metastatic BCC. In this review, we provide a current understanding of the role of the HH signaling pathway in embryonic skin development, skin homeostasis, and skin cancer.

2. The HH Signaling Pathway

The HH pathway is one of the most important signal transduction pathways in tissue development, homeostasis, and repair. It regulates the morphogenesis of various organs during embryogenesis [7]. We show a schematic diagram of the HH signaling pathway in Figure 1. However, HH signaling is complex, and several comprehensive reviews have been published describing the molecular mechanisms in detail [8–11]. In the canonical pathway, HH signaling is initiated by one of three spatiotemporally-onfined ligands: Sonic Hedgehog (SHH), Indian Hedgehog (IHH), and Desert Hedgehog (DHH). HH ligands control numerous developmental outcomes in a concentration- and duration-dependent manner. Each HH ligand has distinct spatial and temporal expression patterns, but all activate HH signaling by binding to Patched (PTCH) 1 or 2, which are 12-pass transmembrane-spanning receptors. PTCH1 is considered to be the primary receptor for HH ligands. In the absence of HH, PTCH1 is localized in primary cilia and constitutively suppresses the activity of SMO, a seven-pass transmembrane-spanning protein that is a member of the G-protein-coupled receptor superfamily [12]. Suppressor of fused (SUFU) is a key negative regulator of the HH signaling pathway [13]. In the absence of HH ligands, SUFU inhibits signaling by sequestration of GLI proteins in the cytoplasm (Figure 1A). In addition to PTCH, other HH-binding cell surface proteins have been identified, such as CAM-related/down-regulated by oncogenes (CDO), brother of Cdo (BOC) and growth-arrest-specific 1 (GAS1) [14,15]. These molecules function as HH co-receptors to facilitate HH signal reception. Following the binding of a HH ligand to PTCH, SMO accumulates at a specialized compartment in the primary cilium, called the EvC zone, mediated by the Ellis van Creveld Syndrome (EVC)–EVC2 complex [16,17]. From here, SMO transmits signals that activate GLI (Figure 1B) [18,19].

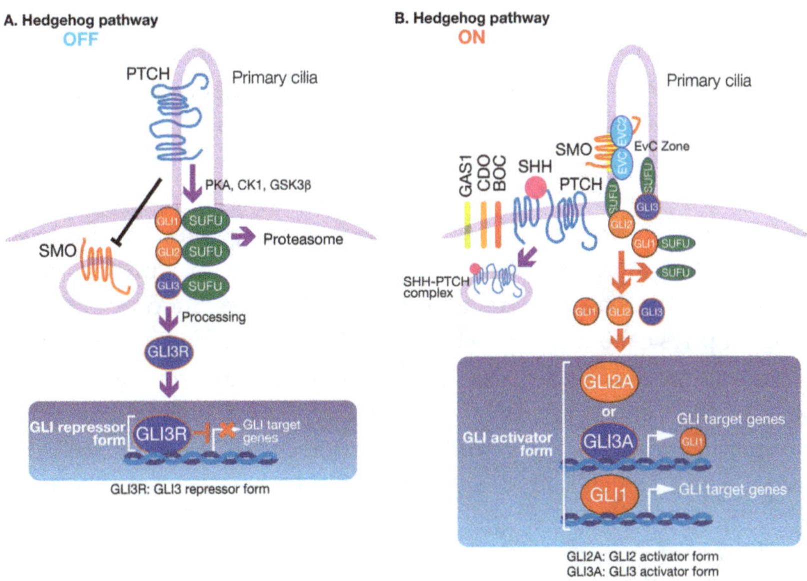

Figure 1. The Hedgehog signaling pathway. (**A**) In the absence of Hedgehog (HH) ligand, Patched (PTCH) blocks the ciliary localization of Smoothened (SMO), and a repressor form of GLI (mainly GLI3R) suppresses the induction of GLI target genes. (**B**) Signaling is activated in the presence of HH ligand. The binding of HH ligand to PTCH prevents PTCH inhibition of SMO, and SMO then interacts with the Ellis van Creveld Syndrome (EVC)–EVC2 complex and translocates into primary cilia to be fully activated. SMO then activates GLI family transcription factors, mainly GLI2. GLI2 up-regulates the expression of GLI1, as well as GLI target genes. GLI1 is also activated downstream of SMO. Activated GLI2 (GLI2A) and GLI1 further upregulate the expression of various GLI target genes. This figure has been adapted from our previous work [20].

Graded levels of HH signaling trigger the expression of different sets of response genes, depending on the ratio of GLI activator (GLI-A) to repressor (GLI-R) forms. In vertebrates, the GLI family consists of three proteins, GLI1, GLI2, and GLI3 [10]. All GLI proteins contain an activator domain at their C-terminus; GLI2 and GLI3 also have an N-terminal repressor domain [21]. Studies in mutant mice suggest that GLI2 is the major activator of HH signaling [22], whereas GLI3 is the major repressor [23,24]. GLI1 most likely serves as a signal amplifier downstream of GLI2 [22,25]. When HH signaling is activated, SUFU–GLI2 complexes dissociate and GLI2 is activated [26]. *Gli2* knockout mice die at birth, whereas *Gli1* knockout mice show normal development, unless one copy of *Gli2* is also defective [27]. Interestingly, experiments in mutant mice further suggest that Gli2 can rescue Gli1 protein function, whereas *Gli1* knock-in into the *Gli2* allele can rescue the *Gli2*-null phenotype [28].

Constitutive activation of HH signaling has been observed in many cancers [29] and promotes cancer cell proliferation, metastasis, and cancer stem cell maintenance. For example, somatic mutations of *PTCH1* and *SMO* have been identified in patients with BCC and medulloblastoma [30–33]. Other mutations in genes encoding HH pathway components have been reported, including *SUFU* mutations in BCC and medulloblastoma [34]. Furthermore, multiple mechanisms of HH pathway activation in cancer have been proposed, and reviewed in detail elsewhere [10,11,35].

HH signaling exhibits crosstalk with other signaling pathways, including Wnt [36–39], TGF-β [40], epidermal growth factor receptor (EGFR) [41], rat sarcoma viral oncogene homolog GTPase (RAS)–extracellular signal-regulated kinase (ERK) [42,43], phosphatidylinositol 3-kinase (PI3K)–Akt–mechanistic target of rapamycin (mTOR) [44], and Notch [45]. Interestingly, the crosstalk between HH signaling and other pathways seem to be different between developmental, homeostatic, and carcinogenic processes.

3. Overview of Epidermal and Hair Follicle Development

3.1. Development of the Epidermis

During embryonic development, the epidermis is derived from the ectoderm. In mice, after gastrulation, the embryo surface consists of a single layer of neuroectoderm, which will form the nervous system and skin epithelium from embryonic day 9.5 (E9.5) to E12.5. Mesenchymal cells from the underlying layer transmit signals that instruct the stratification of the epidermis and dictate the positioning of downgrowths that mark the initiation of hair follicle morphogenesis [1]. In alliance with the mesenchyme, the inner-most, or basal layer of the stratifying epidermis produces and organizes an underlying basement membrane that is rich in extracellular matrix proteins and growth factors. The epidermis adheres to this basement membrane, which serves not only as a growth-promoting platform but also as a physical boundary between the epithelium and the dermis [46]. During the initial steps of stratification (E12.5–E15.5), cell proliferation is almost completely confined to the basal layer, and these cells differentiate immediately. In this period, the single layer of epidermal cells is covered by a transient protective layer called the periderm. Although there is regional variation, the stratification program is largely complete around E17.5 [47].

3.2. Hair Follicle Development

The development of hair follicles is intrinsically related to the stratification of the embryonic epidermis. Hair follicle development is the result of epithelial–mesenchymal interactions. Although hair follicle development occurs in eight morphologically distinct stages, it can more simply be divided into three phases: induction, organogenesis and cytodifferentiation [48]. In the induction phase, which is the initiation of hair follicle development (E13), signaling crosstalk between the epidermis and dermis is induced, and epidermal keratinocytes form clusters that become enlarged and elongated to form hair placodes (E14). Hair follicle development then enters the organogenesis phase. A cluster of specialized fibroblasts, the precursors of the dermal papilla, is formed just under the placode, and the signaling crosstalk between these specialized fibroblasts and the epithelial placode leads to cell proliferation in both structures. The enhanced cell proliferation leads to a downward growth of the epidermal component, shaping the dermal papilla. The resulting structure, called the hair germ, is typically observed at E15.5. Keratinocyte downgrowth continues, whereby the most proximally located keratinocytes begin to wrap around the dermal papilla (E16.5–E17.5). As follicles enter the cytodifferentiation phase, the inner root sheath is formed, triggering the terminal differentiation of keratinocytes above the dermal papilla to generate the hair shaft. Simultaneously, the outer root sheath starts to form a cylinder around the inner root sheath to create a bulbous peg structure (E18.5). Additional appendages such as sebaceous glands and touch domes also develop during this phase. These timelines are summarized in Figure 2. For a more detailed discussion of hair follicle development, please refer to several published reviews [2,4,49].

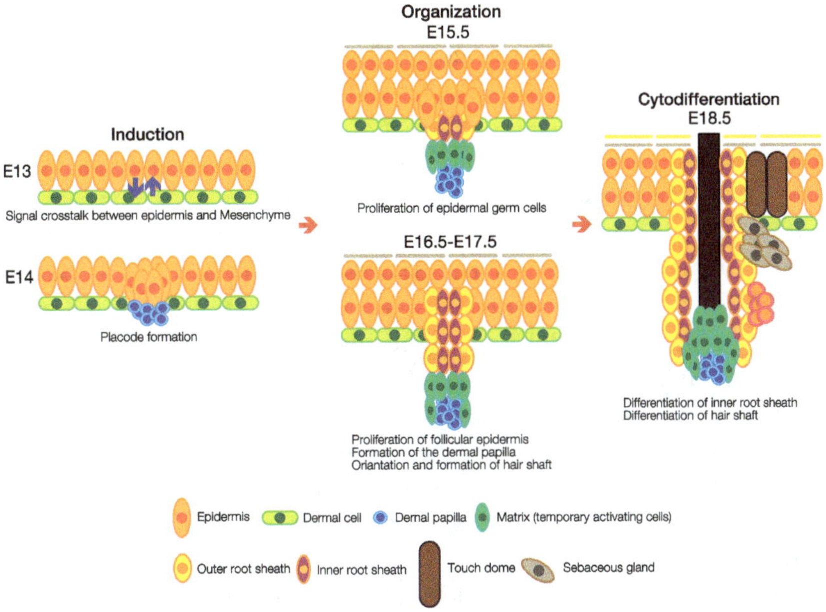

Figure 2. At the beginning of hair follicle development, gradients of activators and inhibitors generate a well-circumscribed inductive field (E13). These gradients induce epithelial and dermal cell proliferation to develop a placode and a dermal papilla (E14). The dermal papilla sends growth signals back to the epithelial cells to form the hair follicle (E15.5–E17.5). In the cytodifferentiation phase, epithelial cells differentiate into inner and outer root sheath, which generates the hair shaft. In addition, melanocytes and hematopoietic cells migrate into the hair follicle, and touch domes and sebaceous glands are formed.

4. The Roles of HH Signaling in Hair Follicle and Touch Dome Development

In mouse skin at E15.5, at the beginning of the organogenesis phase of hair follicle development, *Ptch1*, *Smo*, *Gli1* and *Gli2* are expressed in both the epithelial and dermal components of the early hair follicle, although higher levels of *Ptch1* and *Gli1* are observed in the follicular dermal mesenchyme. In contrast, Ptch2 expression is restricted to epithelial cells of the hair bud where it is co-expressed with *Shh* [47,50]. Therefore, the HH signaling pathway functions in both the epithelium and the mesenchyme. In mice lacking *Shh*, hair follicle formation is initiated and specialized fibroblasts are condensed, indicating that the initial events of hair follicle development are independent of HH signaling. However, in the hair follicle maturation phase, the loss of *Shh* selectively reduces keratinocyte proliferation in the developing follicle epithelium, so that the downgrowth of epithelial cells fails. Therefore, the number of hair follicles is reduced in $Shh^{-/-}$ mice. Thus, HH signaling is necessary for the transition from the induction to the organogenesis phase, and for the downgrowth of epithelial cells after the induction phase. Furthermore, the maturation of the dermal papilla also depends on HH signaling, since $Shh^{-/-}$ mice have smaller dermal papillae [50,51]. It is believed that GLI2 is the primary transducer of HH signaling, whereas GLI1, the expression of which is transcriptionally regulated by GLI2, plays a secondary role in potentiating the HH response [10]. Mill et al. demonstrated that hair follicle maturation fails in *Gli2*$^{-/-}$ skin because of suppressed epithelial cell proliferation, similar to *Shh*$^{-/-}$ skin [52]. Development of the dermal papilla is also severely affected by the failure to respond to SHH in *Gli2*$^{-/-}$ skin. In addition, *Shh*$^{-/-}$, as well as *Gli2*$^{-/-}$, skin fails to induce GLI1. Together, these results show that the SHH–GLI2 axis is required for the proliferation and subsequent downgrowth of the follicular epithelium, and for the maturation of the dermal papilla.

As described above, the expression of *Ptch1*, *Smo*, *Gli1*, and *Gli2* is seen in the dermal papilla [50], indicating that HH signaling also functions in dermal cells. RNAi-mediated SMO knockdown in dermal cells results in the loss of the dermal papilla precursor, the dermal cell condensate, and in hair follicle arrest at the organogenesis phase, similar to *Shh*$^{-/-}$ skin. However, HH signaling in dermal cells does not affect cell survival or epithelial proliferation. Rather, dermal HH signaling regulates the expression of a subset of dermal papilla-specific signature genes. These results suggest that dermal HH signaling regulates genes required to maintain the maturation of the papilla [53]. Sex-determining region Y-box 2 (SOX2) is expressed only in the dermal papilla and determines hair follicle type [54,55]. Recently, Driskel et al. demonstrated that dermal papilla cells are heterogenous and SOX2^{+}/CD133^{+} dermal papilla cells are associated with guard, awl, and auchene hair follicle formation whereas SOX2^{-}/CD133^{+} dermal papilla cells are involved in the formation of zigzag hair follicles. Moreover, they demonstrated the HH signaling pathway is activated in SOX2^{-}/CD133^{+} dermal papilla cells, suggesting that HH signaling regulates hair type fate determination through the maintenance of dermal papilla cell heterogeneity [55].

Touch domes are unique epidermal sensory structures that form exclusively around primary hair follicles [56,57]. Merkel cells are the mechanoreceptors found in touch domes. Merkel cells in the epidermis can be identified based on their expression of atonal bHLH transcription factor 1 (ATOH1), SOX2, and Keratin 8 [58]. Recently, it has been revealed that the hair follicle functions as a niche required for Merkel cells formation. In *Shh*-null skin, Merkel cells are absent, and SHH overexpression results in increased proliferation of Merkel cells. HH signaling pathway activation, which is required for Merkel cell formation, is initiated by the production of SHH in developing hair follicles. Therefore, HH signaling pathway is required for Merkel cell formation, resulting in touch dome development [59,60]. The roles of the HH signaling pathway in hair follicle and touch dome development are summarized in Figure 3.

On the other hand, negative regulators of HH signaling are also important for hair follicle development. PTCH1 and 2 are known to be negative regulators of HH signaling, and Ptch1 is a GLI target gene [47,61]. Basal cell-specific *Ptch1*-null skin exhibits hyperplasia and skin neoplasia formation, although *Ptch2*-null skin is undistinguishable from wild-type skin with respect to embryonic hair follicle development [62,63]. However, the skin of both basal cell-specific *Ptch1*-null and *Ptch2*-null mice exhibits excessive proliferation of epithelial cells, suggesting that epidermal stem and progenitor cells enter a hyperproliferative state. In addition, progenitor cells fail to differentiate, resulting in the loss of epidermal lineage specification. Therefore, PTCH1- and 2-mediated HH signaling suppression is important for regulating epidermal differentiation [64]. Another study reported that increased numbers of immature hair follicles in the induction phase are seen in *Shh*$^{-/-}$/*Gli3*$^{-/-}$ skin, in which the repressor form of GLI3 (GLI3R), another negative regulator of HH signaling, is lost. Furthermore, keratinocyte differentiation fails in *Shh*$^{-/-}$/*Gli3*$^{-/-}$ skin, resulting in the failure of hair follicle maturation [65]. Recently, Adolphe et al. discovered that there is a HH signaling gradient along the proximodistal axis of developing hair follicles [66]. At E17.5, the peak of HH signaling activity is in follicular cells, which are a source of SHH, and HH signaling is endogenously low in basal cells. This HH signaling gradient is established by PTCH1 and PTCH2. These results suggest that HH signaling has the capacity for progenitor cell maintenance, and the intensity of HH signaling is important for both epithelial and mesenchymal cell fate determination during hair follicle development.

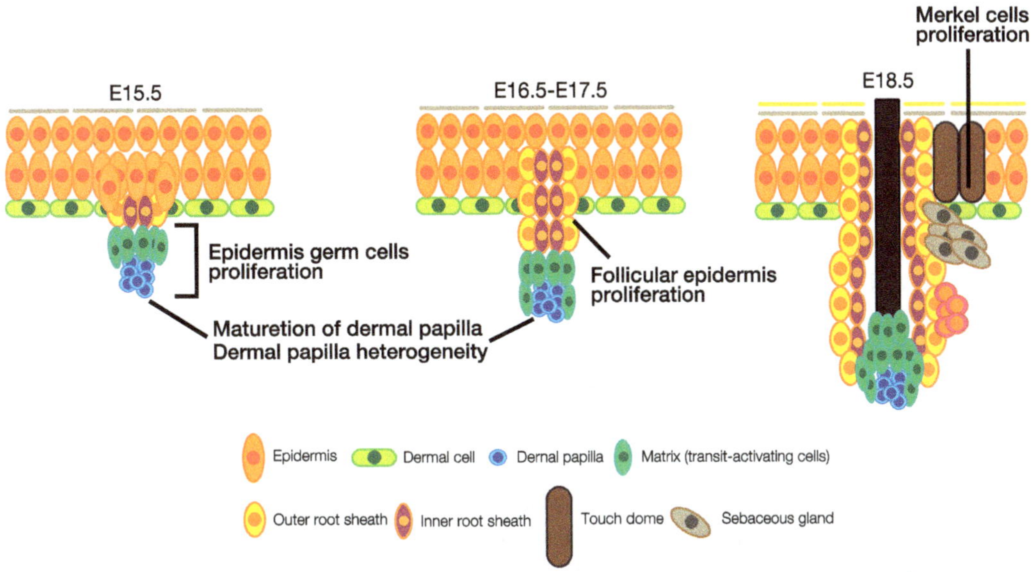

Figure 3. The roles of the HH signaling pathway in embryonic hair follicle development. In epidermal cells, HH signaling is involved in germ cell and follicular cell proliferation. HH signaling is also involved in the maturation of the dermal papilla, as well as dermal papilla heterogeneity. Furthermore, HH signaling is essential for Merkel cell proliferation and touch dome development.

5. Signaling Crosstalk between HH and Other Pathways in Hair Follicle Development

The hair follicle develops as a result of complex reciprocal signaling between epithelial and mesenchymal cells. Various signaling pathways, such as those activated by Wnt, bone morphogenetic protein (BMP), platelet-derived growth factor (PDGF), Notch and ectodysplasin, are involved in hair follicle development, and exhibit signaling crosstalk with the HH pathway. In this section, we discuss the crosstalk between HH and these other signaling pathways in hair follicle development.

5.1. Wnt Signaling

The Wnt signaling pathway has been implicated in many developmental processes [67]. Canonical Wnt signaling is mediated by a transcriptionally active complex between stabilized β-catenin and members of the T-cell factor (TCF)/Lymphoid enhancer-binding factor 1 (LEF-1) family of DNA-binding proteins [68]. One of the first, essential events required to form a follicle placode is the activation of Wnt signaling in the epidermis. Wnt signaling also transmits subsequent signals for hair follicle development and patterning. Wnt ligands are expressed in interfollicular epidermis as well as the hair follicle during all stages of follicle development, suggesting that Wnt signaling is a master regulator of follicle development [36,69,70].

The Wnt and HH signaling pathways are often associated. During hair follicle development in mouse embryos, the canonical Wnt signaling pathway is upstream of HH signaling, since β-catenin activation induces Shh expression in epidermis during skin development [36,37]. On the other hand, Reddy et al. showed that dermal Wnt5a disappears in $Shh^{-/-}$ skin, indicating that Wnt5a is downstream, or even a direct target, of HH signaling [71]. Therefore, SHH-induced dermal Wnt5a, as well as GLI1 expression, is important for dermal papilla maturation. Wnt5a suppresses expression of LEF-1, a well-recognized Wnt/β-catenin signaling target in human dermal papilla cells, indicating suppression of canonical Wnt signaling [72]. Therefore, the HH–Wnt5a axis terminates cell proliferation and induces differentiation to promote mature hair follicle formation. Since PTCH1 and 2, repressors of HH signaling, are also involved in the induction of cell differentiation, cooperation between Wnt5a and negative regulators of HH signaling may be important for cell differentiation during hair follicle development.

5.2. BMP Signaling

BMPs are members of TGF-β family. BMP signaling has been suggested to regulate hair follicle induction and the patterning of follicles within the skin by repressing the placode fate [73,74]. Suzuki et al. demonstrated crosstalk among the Wnt, HH, and BMP signaling pathways during hair follicle development [37]. Expression of constitutively active β-catenin induces aberrant keratinocyte proliferation and differentiation in mouse skin during hair follicle development; the aberrant cell differentiation is suppressed by BMP signaling inhibition, and the aberrant cell proliferation is suppressed by HH signaling inhibition. These results demonstrate that Wnt/β-catenin signaling relayed through HH and BMP signals is the principal regulatory mechanism underlying changes in hair follicle cell fate. Furthermore, activation of the HH signaling pathway induces expression of BMP2 in the organogenesis phase of hair follicle development. A possible mechanism for the regulation of hair follicle development involves cell fate being controlled by the difference in the diffusion area between promoters, such as SHH, and suppressors, such as BMP [75,76]. Therefore, the presence of signaling crosstalk involving the Wnt, HH, and BMP pathways would regulate the diffusion areas of HH and BMP ligands to determine hair follicle cell fate.

The HH signaling pathway also affects dermal cells during hair follicle development. Noggin is known as an antagonist of BMPs, and Noggin signaling is a dermal papilla-specific pathway. In *noggin*$^{-/-}$ mice, hair follicles arrest at the placode formation stage of embryonic development [73,77]. Oro et al. showed that interaction between HH signaling and Noggin regulates dermal papilla maturation during hair follicle development [53]. Dermal cell-specific RNAi-mediated SMO knockdown does not affect dermal cell proliferation in mice. However, dermal cell condensation following placode formation fails in these mice. These authors found that dermal HH signaling induces dermal papilla signature genes such as *Sox2* [54,55], *Sox18* [78,79], and *noggin*. Among these genes, noggin partially rescues dermal SMO knockdown-associated hair regeneration defects by increasing epithelial *Shh* expression. These results suggest that dermal HH signaling regulates specific dermal papilla gene expression signatures to maintain dermal papilla maturation while maintaining a reciprocal HH–Noggin signaling loop to drive embryonic hair follicle development.

5.3. PDGF Signaling

A role for PDGF-A in hair follicle development has been suggested by the injection of PDGF-receptor α (PDGFRα) antibodies into newborn mice, which perturbed hair formation [80]. Karlsson et al. demonstrated the roles of crosstalk between PDGF and HH signaling in hair follicle development [81]. *pdgf-a*-null skin exhibits thinner dermis, smaller dermal papillae and abnormal hair follicle shape. *Shh* and *pdgf-a* expression overlap during hair follicle development. Moreover, PDGFRα is expressed in dermal cells adjacent to sites of SHH expression (unpublished data referred to in [81]). In *Shh*-null skin, the aggregation of PDGFRα-expressing dermal cells beneath the epithelial cells fails, implying the failure of dermal papilla formation [81]. As described in Section 5.2, HH signaling upregulates noggin expression in dermal cells to maintain dermal papilla maturation. Gao et al. reported that HH signaling together with PDGF signaling promotes the expression and secretion of Noggin in mesenchymal cells [82]. These results suggest that the cooperation of HH and PDGF signaling is necessary for dermal papilla formation during hair follicle development.

5.4. Notch Signaling

Notch signaling plays a crucial role in cell fate determination and developmental border formation [83]. During epidermal development, Notch signaling regulates the process of commitment to either the hair follicle or the interfollicular epidermis lineage [84]. Furthermore, Notch signaling terminates the differentiation program that culminates in skin barrier formation [85]. As described above in Section 4, HH signaling is involved in the maintenance of $SOX2^-/CD133^+$ dermal papilla cells [55]. The Notch signaling pathway is also activated in these cells [55]. Previous studies demonstrated that Notch signaling induces SHH [45] and GLI2 [86,87] expression in stem cells. Therefore, Notch signaling may regulate HH signaling for the maintenance of dermal papilla cell heterogeneity.

5.5. The EdaA1–EdaR–NF-κB Axis

Ectodysplasin A1 (EdaA1) and ectodysplasin receptor (EdaR) respectively belong to the tumor necrosis factor (TNF) multigene family of ligands and receptors, and are able to activate the nuclear factor-κB (NF-κB) and Jun amino-terminal kinase (JNK)/activator protein 1 (AP1) pathway in vitro [88–91]. The EdaA1–EdaR pathway is activated at placode formation only in primary guard hair follicles. It has been reported that the EdaA1–EdaR–NF-κB axis is involved in placode formation, since NF-κB is directly activated by EdaA1–EdaR signaling via the IκB kinase (IKK) complex. Schmidt-Ullrich et al. demonstrated an interaction between the EdaA1–EdaR–NF-κB axis and the HH signaling pathway [92]. They found that EdaA1–EdaR signaling-mediated NF-κB activation is essential for the induction of SHH expression and subsequent placode downgrowth. However, NF-κB induces cyclin D1 indirectly, suggesting direct induction by HH (or Wnt) signaling. These results, taken together, suggest that, during hair follicle development, NF-κB transmits EdaA1–EdaR signaling to activate the HH signaling pathway via the direct NF-κB-mediated induction of SHH.

6. Epidermal Homeostasis

The adult epidermis consists of several layers, each with its own characteristics. Cells in the basal layer proliferate to provide new cells that replace those shed from the cornified exterior of the tissue [93]. This renewal process is called homeostasis. Epidermal homeostasis requires a balance between keratinocyte proliferation and differentiation [94,95]. In the skin, multiple populations of epidermal stem cells have a crucial role in maintaining tissue homeostasis, providing new cells to replace those that are constantly lost during tissue turnover or following injury [96]. The epidermis is composed of the interfollicular epidermis, hair follicles, sebaceous glands and sweat glands, and each of these tissues has its resident stem cells. The permanent, lower part of the hair follicle, known as the bulge, is one reservoir of stem cells. Bulge stem cells were originally identified as slow cycling cells [97,98]. A large number of markers have been described for these cells. CD34 cell surface glycoprotein [99,100] and Leu-rich repeat-containing G protein-coupled receptor 5 (LGR5) are considered to be markers of bulge stem cells [101,102]. Recent studies have reported that HH signaling is involved in the maintenance of one type of bulge stem cells. We discuss the role of the HH signaling pathway in bulge stem cell maintenance below.

7. Roles of the HH Signaling Pathway in Epidermal Homeostasis

Treatment of adult mice with an anti-SHH antibody blocks the active phase of the hair growth cycle (anagen) and blocks hair regrowth [103], indicating that SHH is required for the regenerative function of adult bulge stem cells. Conversely, exogenously-administered SHH triggers anagen onset in follicles in the resting phase of the hair cycle (telogen), and stimulates hair growth [104]. These reports suggest that HH signaling is involved in hair follicle maintenance.

In adult skin, stem cells in the hair follicle bulge cyclically regenerate the follicle, whereas a distinct stem cell population maintains the epidermis. In telogen hair follicles, GLI2 and GLI3 continue to be broadly expressed in both the follicle and the surrounding dermis. Throughout the hair cycle, GLI1 expression is restricted to two distinct epidermal stem cell domains [105]. One population of GLI1-expressing ($GLI1^+$) stem cells co-expresses LGR5 [101,102]. Stem cells expressing both GLI1 and LGR5 ($GLI1^+/LGR5^+$) exhibit greater proliferative properties and are localized to the lower portion of the bulge, close to the telogen hair germ. In this phase of the hair cycle, $GLI1^+/LGR5^+$ stem cells only partially overlap with other bulge stem cell markers, such as CD34 and keratin 15. However, in anagen hair follicles, $GLI1^+/LGR5^+$ stem cells are localized to the lower part of the outer root sheath of the reforming hair follicle, where they lose expression of the bulge stem cell markers [106]. This localization change during the hair cycle suggests that GLI1$^+$/LGR5$^+$ stem cells may also be necessary for hair follicle growth in anagen. Another population of GLI1$^+$ stem cells in the dermal papilla is involved in dermal papilla maintenance. These results show that, during anagen, the induction of HH target genes, including *Gli1*, occurs in both epidermal and dermal papilla cells adjacent to SHH-expressing cells, such as those in the follicle matrix.

Another population of GLI1$^+$ stem cells that does not express LGR5, CD34, or keratin 15 exists in the upper fringe of the bulge. GLI1$^+$ upper bulge stem cells are induced by bulgeassociated nerve cells releasing SHH. GLI1$^+$ stem cells from the upper bulge stably remain in a wounded area of the skin long after it has been repaired [105]. Therefore, upper bulge cells might constitute a long-term reservoir of stem cells poised to contribute to the repair of the interfollicular epidermis in a permanent manner. Interestingly, the HH signaling pathway is essential for the contribution of upper bulge stem cells to wound repair, but not for hair follicle cycling. This emphasizes the relevance of the niche for the function of different populations of hair follicle stem cells [105]. Matrix cells are known to be a source of SHH, and secreted SHH regulates epithelial cell proliferation and dermal papilla maturation. Moreover, SHH governs dermal adipocyte precursors in adult hair cycling. In turn, dermal adipocytes and adipocyte precursors influence hair cycle progression [107–109]. In addition, dermal adipocytes secrete antimicrobial peptides to protect the skin from bacterial infection [110]. The roles of the HH signaling pathway in skin homeostasis are summarized in Figure 4.

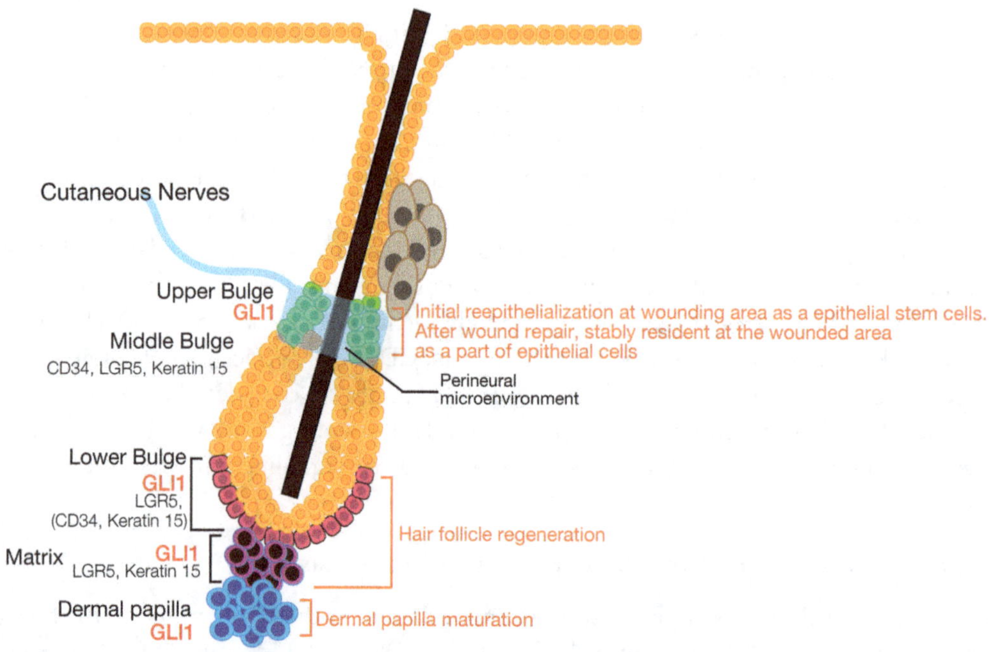

Figure 4. GLI1-expressing hair follicle stem cells and their roles. The expression of GLI1 and other markers define molecularly distinct zones in the telogen hair follicle, including regionalization of the bulge into the upper, middle, and lower bulge. GLI1-expressing cells in the upper bulge receive a HH signal from hair follicle-associated nerve cells, and are functionally distinct from cells in the middle bulge, lower bulge, and matrix in their ability to become epidermal stem cells during wound repair. The roles of each GLI1$^+$ bulge stem cell population are described in orange text on the right.

8. Signaling Crosstalk between HH and Other Signaling Pathways in Epidermal Homeostasis

8.1. Wnt Signaling

The Wnt signaling pathway is also important for the maintenance of hair follicles in adult skin. As described above, activation of canonical Wnt signaling induces SHH expression, indicating HH signaling activation during hair follicle development in mouse embryos. Previous studies have demonstrated that the Wnt–HH signaling axis is also essential for maintaining the hair cycle. Epidermal basal cell-specific β-catenin activation in adult mouse skin also induces SHH expression, resulting in HH signaling activation [38,39], and HH signaling is required for β-catenin-mediated hair follicle formation [39].

8.2. The Hippo–YAP Signaling Pathway

The Hippo signaling pathway is a master regulator of cell proliferation and organ size. Hippo pathway molecules, such as macrophage-stimulating 1/2 (MST1/2) and large tumor suppressor 1/2 (LATS1/2), regulate a transcriptional coactivator, Yes-associated protein (YAP) [111,112]. YAP controls stem or progenitor cell proliferation in the mouse postnatal epidermis [113]. Furthermore, YAP functions in balancing growth and differentiation in the skin [114,115].

Akladios et al. proposed a novel interplay among HH, YAP, Rho-associated protein kinase (ROCK) and β-catenin in the control of skin homeostasis [116]. HH signaling induces YAP activation, and activated YAP in turn induces β-catenin activation. HH, YAP, and β-catenin together induce GLI2 activation to regulate epidermal homeostasis. At the same time, HH signaling activates ROCK in the epidermis. HH signaling and activated ROCK regulate cell proliferation and actin remodeling in the dermis.

8.3. p53/p63

The tumor suppressor, p53, is expressed in the epidermal basal layer and also in the outer root sheath and matrix of the hair follicles. Its expression is increased during the transition from anagen to the regression phase of the hair cycle (catagen), suggesting its role in hair follicle involution [117]. In addition, we previously demonstrated that HH signaling enhances the ubiquitin-mediated degradation of p53 through GLI1 [118]. Therefore, it is possible that HH signaling regulates hair cycling through regulating p53 protein.

A homolog of p53, p63, has an essential role in development. Mice lacking p63 show extensive defects in epidermal development and die at birth [119,120]. In epidermis, an anti-apoptotic protein and GLI target gene product, B-cell lymphoma 2 (BCL2) [121,122], is expressed in basal keratinocytes and the outer root sheath cells of the hair follicle. Its expression is associated with undifferentiated cells having proliferative potential. On the other hand, suprabasal keratinocytes do not express BCL2 [123]. These results suggest that BCL2 has an important role in regulating keratinocyte proliferation for skin homeostasis. Chari et al. demonstrated crosstalk between ΔNp63α, an isoform of p63 that is a crucial regulator of epidermal development, and HH signaling in keratinocytes [124]. HH signaling activation inhibits keratinocyte differentiation via GLI1-mediated upregulation of BCL2. At the same time, HH signaling activation results in ΔNp63α expression. Upregulated p63 then activates transcription in keratinocytes of a negative regulator of HH signaling, SUFU. In vivo analysis indicates that SUFU is also a negative regulator of the Wnt signaling pathway [125]. Overall, these results suggest that the HH–ΔNp63α axis regulates keratinocyte fate determination (proliferation or differentiation) for skin homeostasis.

9. Aberrant HH Signaling Pathway Activation and BCC

BCCs are so named because of their histological resemblance to basal cells of the interfollicular epidermis and epidermal appendages, that is, the keratinocytes adjacent to the stroma [126,127]. Activation of the HH signaling pathway through various mechanisms has been observed in BCCs [5]. Below, we introduce the mechanisms of BCC formation via the HH signaling pathway.

Although in general, BCCs seem to have relatively stable genomes, 67% of BCCs exhibit loss of *PTCH1*, 5% exhibit loss of *SUFU* and 10% have activating mutations of *SMO* [33,128,129]. However, GLI mutations (*GLI1*, *GLI2*, and *GLI3*) are not observed in BCC patients, although GLI overexpression is observed. In addition, *NRAS*, *KRAS*, *HRAS*, *BRAF*, and *CTNNB1* are not known to carry mutations in BCC patients [128].

Based on the molecular genetics analysis of human BCCs, several studies have been performed using genetically-modified mice in which the HH signaling pathway is constitutively activated. Constitutive or conditional overexpression of either GLI1 [130] or GLI2 [131] in keratinocytes can produce BCC-like lesions in the skin. Similarly, expression of mutant SMO carrying the activating mutations identified in human BCCs [33] can also produce murine BCCs. Furthermore, $Ptch1^{+/-}$ mice develop BCCs, and those BCCs often exhibit a deletion of the wild-type copy of *Ptch1*, as well as up-regulation of HH signaling. Similarly, mice carrying one inactivated, mutant allele of the HH suppressor, *SUFU* are also susceptible to BCC development [132].

Although these results show that the HH signaling pathway has important roles in BCC formation, the actual cell of origin remains to be elucidated. X-ray-induced BCCs in $Ptch1^{+/-}$ mice arise predominantly, and likely exclusively, from hair follicle bulge stem cells expressing keratin 15 (a bulge cell marker). Furthermore, upregulation of SMO expression is necessary for BCC formation in X-irradiated $Ptch1^{+/-}$ mice [133]. Using an inducible mouse model enabling expression of a constitutively active SMO mutant (SMO-M2) [33] in the adult epidermis, HH pathway activation was found to reprogram adult interfollicular cells to an embryonic hair follicle progenitor fate. Furthermore, HH pathway activation induces the activation of Wnt/β-catenin signaling. Deletion of β-catenin in adult SMO-M2-expressing cells prevents reprogramming to an embryonic hair follicle progenitor fate and also prevents tumor initiation [134]. Recently, deletion of *Ptch1* in different cellular compartments revealed that tumors can arise from multiple hair follicle stem cells and from innervated GLI1-expressing progenitor cells in touch domes. In contrast, tumors do not arise from interfollicular epidermal stem cells [135].

BCC-specific signaling crosstalk between the HH and other signaling pathways has been reported. Eberl et al. reported crosstalk between the HH and EGFR signaling pathways in BCCs [41]. Skin-specific SMO-M2-expressing transgenic mice develop BCC-like lesions. However, skin-specific knockout of EGFR in SMO-M2-expressing mice reduces the number and size of BCCs. Furthermore, cooperation of HH and EGFR signaling induces specific GLI target genes such as *FGF19*, *TGFA*, *CXCR4*, *SOX9*, and *SOX2*. These genes are not induced by activation of the canonical HH signaling pathway alone. These results suggest that cooperation between HH and EGFR signaling is important for BCC development, which is mediated by HH–EGFR cooperation-responsive GLI target genes.

Vismodegib [6] is the first SMO inhibitor to receive approval from the USA FDA as a therapeutic for treating advanced or metastatic BCC. However, some studies have reported the emergence of SMO inhibitor-resistant tumors [136,137]. It will be important to understand the mechanisms by which SMO inhibitor-resistant cancer cells emerge, and to find novel therapeutic targets for these tumors. As a therapeutic strategy for SMO inhibitor-resistant tumors, several GLI1 inhibitors, such as JQ1 [138], Glabrescione B (GlaB) [139], and arsenic trioxide (ATO) [140], have been developed and tested using in vitro or in vivo models of BCC. Tang et al. demonstrated that bromodomain containing 4 (BRD4) directly occupies the *GLI1* and *GLI2* promoters and regulates *GLI1* and *GLI2* transcription downstream of SMO [139]. Treatment with JQ1, a small-molecule inhibitor targeting BRD4 [141], attenuates mouse BCC cell-derived tumor growth through the suppression of *GLI1* and *GLI2* transcription. In addition, JQ1-mediated downregulation of *GLI1* and *GLI2* transcription is also observed in SMO inhibitor-resistant BCC cells [138]. Infante et al. developed a novel GLI1 inhibitor (GlaB) that inhibits the interaction of GLI1 with DNA. They then demonstrated that GlaB inhibits GLI1-dependent BCC cell growth in vitro and BCC tumor growth in ASZ001 BCC allografts in vivo [139]. Ally et al. tried a combined treatment with ATO plus itraconazole for metastatic BCC patients [140]. Although tumor shrinkage was not observed, tumor growth was arrested for three months, and GLI1 mRNA levels were dramatically decreased. Improvement of the drug dose or dosing schedule may yield a better clinical response. Finally, we review another therapeutic target for SMO inhibitor-resistant BCC. Atwood et al. reported that atypical protein kinase Ci/λ (aPKC-i/λ) is a novel GLI1 target gene product, and a novel therapeutic target for SMO inhibitor-resistant BCC [142]. aPKC-i/λ functions downstream of SMO to phosphorylate and activate GLI1, resulting in maximal DNA binding and transcriptional activation. Activated aPKC-i/λ is upregulated in SMO inhibitor-resistant tumors, and targeting aPKC-i/λ suppresses GLI1 activation and the growth of SMO inhibitor-resistant BCC cell lines. These results show the potential of aPKC-i/λ as a therapeutic target for SMO inhibitor-resistant BCC.

10. Conclusions

The HH signaling pathway is one of the most fundamental in epidermal development and epidermal stem cell maintenance. However, deregulated HH signaling induces BCC. Hence, it is important that the intensity of the HH signal is strictly regulated during skin generation and homeostasis. To achieve precise regulation of HH signaling intensity, negative regulators of HH signaling are necessary, as well as crosstalk between the HH and other signal transduction pathways such as those activated by Wnt and BMP. Recently, it has been reported that low HH signaling activity is important for lung homeostasis [143]. Therefore, it is possible that fine-tuning the intensity of a HH signal may be involved in the homeostasis of multiple organs, including the skin.

As described above, aberrant activation of HH signaling causes BCC formation. A tumor microenvironment would exhibit not only the HH pathway regulatory mechanisms that are common with skin development and homeostasis, but also cancer cell-specific regulatory mechanisms. Therefore, understanding the precise mechanisms of HH pathway regulation in the tumor microenvironment is important to advance new therapeutic strategies for SMO inhibitor-resistant BCC. Moreover, such knowledge may also help to provide new therapeutic strategies for other HH pathway-activated cancers.

Acknowledgments: We thank Nicholas Rufaut, from Edanz Group (www.edanzediting.com/ac) for editing a draft of this manuscript.

Author Contributions: Yoshinori Abe and Nobuyuki Tanaka contributed equally to this work.

Conflicts of Interest: The authors declare no conflict of interest.

References

1. Koster, M.I.; Roop, D.R. Mechanisms Regulating Epithelial Stratification. *Annu. Rev. Cell Dev. Biol.* **2007**, *23*, 93–113. [CrossRef] [PubMed]
2. Fuchs, E. Scratching the surface of skin development. *Nature* **2007**, *445*, 834–842. [CrossRef] [PubMed]
3. Blanpain, C.; Fuchs, E. Epidermal homeostasis: A balancing act of stem cells in the skin. *Nat. Rev. Mol. Cell Biol.* **2009**, *10*, 207–217. [CrossRef] [PubMed]
4. Schmidt-Ullrich, R.; Paus, R. Molecular principles of hair follicle induction and morphogenesis. *BioEssays* **2005**, *27*, 247–261. [CrossRef] [PubMed]
5. Epstein, E.H. Basal cell carcinomas: Attack of the hedgehog. *Nat. Rev. Cancer* **2008**, *8*, 743–754. [CrossRef] [PubMed]
6. Dlugosz, A.; Agrawal, S.; Kirkpatrick, P. Vismodegib. *Nat. Rev. Drug Discov.* **2012**, *11*, 437–438. [CrossRef] [PubMed]
7. Ingham, P.W.; McMahon, A.P. Hedgehog signaling in animal development: Paradigms and principles. *Genes Dev.* **2001**, *15*, 3059–3087. [CrossRef] [PubMed]
8. Ng, J.M.Y.; Curran, T. The Hedgehog's tale: Developing strategies for targeting cancer. *Nat. Rev. Cancer* **2011**, *11*, 493–501. [CrossRef] [PubMed]
9. Brechbiel, J.; Miller-Moslin, K.; Adjei, A.A. Crosstalk between hedgehog and other signaling pathways as a basis for combination therapies in cancer. *Cancer Treat. Rev.* **2014**, *40*, 750–759. [CrossRef] [PubMed]
10. Aberger, F.; i Altaba, A.R. Context-dependent signal integration by the GLI code: The oncogenic load, pathways, modifiers and implications for cancer therapy. *Semin. Cell Dev. Biol.* **2014**, *33*, 93–104. [CrossRef] [PubMed]
11. Pak, E.; Segal, R.A. Hedgehog Signal Transduction: Key Players, Oncogenic Drivers, and Cancer Therapy. *Dev. Cell* **2016**, *38*, 333–344. [CrossRef] [PubMed]

12. Rohatgi, R.; Milenkovic, L.; Scott, M.P. Patched1 Regulates Hedgehog Signaling at the Primary Cilium. *Science* **2007**, *317*, 372–376. [CrossRef] [PubMed]
13. Lee, Y.; Kawagoe, R.; Sasai, K.; Li, Y.; Russell, H.R.; Curran, T.; McKinnon, P.J. Loss of suppressor-of-fused function promotes tumorigenesis. *Oncogene* **2007**, *26*, 6442–6447. [CrossRef] [PubMed]
14. Allen, B.L.; Tenzen, T.; McMahon, A.P. The Hedgehog-binding proteins Gas1 and Cdo cooperate to positively regulate Shh signaling during mouse development. *Genes Dev.* **2007**, *21*, 1244–1257. [CrossRef] [PubMed]
15. Okada, A.; Charron, F.; Morin, S.; Shin, D.S.; Wong, K.; Fabre, P.J.; Tessier-Lavigne, M.; McConnell, S.K. Boc is a receptor for sonic hedgehog in the guidance of commissural axons. *Nature* **2006**, *444*, 369–373. [CrossRef] [PubMed]
16. Yang, C.; Chen, W.; Chen, Y.; Jiang, J. Smoothened transduces Hedgehog signal by forming a complex with Evc/Evc2. *Cell Res.* **2012**, *22*, 1593–1604. [CrossRef] [PubMed]
17. Dorn, K.V.; Hughes, C.E.; Rohatgi, R. A Smoothened-Evc2 Complex Transduces the Hedgehog Signal at Primary Cilia. *Dev. Cell* **2012**, *23*, 823–835. [CrossRef] [PubMed]
18. Corbit, K.C.; Aanstad, P.; Singla, V.; Norman, A.R.; Stainier, D.Y.R.; Reiter, J.F. Vertebrate Smoothened functions at the primary cilium. *Nature* **2005**, *437*, 1018–1021. [CrossRef] [PubMed]
19. Haycraft, C.J.; Banizs, B.; Aydin-Son, Y.; Zhang, Q.; Michaud, E.J.; Yoder, B.K. Gli2 and Gli3 localize to cilia and require the intraflagellar transport protein polaris for processing and function. *PLoS Genet.* **2005**, *1*, e53. [CrossRef] [PubMed]
20. Abe, Y.; Tanaka, N. The Hedgehog Signaling Networks in Lung Cancer: The Mechanisms and Roles in Tumor Progression and Implications for Cancer Therapy. *BioMed Res. Int.* **2016**, *2016*, 1–11. [CrossRef] [PubMed]
21. Sasaki, H.; Nishizaki, Y.; Hui, C.; Nakafuku, M.; Kondoh, H. Regulation of Gli2 and Gli3 activities by an amino-terminal repression domain: Implication of Gli2 and Gli3 as primary mediators of Shh signaling. *Development (Camb. Engl.)* **1999**, *126*, 3915–3924.
22. Bai, C.B.; Auerbach, W.; Lee, J.S.; Stephen, D.; Joyner, A.L. Gli2, but not Gli1, is required for initial Shh signaling and ectopic activation of the Shh pathway. *Development (Camb. Engl.)* **2002**, *129*, 4753–4761.
23. Masuya, H.; Sagai, T.; Moriwaki, K.; Shiroishi, T. Multigenic control of the localization of the zone of polarizing activity in limb morphogenesis in the mouse. *Dev. Biol.* **1997**, *182*, 42–51. [CrossRef] [PubMed]
24. Büscher, D.; Bosse, B.; Heymer, J.; Rüther, U. Evidence for genetic control of Sonic hedgehog by Gli3 in mouse limb development. *Mech. Dev.* **1997**, *62*, 175–182. [CrossRef]
25. Ding, Q.; Motoyama, J.; Gasca, S.; Mo, R.; Sasaki, H.; Rossant, J.; Hui, C.C. Diminished Sonic hedgehog signaling and lack of floor plate differentiation in Gli2 mutant mice. *Development (Camb. Engl.)* **1998**, *125*, 2533–2543.
26. Tukachinsky, H.; Lopez, L.V.; Salic, A. A mechanism for vertebrate Hedgehog signaling: Recruitment to cilia and dissociation of SuFu-Gli protein complexes. *J. Cell Biol.* **2010**, *191*, 415–428. [CrossRef] [PubMed]
27. Park, H.L.; Bai, C.; Platt, K.A.; Matise, M.P.; Beeghly, A.; Hui, C.C.; Nakashima, M.; Joyner, A.L. Mouse Gli1 mutants are viable but have defects in SHH signaling in combination with a Gli2 mutation. *Development (Camb. Engl.)* **2000**, *127*, 1593–1605. [CrossRef]
28. Bai, C.B.; Joyner, A.L. Gli1 can rescue the in vivo function of Gli2. *Development (Camb. Engl.)* **2001**, *128*, 5161–5172.
29. Rubin, L.L.; de Sauvage, F.J. Targeting the Hedgehog pathway in cancer. *Nat. Rev. Drug Discov.* **2006**, *5*, 1026–1033. [CrossRef] [PubMed]
30. Raffel, C.; Jenkins, R.B.; Frederick, L.; Hebrink, D.; Alderete, B.; Fults, D.W.; James, C.D. Sporadic medulloblastomas contain PTCH mutations. *Cancer Res.* **1997**, *57*, 842–845. [PubMed]
31. Wolter, M.; Reifenberger, J.; Sommer, C.; Ruzicka, T.; Reifenberger, G. Mutations in the human homologue of the Drosophila segment polarity gene patched (PTCH) in sporadic basal cell carcinomas of the skin and primitive neuroectodermal tumors of the central nervous system. *Cancer Res.* **1997**, *57*, 2581–2585. [PubMed]
32. Reifenberger, J.; Wolter, M.; Weber, R.G.; Megahed, M.; Ruzicka, T.; Lichter, P.; Reifenberger, G. Missense mutations in SMOH in sporadic basal cell carcinomas of the skin and primitive neuroectodermal tumors of the central nervous system. *Cancer Res.* **1998**, *58*, 1798–1803. [PubMed]
33. Xie, J.; Murone, M.; Luoh, S.M.; Ryan, A.; Gu, Q.; Zhang, C.; Bonifas, J.M.; Lam, C.W.; Hynes, M.; Goddard, A.; et al. Activating Smoothened mutations in sporadic basal-cell carcinoma. *Nature* **1998**, *391*, 90–92. [CrossRef] [PubMed]

34. Taylor, M.D.; Liu, L.; Raffel, C.; Hui, C.; Mainprize, T.G.; Zhang, X.; Agatep, R.; Chiappa, S.; Gao, L.; Lowrance, A.; et al. Mutations in SUFU predispose to medulloblastoma. *Nat. Genet.* **2002**, *31*, 306–310. [CrossRef] [PubMed]

35. Pandolfi, S.; Stecca, B. Cooperative integration between HEDGEHOG-GLI signalling and other oncogenic pathways: Implications for cancer therapy. *Expert Rev. Mol. Med.* **2015**, *17*, e5. [CrossRef] [PubMed]

36. Huelsken, J.; Vogel, R.; Erdmann, B.; Cotsarelis, G.; Birchmeier, W. β-Catenin Controls Hair Follicle Morphogenesis and Stem Cell Differentiation in the Skin. *Cell* **2001**, *105*, 533–545. [CrossRef]

37. Suzuki, K.; Yamaguchi, Y.; Villacorte, M.; Mihara, K.; Akiyama, M.; Shimizu, H.; Taketo, M.M.; Nakagata, N.; Tsukiyama, T.; Yamaguchi, T.P.; et al. Embryonic hair follicle fate change by augmented -catenin through Shh and Bmp signaling. *Development* **2009**, *136*, 367–372. [CrossRef] [PubMed]

38. Gat, U.; DasGupta, R.; Degenstein, L.; Fuchs, E. De Novo hair follicle morphogenesis and hair tumors in mice expressing a truncated beta-catenin in skin. *Cell* **1998**, *95*, 605–614. [CrossRef]

39. Silva-Vargas, V.; Lo Celso, C.; Giangreco, A.; Ofstad, T.; Prowse, D.M.; Braun, K.M.; Watt, F.M. β-Catenin and Hedgehog Signal Strength Can Specify Number and Location of Hair Follicles in Adult Epidermis without Recruitment of Bulge Stem Cells. *Dev. Cell* **2005**, *9*, 121–131. [CrossRef] [PubMed]

40. Javelaud, D.; Pierrat, M.-J.; Mauviel, A. Crosstalk between TGF-β and hedgehog signaling in cancer. *FEBS Lett.* **2012**, *586*, 2016–2025. [CrossRef] [PubMed]

41. Eberl, M.; Klingler, S.; Mangelberger, D.; Loipetzberger, A.; Damhofer, H.; Zoidl, K.; Schnidar, H.; Hache, H.; Bauer, H.-C.; Solca, F.; et al. Hedgehog-EGFR cooperation response genes determine the oncogenic phenotype of basal cell carcinoma and tumour-initiating pancreatic cancer cells. *EMBO Mol. Med.* **2012**, *4*, 218–233. [CrossRef] [PubMed]

42. Nolan-Stevaux, O.; Lau, J.; Truitt, M.L.; Chu, G.C.; Hebrok, M.; Fernández-Zapico, M.E.; Hanahan, D. GLI1 is regulated through Smoothened-independent mechanisms in neoplastic pancreatic ducts and mediates PDAC cell survival and transformation. *Genes Dev.* **2009**, *23*, 24–36. [CrossRef] [PubMed]

43. Mazumdar, T.; DeVecchio, J.; Agyeman, A.; Shi, T.; Houghton, J.A. The GLI genes as the molecular switch in disrupting Hedgehog signaling in colon cancer. *Oncotarget* **2011**, *2*, 638–645. [CrossRef] [PubMed]

44. Wang, Y.; Ding, Q.; Yen, C.-J.; Xia, W.; Izzo, J.G.; Lang, J.-Y.; Li, C.-W.; Hsu, J.L.; Miller, S.A.; Wang, X.; et al. The crosstalk of mTOR/S6K1 and Hedgehog pathways. *Cancer Cell* **2012**, *21*, 374–387. [CrossRef] [PubMed]

45. Androutsellis-Theotokis, A.; Leker, R.R.; Soldner, F.; Hoeppner, D.J.; Ravin, R.; Poser, S.W.; Rueger, M.A.; Bae, S.-K.; Kittappa, R.; McKay, R.D.G. Notch signalling regulates stem cell numbers in vitro and in vivo. *Nature* **2006**, *442*, 823–826. [CrossRef] [PubMed]

46. Mikkola, M.L. Genetic basis of skin appendage development. *Semin. Cell Dev. Biol.* **2007**, *18*, 225–236. [CrossRef] [PubMed]

47. Motoyama, J.; Takabatake, T.; Takeshima, K.; Hui, C. Ptch2, a second mouse Patched gene is co-expressed with Sonic hedgehog. *Nat. Genet.* **1998**, *18*, 104–106. [CrossRef] [PubMed]

48. Paus, R.; Müller-Röver, S.; Van Der Veen, C.; Maurer, M.; Eichmüller, S.; Ling, G.; Hofmann, U.; Foitzik, K.; Mecklenburg, L.; Handjiski, B. A comprehensive guide for the recognition and classification of distinct stages of hair follicle morphogenesis. *J. Investig. Dermatol.* **1999**, *113*, 523–532. [PubMed]

49. Forni, M.F.; Trombetta-Lima, M.; Sogayar, M.C. Stem cells in embryonic skin development. *Biol. Res.* **2012**, *45*, 215–222. [CrossRef] [PubMed]

50. St-Jacques, B.; Dassule, H.R.; Karavanova, I.; Botchkarev, V.A.; Li, J.; Danielian, P.S.; McMahon, J.A.; Lewis, P.M.; Paus, R.; McMahon, A.P. Sonic hedgehog signaling is essential for hair development. *Curr. Biol.* **1998**, *8*, 1058–1068. [CrossRef]

51. Chiang, C.; Swan, R.Z.; Grachtchouk, M.; Bolinger, M.; Litingtung, Y.; Robertson, E.K.; Cooper, M.K.; Gaffield, W.; Westphal, H.; Beachy, P.A.; et al. Essential role for Sonic hedgehog during hair follicle morphogenesis. *Dev. Biol.* **1999**, *205*, 1–9. [CrossRef] [PubMed]

52. Mill, P. Sonic hedgehog-dependent activation of Gli2 is essential for embryonic hair follicle development. *Genes Dev.* **2003**, *17*, 282–294. [CrossRef] [PubMed]

53. Woo, W.-M.; Zhen, H.H.; Oro, A.E. Shh maintains dermal papilla identity and hair morphogenesis via a Noggin-Shh regulatory loop. *Genes Dev.* **2012**, *26*, 1235–1246. [CrossRef] [PubMed]

54. Biernaskie, J.; Paris, M.; Morozova, O.; Fagan, B.M.; Marra, M.; Pevny, L.; Miller, F.D. SKPs Derive from

Hair Follicle Precursors and Exhibit Properties of Adult Dermal Stem Cells. *Cell Stem Cell* **2009**, *5*, 610–623. [CrossRef] [PubMed]

55. Driskell, R.R.; Giangreco, A.; Jensen, K.B.; Mulder, K.W.; Watt, F.M. Sox2-positive dermal papilla cells specify hair follicle type in mammalian epidermis. *Development (Camb. Engl.)* **2009**, *136*, 2815–2823. [CrossRef] [PubMed]
56. Morrison, K.M.; Miesegaes, G.R.; Lumpkin, E.A.; Maricich, S.M. Mammalian Merkel cells are descended from the epidermal lineage. *Dev. Biol.* **2009**, *336*, 76–83. [CrossRef] [PubMed]
57. Van Keymeulen, A.; Mascre, G.; Youseff, K.K.; Harel, I.; Michaux, C.; De Geest, N.; Szpalski, C.; Achouri, Y.; Bloch, W.; Hassan, B.A.; et al. Epidermal progenitors give rise to Merkel cells during embryonic development and adult homeostasis. *J. Cell Biol.* **2009**, *187*, 91–100. [CrossRef] [PubMed]
58. Xiao, Y.; Williams, J.S.; Brownell, I. Merkel cells and touch domes: More than mechanosensory functions? *Exp. Dermatol.* **2014**, *23*, 692–695. [CrossRef] [PubMed]
59. Xiao, Y.; Thoresen, D.T.; Miao, L.; Williams, J.S.; Wang, C.; Atit, R.P.; Wong, S.Y.; Brownell, I. A Cascade of Wnt, Eda, and Shh Signaling Is Essential for Touch Dome Merkel Cell Development. *PLoS Genet.* **2016**, *12*, 1–19. [CrossRef] [PubMed]
60. Perdigoto, C.N.; Dauber, K.L.; Bar, C.; Tsai, P.C.; Valdes, V.J.; Cohen, I.; Santoriello, F.J.; Zhao, D.; Zheng, D.; Hsu, Y.C.; et al. Polycomb-Mediated Repression and Sonic Hedgehog Signaling Interact to Regulate Merkel Cell Specification during Skin Development. *PLoS Genet.* **2016**, *12*, 1–27. [CrossRef] [PubMed]
61. Chen, Y.; Struhl, G. Dual roles for patched in sequestering and transducing Hedgehog. *Cell.* **1996**, *87*, 553–563. [CrossRef]
62. Adolphe, C.; Hetherington, R.; Ellis, T.; Wainwright, B. Patched1 functions as a gatekeeper by promoting cell cycle progression. *Cancer Res.* **2006**, *66*, 2081–2088. [CrossRef] [PubMed]
63. Nieuwenhuis, E.; Motoyama, J.; Barnfield, P.C.; Yoshikawa, Y.; Zhang, X.; Mo, R.; Crackower, M.A.; Hui, C.-C. Mice with a Targeted Mutation of Patched2 Are Viable but Develop Alopecia and Epidermal Hyperplasia. *Mol. Cell. Biol.* **2006**, *26*, 6609–6622. [CrossRef] [PubMed]
64. Adolphe, C.; Nieuwenhuis, E.; Villani, R.; Li, Z.J.; Kaur, P.; Hui, C.; Wainwright, B.J. Patched 1 and Patched 2 Redundancy Has a Key Role in Regulating Epidermal Differentiation. *J. Investig. Dermatol.* **2014**, *134*, 1981–1990. [CrossRef] [PubMed]
65. Mill, P.; Mo, R.; Hu, M.C.; Dagnino, L.; Rosenblum, N.D.; Hui, C.-C. Shh controls epithelial proliferation via independent pathways that converge on N-Myc. *Dev. Cell* **2005**, *9*, 293–303. [CrossRef] [PubMed]
66. Adolphe, C.; Junker, J.P.; Lyubimova, A.; van Oudenaarden, A.; Wainwright, B. Patched Receptors Sense, Interpret, and Establish an Epidermal Hedgehog Signaling Gradient. *J. Investig. Dermatol.* **2017**, *137*, 179–186. [CrossRef] [PubMed]
67. Clevers, H. Wnt/beta-catenin signaling in development and disease. *Cell* **2006**, *127*, 469–480. [CrossRef] [PubMed]
68. Van Camp, J.K.; Beckers, S.; Zegers, D.; Van Hul, W. Wnt Signaling and the Control of Human Stem Cell Fate. *Stem Cell Rev. Rep.* **2014**, *10*, 207–229. [CrossRef] [PubMed]
69. Fuchs, E.; Merrill, B.J.; Jamora, C.; Dasgupta, R. At the Roots of a Never-Ending Cycle. *Dev. Cell* **2001**, *1*, 13–25. [CrossRef]
70. Widelitz, R.B. Wnt signaling in skin organogenesis. *Organogenesis* **2008**, *4*, 123–133. [CrossRef] [PubMed]
71. Reddy, S.; Andl, T.; Bagasra, A.; Lu, M.M.; Epstein, D.J.; Morrisey, E.E.; Millar, S.E. Characterization of Wnt gene expression in developing and postnatal hair follicles and identification of Wnt5a as a target of Sonic hedgehog in hair follicle morphogenesis. *Mech. Dev.* **2001**, *107*, 69–82. [CrossRef]
72. Kwack, M.H.; Kim, M.K.; Kim, J.C.; Sung, Y.K. Wnt5a attenuates Wnt/β-catenin signalling in human dermal papilla cells. *Exp. Dermatol.* **2013**, *22*, 229–231. [CrossRef] [PubMed]
73. Jamora, C.; DasGupta, R.; Kocieniewski, P.; Fuchs, E. Links between signal transduction, transcription and adhesion in epithelial bud development. *Nature* **2003**, *422*, 317–322. [CrossRef] [PubMed]
74. Rendl, M.; Polak, L.; Fuchs, E. BMP signaling in dermal papilla cells is required for their hair follicle-inductive properties. *Genes Dev.* **2008**, *22*, 543–557. [CrossRef] [PubMed]
75. Millar, S.E. Molecular mechanisms regulating hair follicle development. *J. Investig. Dermatol.* **2002**, *118*, 216–225. [CrossRef] [PubMed]
76. Mikkola, M.L.; Millar, S.E. The mammary bud as a skin appendage: Unique and shared aspects of development. *J. Mammary Gland Biol. Neoplasia* **2006**, *11*, 187–203. [CrossRef] [PubMed]

77. McMahon, A.P.; Paus, R. Noggin is a mesenchymally derived stimulator of hair-follicle induction. *Nat. Cell Biol.* **1999**, *1*, 158–164. [CrossRef] [PubMed]
78. Irrthum, A.; Devriendt, K.; Chitayat, D.; Matthijs, G.; Glade, C.; Steijlen, P.M.; Fryns, J.-P.; Van Steensel, M.A.M.; Vikkula, M. Mutations in the Transcription Factor Gene SOX18 Underlie Recessive and Dominant Forms of Hypotrichosis-Lymphedema-Telangiectasia. *Am. J. Hum. Genet.* **2003**, *72*, 1470–1478. [CrossRef] [PubMed]
79. Pennisi, D.; Bowles, J.; Nagy, A.; Muscat, G.; Koopman, P. Mice null for sox18 are viable and display a mild coat defect. *Mol. Cell. Biol.* **2000**, *20*, 9331–9336. [CrossRef] [PubMed]
80. Takakura, N.; Yoshida, H.; Kunisada, T.; Nishikawa, S.; Nishikawa, S.I. Involvement of platelet-derived growth factor receptor-alpha in hair canal formation. *J. Investig. Dermatol.* **1996**, *107*, 770–777. [CrossRef] [PubMed]
81. Karlsson, L.; Bondjers, C.; Betsholtz, C. Roles for PDGF-A and sonic hedgehog in development of mesenchymal components of the hair follicle. *Development (Camb. Engl.)* **1999**, *126*, 2611–2621.
82. Gao, J.; DeRouen, M.C.; Chen, C.-H.; Nguyen, M.; Nguyen, N.T.; Ido, H.; Harada, K.; Sekiguchi, K.; Morgan, B.A.; Miner, J.H.; et al. Laminin-511 is an epithelial message promoting dermal papilla development and function during early hair morphogenesis. *Genes Dev.* **2008**, *22*, 2111–2124. [CrossRef] [PubMed]
83. Artavanis-Tsakonas, S. Notch Signaling: Cell Fate Control and Signal Integration in Development. *Science* **1999**, *284*, 770–776. [CrossRef] [PubMed]
84. Yamamoto, N.; Tanigaki, K.; Han, H.; Hiai, H.; Honjo, T. Notch/RBP-J signaling regulates epidermis/hair fate determination of hair follicular stem cells. *Curr. Biol.* **2003**, *13*, 333–338. [CrossRef]
85. Nguyen, B.C.; Lefort, K.; Mandinova, A.; Antonini, D.; Devgan, V.; Della Gatta, G.; Koster, M.I.; Zhang, Z.; Wang, J.; Di Vignano, A.T.; et al. Cross-regulation between Notch and p63 in keratinocyte commitment to differentiation. *Genes Dev.* **2006**, *20*, 1028–1042. [CrossRef] [PubMed]
86. Li, Y.; Hibbs, M.A.; Gard, A.L.; Shylo, N.A.; Yun, K. Genome-wide analysis of N1ICD/RBPJ targets in vivo reveals direct transcriptional regulation of Wnt, SHH, and Hippo pathway effectors by Notch1. *Stem Cells* **2012**, *30*, 741–752. [CrossRef] [PubMed]
87. Ringuette, R.; Atkins, M.; Lagali, P.S.; Bassett, E.A.; Campbell, C.; Mazerolle, C.; Mears, A.J.; Picketts, D.J.; Wallace, V.A. A Notch-Gli2 axis sustains Hedgehog responsiveness of neural progenitors and Müller glia. *Dev. Biol.* **2016**, *411*, 85–100. [CrossRef] [PubMed]
88. Headon, D.J.; Overbeek, P.A. Involvement of a novel Tnf receptor homologue in hair follicle induction. *Nat. Genet.* **1999**, *22*, 370–374. [CrossRef] [PubMed]
89. Aggarwal, B.B. Signalling pathways of the TNF superfamily: A double-edged sword. *Nat. Rev. Immunol.* **2003**, *3*, 745–756. [CrossRef] [PubMed]
90. Kumar, A.; Eby, M.T.; Sinha, S.; Jasmin, A.; Chaudhary, P.M. The ectodermal dysplasia receptor activates the nuclear factor-kappaB, JNK, and cell death pathways and binds to ectodysplasin A. *J. Biol. Chem.* **2001**, *276*, 2668–2677. [CrossRef] [PubMed]
91. Mikkola, M.L.; Pispa, J.; Pekkanen, M.; Paulin, L.; Nieminen, P.; Kere, J.; Thesleff, I. Ectodysplasin, a protein required for epithelial morphogenesis, is a novel TNF homologue and promotes cell-matrix adhesion. *Mech. Dev.* **1999**, *88*, 133–146. [CrossRef]
92. Schmidt-Ullrich, R.; Tobin, D.J.; Lenhard, D.; Schneider, P.; Paus, R.; Scheidereit, C. NF-kappaB transmits Eda A1/EdaR signalling to activate Shh and cyclin D1 expression, and controls post-initiation hair placode down growth. *Development (Camb. Engl.)* **2006**, *133*, 1045–1057. [CrossRef] [PubMed]
93. Dai, X.; Segre, J.A. Transcriptional control of epidermal specification and differentiation. *Curr. Opin. Genet. Dev.* **2004**, *14*, 485–491. [CrossRef] [PubMed]
94. Fuchs, E.; Byrne, C. The epidermis: Rising to the surface. *Curr. Opin. Genet. Dev.* **1994**, *4*, 725–736. [CrossRef]
95. Simpson, C.L.; Patel, D.M.; Green, K.J. Deconstructing the skin: Cytoarchitectural determinants of epidermal morphogenesis. *Nat. Rev. Mol. Cell Biol.* **2011**, *12*, 565–580. [CrossRef] [PubMed]
96. Blanpain, C.; Horsley, V.; Fuchs, E. Epithelial Stem Cells: Turning over New Leaves. *Cell* **2007**, *128*, 445–458. [CrossRef] [PubMed]
97. Cotsarelis, G.; Sun, T.T.; Lavker, R.M. Label-retaining cells reside in the bulge area of pilosebaceous unit: Implications for follicular stem cells, hair cycle, and skin carcinogenesis. *Cell* **1990**, *61*, 1329–1337. [CrossRef]
98. Braun, K.M.; Niemann, C.; Jensen, U.B.; Sundberg, J.P.; Silva-Vargas, V.; Watt, F.M. Manipulation of stem cell proliferation and lineage commitment: Visualisation of label-retaining cells in wholemounts of mouse epidermis. *Development (Camb. Engl.)* **2003**, *130*, 5241–5255. [CrossRef] [PubMed]

99. Blanpain, C.; Lowry, W.E.; Geoghegan, A.; Polak, L.; Fuchs, E. Self-renewal, multipotency, and the existence of two cell populations within an epithelial stem cell niche. *Cell* **2004**, *118*, 635–648. [CrossRef] [PubMed]
100. Trempus, C.S.; Morris, R.J.; Bortner, C.D.; Cotsarelis, G.; Faircloth, R.S.; Reece, J.M.; Tennant, R.W. Enrichment for living murine keratinocytes from the hair follicle bulge with the cell surface marker CD34. *J. Investig. Dermatol.* **2003**, *120*, 501–511. [CrossRef] [PubMed]
101. Panteleyev, A.A.; Jahoda, C.A.; Christiano, A.M. Hair follicle predetermination. *J. Cell Sci.* **2001**, *114*, 3419–3431. [PubMed]
102. Barker, N.; van Es, J.H.; Jaks, V.; Kasper, M.; Snippert, H.; Toftgård, R.; Clevers, H. Very long-term self-renewal of small intestine, colon, and hair follicles from cycling Lgr5+ve stem cells. *Cold Spring Harb. Symp. Quant. Biol.* **2008**, *73*, 351–356. [CrossRef] [PubMed]
103. Wang, L.C.; Liu, Z.Y.; Gambardella, L.; Delacour, A.; Shapiro, R.; Yang, J.; Sizing, I.; Rayhorn, P.; Garber, E.A.; Benjamin, C.D.; et al. Conditional disruption of hedgehog signaling pathway defines its critical role in hair development and regeneration. *J. Investig. Dermatol.* **2000**, *114*, 901–908. [CrossRef] [PubMed]
104. Sato, N.; Leopold, P.L.; Crystal, R.G. Induction of the hair growth phase in postnatal mice by localized transient expression of Sonic hedgehog. *J. Clin. Investig.* **1999**, *104*, 855–864. [CrossRef] [PubMed]
105. Brownell, I.; Guevara, E.; Bai, C.B.; Loomis, C.A.; Joyner, A.L. Nerve-derived sonic hedgehog defines a niche for hair follicle stem cells capable of becoming epidermal stem cells. *Cell Stem Cell* **2011**, *8*, 552–565. [CrossRef] [PubMed]
106. Hsu, Y.C.; Pasolli, H.A.; Fuchs, E. Dynamics between stem cells, niche, and progeny in the hair follicle. *Cell* **2011**, *144*, 92–105. [CrossRef] [PubMed]
107. Plikus, M.V.; Mayer, J.A.; de la Cruz, D.; Baker, R.E.; Maini, P.K.; Maxson, R.; Chuong, C.-M. Cyclic dermal BMP signalling regulates stem cell activation during hair regeneration. *Nature* **2008**, *451*, 340–344. [CrossRef] [PubMed]
108. Festa, E.; Fretz, J.; Berry, R.; Schmidt, B.; Rodeheffer, M.; Horowitz, M.; Horsley, V. Adipocyte lineage cells contribute to the skin stem cell niche to drive hair cycling. *Cell* **2011**, *146*, 761–771. [CrossRef] [PubMed]
109. Keyes, B.E.; Segal, J.P.; Heller, E.; Lien, W.-H.; Chang, C.-Y.; Guo, X.; Oristian, D.S.; Zheng, D.; Fuchs, E. Nfatc1 orchestrates aging in hair follicle stem cells. *Proc. Natl. Acad. Sci. USA* **2013**, *110*, E4950–E4959. [CrossRef] [PubMed]
110. Zhang, L.-J.; Guerrero-Juarez, C.F.; Hata, T.; Bapat, S.P.; Ramos, R.; Plikus, M.V.; Gallo, R.L. Dermal adipocytes protect against invasive Staphylococcus aureus skin infection. *Science* **2015**, *347*, 67–71. [CrossRef] [PubMed]
111. Halder, G.; Johnson, R.L. Hippo signaling: Growth control and beyond. *Development (Camb. Engl.)* **2011**, *138*, 9–22. [CrossRef] [PubMed]
112. Zhao, B.; Li, L.; Lei, Q.; Guan, K.L. The Hippo-YAP pathway in organ size control and tumorgenesis: An updated version. *Genes Dev.* **2010**, *24*, 862–874. [CrossRef] [PubMed]
113. Akladios, B.; Mendoza-Reinoso, V.; Samuel, M.S.; Hardeman, E.C.; Khosrotehrani, K.; Key, B.; Beverdam, A. Epidermal YAP2-5SA-ΔC Drives β-Catenin Activation to Promote Keratinocyte Proliferation in Mouse Skin In Vivo. *J. Investig. Dermatol.* **2017**, *137*, 716–726. [CrossRef] [PubMed]
114. Schlegelmilch, K.; Mohseni, M.; Kirak, O.; Pruszak, J.; Rodriguez, J.R.; Zhou, D.; Kreger, B.T.; Vasioukhin, V.; Avruch, J.; Brummelkamp, T.R.; et al. Yap1 acts downstream of α-catenin to control epidermal proliferation. *Cell* **2011**, *144*, 782–795. [CrossRef] [PubMed]
115. Zhang, H.; Pasolli, H.A.; Fuchs, E. Yes-associated protein (YAP) transcriptional coactivator functions in balancing growth and differentiation in skin. *Proc. Natl. Acad. Sci. USA* **2011**, *108*, 2270–2275. [CrossRef] [PubMed]
116. Akladios, B.; Mendoza Reinoso, V.; Cain, J.E.; Wang, T.; Lambie, D.L.; Watkins, D.N.; Beverdam, A. Positive regulatory interactions between YAP and Hedgehog signalling in skin homeostasis and BCC development in mouse skin in vivo. *PLoS ONE* **2017**, *12*, e0183178. [CrossRef] [PubMed]
117. Botchkarev, V.A.; Flores, E.R. p53/p63/p73 in the epidermis in health and disease. *Cold Spring Harb. Perspect. Med.* **2014**, *4*, 466–472. [CrossRef] [PubMed]
118. Abe, Y.; Oda-Sato, E.; Tobiume, K.; Kawauchi, K.; Taya, Y.; Okamoto, K.; Oren, M.; Tanaka, N. Hedgehog signaling overrides p53-mediated tumor suppression by activating Mdm2. *Proc. Natl. Acad. Sci. USA* **2008**, *105*, 4838–4843. [CrossRef] [PubMed]
119. Mills, A.A.; Zheng, B.; Wang, X.J.; Vogel, H.; Roop, D.R.; Bradley, A. P63 Is a P53 Homologue Required for Limb and Epidermal Morphogenesis. *Nature* **1999**, *398*, 708–713. [CrossRef] [PubMed]

120. Yang, A.; Schweitzer, R.; Sun, D.; Kaghad, M.; Walker, N.; Bronson, R.T.; Tabin, C.; Sharpe, A.; Caput, D.; Crum, C.; et al. p63 is essential for regenerative proliferation in limb, craniofacial and epithelial development. *Nature* **1999**, *398*, 714–718. [CrossRef] [PubMed]

121. Bigelow, R.L.H.; Chari, N.S.; Undén, A.B.; Spurgers, K.B.; Lee, S.; Roop, D.R.; Toftgård, R.; McDonnell, T.J. Transcriptional Regulation of bcl-2 Mediated by the Sonic Hedgehog Signaling Pathway through gli-1. *J. Biol. Chem.* **2004**, *279*, 1197–1205. [CrossRef] [PubMed]

122. Regl, G.; Kasper, M.; Schnidar, H.; Eichberger, T.; Neill, G.W.; Philpott, M.P.; Esterbauer, H.; Hauser-Kronberger, C.; Frischauf, A.-M.; Aberger, F. Activation of the BCL2 promoter in response to Hedgehog/GLI signal transduction is predominantly mediated by GLI2. *Cancer Res.* **2004**, *64*, 7724–7731. [CrossRef] [PubMed]

123. Polakowska, R.R.; Piacentini, M.; Bartlett, R.; Goldsmith, L.A.; Haake, A.R. Apoptosis in human skin development: Morphogenesis, periderm, and stem cells. *Dev. Dyn.* **1994**, *199*, 176–188. [CrossRef] [PubMed]

124. Chari, N.S.; Romano, R.A.; Koster, M.I.; Jaks, V.; Roop, D.; Flores, E.R.; Teglund, S.; Sinha, S.; Gruber, W.; Aberger, F.; et al. Interaction between the TP63 and SHH pathways is an important determinant of epidermal homeostasis. *Cell Death Differ.* **2013**, *20*, 1080–1088. [CrossRef] [PubMed]

125. Taylor, M.D.; Zhang, X.; Liu, L.; Hui, C.-C.; Mainprize, T.G.; Scherer, S.W.; Wainwright, B.; Hogg, D.; Rutka, J.T. Failure of a medulloblastoma-derived mutant of SUFU to suppress WNT signaling. *Oncogene* **2004**, *23*, 4577–4583. [CrossRef] [PubMed]

126. Rubin, A.I.; Chen, E.H.; Ratner, D. Basal-Cell Carcinoma. *N. Engl. J. Med.* **2005**, *353*, 2262–2269. [CrossRef] [PubMed]

127. Donovan, J. Review of the hair follicle origin hypothesis for basal cell Carcinoma. *Dermatol. Surg.* **2009**, *35*, 1311–1323. [CrossRef] [PubMed]

128. Reifenberger, J.; Wolter, M.; Knobbe, C.B.; Köhler, B.; Schönicke, A.; Scharwächter, C.; Kumar, K.; Blaschke, B.; Ruzicka, T.; Reifenberger, G. Somatic mutations in the PTCH, SMOH, SUFUH and TP53 genes in sporadic basal cell carcinomas. *Br. J. Dermatol.* **2005**, *152*, 43–51. [CrossRef] [PubMed]

129. Khavari, P.A. Modelling cancer in human skin tissue. *Nat. Rev. Cancer* **2006**, *6*, 270–280. [CrossRef] [PubMed]

130. Nilsson, M.; Undèn, A.B.; Krause, D.; Malmqwist, U.; Raza, K.; Zaphiropoulos, P.G.; Toftgård, R. Induction of basal cell carcinomas and trichoepitheliomas in mice overexpressing GLI-1. *Proc. Natl. Acad. Sci. USA* **2000**, *97*, 3438–3443. [CrossRef] [PubMed]

131. Grachtchouk, M.; Mo, R.; Yu, S.; Zhang, X.; Sasaki, H.; Hui, C.C.; Dlugosz, A.A. Basal cell carcinomas in mice overexpressing Gli2 in skin. *Nat. Genet.* **2000**, *24*, 216–217. [CrossRef] [PubMed]

132. Svärd, J.; Heby-Henricson, K.; Henricson, K.H.; Persson-Lek, M.; Rozell, B.; Lauth, M.; Bergström, A.; Ericson, J.; Toftgård, R.; Teglund, S. Genetic elimination of Suppressor of fused reveals an essential repressor function in the mammalian Hedgehog signaling pathway. *Dev. Cell* **2006**, *10*, 187–197. [CrossRef] [PubMed]

133. Wang, G.Y.; Wang, J.; Mancianti, M.L.; Epstein, E.H. Basal cell carcinomas arise from hair follicle stem cells in Ptch1+/− mice. *Cancer Cell* **2011**, *19*, 114–124. [CrossRef] [PubMed]

134. Youssef, K.K.; Lapouge, G.; Bouvrée, K.; Rorive, S.; Brohée, S.; Appelstein, O.; Larsimont, J.-C.; Sukumaran, V.; Van de Sande, B.; Pucci, D.; et al. Adult interfollicular tumour-initiating cells are reprogrammed into an embryonic hair follicle progenitor-like fate during basal cell carcinoma initiation. *Nat. Cell Biol.* **2012**, *14*, 1282–1294. [CrossRef] [PubMed]

135. Peterson, S.C.; Eberl, M.; Vagnozzi, A.N.; Belkadi, A.; Veniaminova, N.A.; Verhaegen, M.E.; Bichakjian, C.K.; Ward, N.L.; Dlugosz, A.A.; Wong, S.Y. Basal cell carcinoma preferentially arises from stem cells within hair follicle and mechanosensory niches. *Cell Stem Cell* **2015**, *16*, 400–412. [CrossRef] [PubMed]

136. Sharpe, H.J.; Pau, G.; Dijkgraaf, G.J.; Basset-Seguin, N.; Modrusan, Z.; Januario, T.; Tsui, V.; Durham, A.B.; Dlugosz, A.A.; Haverty, P.M.; et al. Genomic Analysis of Smoothened Inhibitor Resistance in Basal Cell Carcinoma. *Cancer Cell* **2015**, *27*, 327–341. [CrossRef] [PubMed]

137. Atwood, S.X.; Sarin, K.Y.; Li, J.R.; Yao, C.Y.; Urman, N.M.; Chang, A.L.S.; Tang, J.Y.; Oro, A.E. Rolling the Genetic Dice: Neutral and Deleterious Smoothened Mutations in Drug-Resistant Basal Cell Carcinoma. *J. Investig. Dermatol.* **2015**, *135*, 2138–2141. [CrossRef] [PubMed]

138. Tang, Y.; Gholamin, S.; Schubert, S.; Willardson, M.I.; Lee, A.; Bandopadhayay, P.; Bergthold, G.; Masoud, S.; Nguyen, B.; Vue, N.; et al. Epigenetic targeting of Hedgehog pathway transcriptional output through BET bromodomain inhibition. *Nat. Med.* **2014**, *20*, 732–740. [CrossRef] [PubMed]

139. Infante, P.; Mori, M.; Alfonsi, R.; Ghirga, F.; Aiello, F.; Toscano, S.; Ingallina, C.; Siler, M.; Cucchi, D.; Po, A.; et al. Gli1/DNA interaction is a druggable target for Hedgehog-dependent tumors. *EMBO J.* **2015**, *34*, 200–217. [CrossRef] [PubMed]
140. Ally, M.S.; Ransohoff, K.; Sarin, K.; Atwood, S.X.; Rezaee, M.; Bailey-Healy, I.; Kim, J.; Beachy, P.A.; Chang, A.L.S.; Oro, A.; et al. Effects of Combined Treatment With Arsenic Trioxide and Itraconazole in Patients With Refractory Metastatic Basal Cell Carcinoma. *JAMA Dermatol.* **2016**, *152*, 452. [CrossRef] [PubMed]
141. Filippakopoulos, P.; Qi, J.; Picaud, S.; Shen, Y.; Smith, W.B.; Fedorov, O.; Morse, E.M.; Keates, T.; Hickman, T.T.; Felletar, I.; et al. Selective inhibition of BET bromodomains. *Nature* **2010**, *468*, 1067–1073. [CrossRef] [PubMed]
142. Atwood, S.X.; Li, M.; Lee, A.; Tang, J.Y.; Oro, A.E. GLI activation by atypical protein kinase C ι/λ regulates the growth of basal cell carcinomas. *Nature* **2013**, *494*, 484–488. [CrossRef] [PubMed]
143. Peng, T.; Frank, D.B.; Kadzik, R.S.; Morley, M.P.; Rathi, K.S.; Wang, T.; Zhou, S.; Cheng, L.; Lu, M.M.; Morrisey, E.E. Hedgehog actively maintains adult lung quiescence and regulates repair and regeneration. *Nature* **2015**, *526*, 578–582. [CrossRef] [PubMed]

3

Emerging Roles of DYRK Kinases in Embryogenesis and Hedgehog Pathway Control

Rajeev Singh and Matthias Lauth *

Philipps University Marburg, Institute of Molecular Biology and Tumor Research (IMT), Center for Tumor and Immune Biology (ZTI), Hans-Meerwein-Str. 3, 35043 Marburg, Germany; singhr@staff.uni-marburg.de
* Correspondence: lauth@imt.uni-marburg.de

Abstract: Hedgehog (Hh)/GLI signaling is an important instructive cue in various processes during embryonic development, such as tissue patterning, stem cell maintenance, and cell differentiation. It also plays crucial roles in the development of many pediatric and adult malignancies. Understanding the molecular mechanisms of pathway regulation is therefore of high interest. Dual-specificity tyrosine phosphorylation-regulated kinases (DYRKs) comprise a group of protein kinases which are emerging modulators of signal transduction, cell proliferation, survival, and cell differentiation. Work from the last years has identified a close regulatory connection between DYRKs and the Hh signaling system. In this manuscript, we outline the mechanistic influence of DYRK kinases on Hh signaling with a focus on the mammalian situation. We furthermore aim to bring together what is known about the functional consequences of a DYRK-Hh cross-talk and how this might affect cellular processes in development, physiology, and pathology.

Keywords: hedgehog; GLI1; dual-specificity tyrosine-regulated kinase; DYRK; MIRK; Down syndrome

1. Introduction

1.1. The Hedgehog Signaling Pathway

The *Hedgehog (Hh)* gene was first identified in genetic screens for mutations that disrupt the larval body plan in *Drosophila melanogaster* [1]. The name Hedgehog originates from the short and 'spiked' phenotype of the cuticle of the Hh mutant *Drosophila* larvae, which resembles the spikes of a hedgehog [1,2]. The members of the Hh family of proteins have since been recognized as key mediators of many fundamental processes in embryonic development, playing a crucial role in controlling cell fate, patterning, proliferation, survival, and differentiation. Furthermore, Hh signaling also regulates the maintenance of tissue stem cells and affects oncogenic transformation and the development of tumors [3–5]. Vertebrates possess three Hedgehog homologues: Desert (*DHH*), Indian (*IHH*), and Sonic (*SHH*). All three genes have evolutionary conserved roles in body plan organization and development [2,6–8]. The polarizing activity of the organizing centers located in the limb bud, the notochord, or the floor plate of the neural tube is regulated by SHH [9,10]. IHH regulates the coordination of multiple cellular events during endochondral bone development including osteoblast differentiation [11,12], while DHH is required for the development of germ cells in testes and peripheral nerve sheath formation [13].

The Hh signaling cascade has been discussed in depth by other excellent reviews of this special issue on embryogenesis (e.g., [14–17]). Briefly, the canonical Hh signaling cascade is initiated in the target cell by the Hh ligand binding to the Patched1 receptor (PTCH1) [18,19], a 12-span transmembrane protein located in the ciliary membrane relieving the repression of Smoothened (SMO) [20,21], a 7-span transmembrane protein, which is a member of the G protein-coupled receptor (GPCR) superfamily.

This de-repression results in the activation of the Hh transcriptional effectors, the zinc finger proteins of the GLI (Cubitus interruptus (Ci) in *Drosophila melanogaster*) family [22].

Several studies have reported the modulation of Hh signaling through protein kinases, amongst others PKA, PKC, GRK2, MEK, ERK, AKT, S6K, and GSK3β all of which have been documented to play a role in Hh signal transduction [23–28]. Moreover, recent studies have outlined the importance of dual-specificity tyrosine phosphorylation-regulated kinases (DYRKs) in the positive and negative regulation of Hh pathway activity [29–36]. This review centers on the DYRK family of kinases and their role in regulating the developmentally important Hh signaling pathway.

1.2. Protein Kinases: An Introduction

Protein kinases are central for the regulation of major cellular processes. Kinases play particularly prominent roles in signal transduction as they direct the cellular activities by the addition or removal of a phosphate group. As abnormal levels of protein phosphorylation are associated with the development of several diseases [37], it is crucial to delve deeper into the understanding of the varying mechanisms that control these phosphorylation events [38]. Eukaryotic protein kinases (ePKs) are divided into nine large groups (plus one atypical group which does not show similarity to ePKs), which are further divided into families and subfamilies [39,40]. These groups are: (1) Tyrosine kinases (TK); (2) Tyrosine kinase-like (TKL); (3) cAMP-dependent protein kinase, cGMP-dependent protein kinase and protein kinase C (AGC); (4) Calcium/calmodulin-dependent kinases (CAMK); (5) Casein kinase 1 (CK1); (6) Cyclin-dependent kinases (CDK), Mitogen-activated protein kinases (MAPK), Glycogen synthase kinase (GSK3) and CDC-like kinase (CLK) group of protein kinases (CMGC); (7) Homologs of the yeast STE7, STE11 and STE20 genes (STE); (8) Receptor Guanylate Cyclases (RGC); and (9) Others (kinases that do not fit within any of the other main kinase groups) [40–42].

2. The CMGC Group of Kinases

Due to sequence homologies in their kinase domains, CDKs, MAPKs, GSK3s, CLKs, and related kinases (CMGCs) form one big group of eukaryotic protein kinases [40]. The CMGC group consists of 62 members in total, which are subdivided further into nine families (CDK-, CDKL-, GSK-, CLK-, MAPK-, HIPK-, DYRK-, RCK-, and SRPK-families) [43]. This group is highly conserved during evolution, arguing that its members fulfill important functions from nematodes to humans. Given their involvement in cell proliferation, MAPKs and CDKs are the most studied kinases within the CMGC group and are the subject of intense research efforts in oncological research. Less studied candidates include the dual-specificity tyrosine regulated kinases (DYRKs) and the serine-arginine protein kinases (SRPK). In general, the kinases in the CMGC group have a broad spectrum of functional roles ranging from signal transduction to cell cycle regulation, RNA related processing, and intracellular communication [43].

3. The DYRK Family of Kinases

Dual-specificity tyrosine phosphorylation-regulated kinases belong to the CMGC group of kinases and contain a characteristic sequence motif called the DYRK-homology box (DH box) (Figure 1). YAK1 from budding yeast was the first member of the DYRK family to be discovered [44,45]. There are five members within the mammalian DYRK subfamily and they are categorized into two classes. Class I consists of DYRK1A and DYRK1B (the latter is also known as Minibrain-related kinase (Mirk)), while class II is made up of DYRK2, DYRK3, and DYRK4 [45,46]. The assortment of mammalian DYRKs in the corresponding classes is based on sequence homologies within the conserved kinase domain [46,47]. Certain sequence motifs can only be found in class I DYRKs, such as a C-terminal PEST domain (a region rich in proline (P), glutamic acid (E), serine (S), and threonine (T)) (Figure 1). The PEST sequence is known to act as a signal for rapid protein degradation [48]. However, to our knowledge, this function has not been formally proven in DYRKs. DYRK1A protein stability is regulated through the ubiquitin/proteasome system, but involves an N-terminal region [49]. DYRK1A is the only family member containing a poly-histidine stretch (13 consecutive histidine residues) and a region enriched in serine/threonine residues (S/T-rich region) [50,51]. The poly-histidine stretch promotes the targeting of DYRK1A to nuclear speckles which are enriched with pre-mRNA splicing factors regulating the splicing machinery [51,52]. Other elements, such as nuclear localization signals (NLS), can be found in many DYRKs. On the other hand, only class II DYRK kinases contain a N-terminal auto-phosphorylation accessory region (NAPA) domain, which is thought to be required for tyrosine auto-phosphorylation specifically in class II DYRKs, although DYRK2 lacking the NAPA domain has been shown to auto-phosphorylate itself under in vitro conditions [47,53–55]. Further differences include the extent of the respective N- and C-termini (Figure 1). In general, DYRK family members are known to regulate protein stability, cell proliferation, and differentiation. These events are mediated by the phosphorylation of DYRK recognition sites in target proteins. The consensus sequence motif consists of Ser or Thr followed by Pro in position +1. Furthermore, an arginine residue at position −2 or −3 relative to Ser/Thr seems to be preferred (RxxS/TP or RPxS/TP), although a considerable degree of divergence to this consensus has also been noted [47,56].

The activation loop of DYRK kinases contains a conserved YXY sequence, the phosphorylation of which leads to the activation of full enzymatic activity. Members of the DYRK family auto-phosphorylate the second tyrosine residue in order to be fully activated and then phosphorylate substrates in trans on Ser/Thr residues, hence they are known as Dual-specificity tyrosine-regulated kinases [45,47,56]. The auto-activation mode of DYRKs is in contrast to MAP kinases, where an upstream kinase is needed for the phosphorylation of the activation loop [47]. The tyrosine specificity of DYRK kinases is thought to be lost once the protein is fully translated and only the Ser/Thr specificity on target proteins remains [53,57].

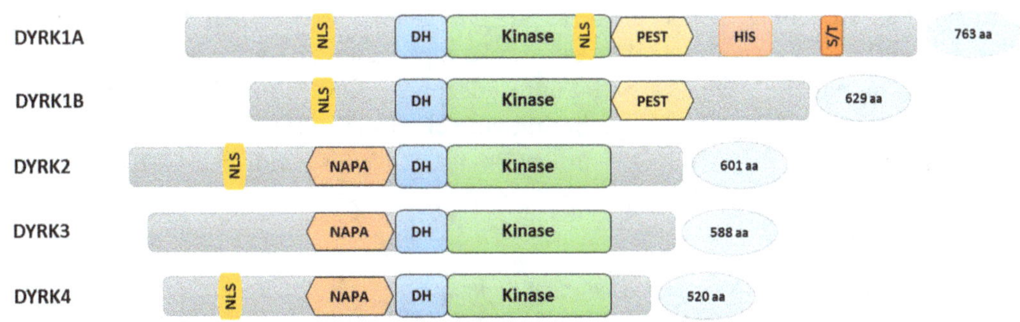

Figure 1. Schematic representation of the DYRK family of proteins: Distinct sequence motifs such as the nuclear localization signal (NLS); DYRK-homology box (DH); a motif rich in proline, glutamic acid, serine, and threonine residues (PEST); a poly-histidine stretch (HIS); a serine/threonine rich region (S/T); a N-terminal auto-phosphorylation accessory region (NAPA); and a conserved kinase domain comprising the structural and functional features of DYRKs.

4. Class I DYRKs: DYRK1A and DYRK1B

4.1. The DYRK1A Kinase

DYRK1A is a nuclear kinase, but can also be found in the cytosol. It represents the most studied member of the DYRK family, which is due to its presumed involvement in the Down syndrome (DS, OMIM #190685). DS is one of the most common genetic defects in humans with an estimated incidence of about 1 in 1000 live births worldwide and is caused by the complete or partial duplication of human chromosome 21 (trisomy 21) [58,59]. In humans, the *DYRK1A* gene is located on chromosome 21 (21q22.13), which is part of the so-called *Down-Syndrome Critical Region* (DSCR) [60]. Genes present within the DSCR (21q22.1–22.3 encompassing 33 genes) are thought to account for the development of DS, characterized by a general intellectual impairment, characteristic craniofacial dysmorphologies, and congenital heart disease [60–62].

Despite the fact that upstream modulators of DYRK1A kinase activity exist [63,64], the prime determinant of DYRK1A protein function is considered to be its overall protein amount making it very sensitive to gene dosage. An altered copy number of the *DYRK1A* gene in mammals or of its orthologous gene, *minibrain (mnb)* in *Drosophila*, impedes with the proper development of the central nervous system [65]. Different studies with the trisomic DS mouse model Ts65Dn or cells derived from Down syndrome patients [66] have shown that an increased kinase expression affects neurogenesis and neuroblast proliferation, and results in impaired behavioral phenotypes. Genetic overexpression of *Dyrk1A* in mice leads to behavioral and cognitive impairment and neuronal alterations [67–69]. In contrast, loss of function of *Dyrk1A* or *mnb* results in significant brain size reduction in mice [70], flies [65,71], and men [72].

Intriguingly, a recent study identified *DYRK1A* loss-of-function mutations which are associated with impaired dendritic and spine growth, cortical development, and the pathophysiology of autism [73]. The exact mechanisms underlying DYRK1A's effects on dendritogenesis and neurogenesis remain open, but might involve its role in actin regulation [36,74–77]. Furthermore, DYRK1A has functions in synaptogenesis and synaptic vesicle endocytosis [74,78]. Haplo-insufficiency of *DYRK1A* is associated with the development of autosomal dominant mental retardation-7 (MRD7) (OMIM #614104), a syndrome characterized by primary microcephaly, facial dysmorphism, and behavioral problems [79]. Also, *DYRK1A* expression might be epigenetically misregulated in the William-Beuren region duplication syndrome (WBS) (OMIM #609757). WBS phenotypes commonly include craniofacial anomalies and cognitive deficits ranging from mental retardation to autism [80,81].

The documented evidence of DYRK1A functioning in brain development suggests that it interacts with embryonic signaling pathways such as Hedgehog, which is known to be crucial for neuronal specification in the neural tube, hippocampal neural stem cell maintenance, and the development of the cerebellar cortex [82,83]. Mice with a genetic *Shh* knockout present with Cyclopia [84], and inactivating mutations in the human *SHH* gene cause holoprosencephaly (OMIM #236100), a common form of structural malformation of the developing brain hemispheres [85–87]. In contrast, human patients suffering from the Hh-activating Gorlin syndrome (Basal Cell Nevus Syndrome, OMIM #109400) have an increased brain size [88].

4.2. DYRK1A as a Regulator of (Neuronal) Hedgehog Signaling

The fact that both Hh signaling and the DYRK1A kinase have such important roles in embryonic brain development suggests that they might be functionally linked. The exact interplay between DYRK1A and Hh signaling seems to be complex and stimulatory, and inhibitory functions have also been described (Figure 2). Indeed, suppression of Hh pathway activity was seen in cerebellar cells derived from a Down syndrome mouse model [89]. Furthermore, some morphological as well as functional deficits could be ameliorated by the application of a synthetic SMO agonist or by genetic Hh pathway activation [90,91], suggesting a too low level of Hh signaling in DS. Subsequent mechanistic studies revealed that increased levels of the DSCR-localized DYRK1A kinase can dampen Hh signaling [92], most likely through its effect on the actin cytoskeleton and on actin-regulated transcriptional regulators [36]. For instance, DYRK1A can phosphorylate the F-actin stabilizing ABLIM proteins and thereby functionally exert a negative impact on the actin cytoskeleton and on actin-modulated transcriptional co-factors such as MAL (MKL1, MRTF), which also modulate the Hh pathway [36]. This mechanism might explain why DS cerebellar cells display a limited response to Purkinje cell-derived SHH (Figure 2). However, this finding was unexpected as previous reports had proposed a direct activating function of DYRK1A on GLI1. Specifically, DYRK1A can phosphorylate amino acid residues in the N-terminus critical for the nuclear import of GLI1 [31,32,36,92] (Figure 2). As a result, the impact of DYRK1A on Hh signaling might be context-dependent and might also be dictated by the exact mode of pathway activation (ligand/receptor-triggered versus direct GLI1 activation). Further investigations are certainly needed to clarify this point.

Hypothetically, a physiological connection between Hh signaling and DYRK1A might also exist in the case of neural stem cell (NSC) division. In general, stem or progenitor cells can undergo symmetric or asymmetric types of cell division in order to generate progeny [93]. Hh signaling preferentially supports symmetric cell divisions [94,95]. DYRK1A has also been implicated in signaling aspects during asymmetric versus symmetric neural stem cell division, although the details await further investigation [96,97]. In general, Hh pathway activity has been associated with brain size (see above), which might be caused by its positive effects on neural stem cell pools. Intriguingly, truncation of DYRK1A results in the stimulation of kinase activity [98] and *DYRK1A* gene truncations have been found in human microcephaly [72]. Finally, it is interesting to note that Down syndrome patients have a reduced risk of developing solid cancer. The fact that DYRK1A is capable of suppressing canonical Hh signaling might contribute to its described potential as a tumor suppressor [99–103], in addition to other Hh-independent proposed mechanisms [104,105].

In addition to the neuronal effects of DYRK1A, one report exists describing the Hh-related impact of DYRK1A and its physical interactors HAN11 and mDia1 in cultured sebocytes [106]. Overexpression of either HAN11 or mDia1 suppressed GLI1 nuclear localization and activity in reporter assays and slowed the growth of these cells. In the murine embryo, *Han11* is expressed in the developing limb bud (E10.5), together with *Gli1* and *Ptch1*. It is possible that the actin-regulating formin mDia1 is functionally linked to the aforementioned DYRK1A-ABLIM-Actin-MAL-GLI axis, but experimental proof is lacking at this point.

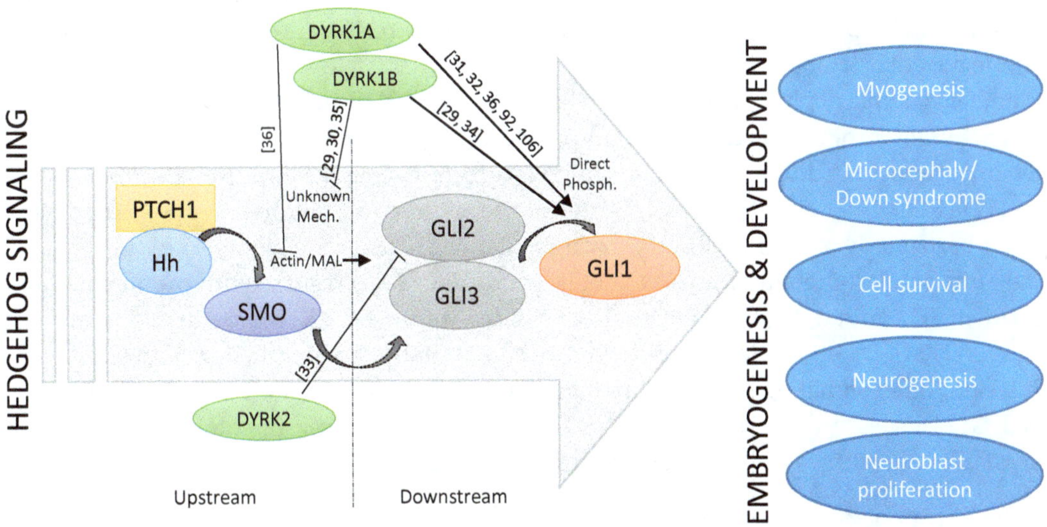

Figure 2. Schematic depiction of the cross-talk between Hh signaling and DYRK kinases.

Three DYRK kinases (DYRK1A, DYRK1B, DYRK2) have been described as regulators of upstream (above GLI transcription factors) and downstream (at the level of GLI transcription factors) Hh signaling. As Hh pathway modulators, they might control important embryogenic and developmental processes, such as myogenesis, neurogenesis, or the pathophysiology of Down syndrome. Literature references are given in square brackets.

4.3. The DYRK1B Kinase

The closest relative of mammalian DYRK1A is the DYRK1B kinase, also referred to as MIRK (Minibrain-related kinase). The human *DYRK1B* gene is located on chromosome 19q13.2, a region often amplified in ovarian and pancreatic cancer [107,108]. This kinase has three splice variants (629aa (p69), 601aa (p66), and 589aa (p65)) and is expressed in abundance in human skeletal muscle and testes [109]. Human DYRK1A and DYRK1B proteins are 84% identical in the N-terminal and catalytic domains but show no extended similarity in the C-terminal domain. Human and mouse DYRK1B proteins share

97% sequence similarity [46,110–113]. In many different cell types, DYRK1B can be found both in the nucleus and in the cytoplasm [114].

4.4. DYRK1B in Developmental and Physiological Processes

The observation that DYRK1B is strongly expressed in skeletal muscle argues for a physiological role in muscle function and/or development. Indeed, DYRK1B levels have been shown to be comparatively low in myoblasts, but to increase significantly upon the induction of differentiation [115]. DYRK1B favors myoblast fusion and the subsequent expression of differentiation markers [116]. Furthermore, DYRK1B supports the survival of muscle progenitor (C2C12) cells in culture and of cells from muscle-related tumors such as rhabdomyosarcoma [117,118]. The pro-differentiating effects of DYRK1B on myoblasts are opposite to the effects which Hh signaling exerts on muscle stem cells (satellite cells) and on C2C12 progenitor cells [119]. Here, Hh promotes cell division and blocks differentiation along the myogenic lineage, thereby maintaining the progenitor cell pool. Although it is not clear whether the influence of Hh or DYRK1B occurs exactly at the same developmental stage, currently available data would suggest a primarily antagonistic relationship between these two pathways. Recent work has identified a complex regulatory relationship between DYRK1B and Hh. While DYRK1B dampens Hh signaling initiated by SMO, it also promotes the stability of the GLI1 transcription factor on the other side [29,30,34]. The latter might be mediated by DYRK1B-induced stimulation of the pro-survival PI3K-AKT signaling pathway, a known positive regulator of GLI stability [23,120] (Figure 2). In addition, at least in cultured fibroblasts, Hh pathway stimulation increases DYRK1B protein levels by currently unknown post-transcriptional mechanisms [29], suggesting a feedback loop.

Another example of physiological cross-talk between DYRK1B and Hh might be the differentiation of mesenchymal progenitor cells into adipocytes. Hh signaling has a generally inhibiting impact on adipocytic differentiation, usually redirecting cellular fate towards the osteogenic lineage [121–123]. In contrast, DYRK1B favors the in vitro differentiation into adipocytes [35]. This holds particularly true for DYRK1B carrying mutations which were identified in families suffering from an autosomal-dominant form of metabolic syndrome [35], a disease with prominent adipocyte involvement. The mutations found result in misfolding of the DYRK1B protein and in intracellular aggregation [124]. It remains to be clarified how these mutations affect the functional integration of DYRK1B into other signaling pathways, but it is intriguing to note that mutant DYRK1B expression reduced GLI2 levels in cultured adipocytes [35]. It is therefore reasonable to speculate that the suppression of Hh pathway activity contributes to these effects.

5. The Class II DYRKs

5.1. The DYRK2, DYRK3, and DYRK4 Kinases

Compared to the DYRK class I members, the class II DYRKs (DYRK2, DYRK3, DYRK4) contain a larger N-terminal region and a shorter C-terminal domain. DYRK2 and DYRK4, but not DYRK3, possess an NLS sequence and all three contain an NAPA (N-terminal autophosphorylation accessory region) domain which is absent in class I DYRKs [47]. The NAPA domain provides a chaperone-like function and transiently converts class II DYRKs into intramolecular tyrosine kinases [54]. Despite lacking an apparent NLS, DYRK3 (also named REDK) is localized in the nucleus in hematopoetic cells [125], whereas DYRK2 is mostly cytosolic, but under conditions of genotoxic stress, it accumulates in the nucleus regulating p53 [126]. DYRK4, which is currently the least studied DYRK family member, displays splice variant-dependent subcellular localization [47].

In contrast to DYRK1B, which has been described as an oncogenic kinase in numerous cancer types, DYRK2 can also exert opposite functions and can display tumor suppressive traits. This is brought about by DYRK2's ability to activate p53-dependent apoptosis following DNA damage [126,127] and by negatively controlling the protein stability of well-established oncogenes such as c-MYC or c-JUN [128]. Phosphorylation-dependent regulation of proteasomal degradation seems to be a recurrent mechanism employed by many if not all DYRK kinases [129–134].

5.2. Class II DYRKs in Development

In zebrafish, *DYRK2* has been shown to be expressed in lateral somites (mesodermal blocks around the anterior-posterior axis of the developing embryo) and adaxial cells (muscle precursor cells that are adjacent to the notochord and part of the presomitic mesoderm) at an early stage of embryogenesis [135]. Co-localization of *Dyrk2* mRNA and *myogenic differentiation factor D (MyoD)* mRNA was seen in muscle progenitor cells in the posterior compartment of somites. Here, DYRK2 might positively regulate fast twitch muscle differentiation in the early stages of embryonic development [135]. Although the link has not yet been experimentally verified, it is intriguing to note that in contrast to DYRK2, Hh signaling promotes the formation of slow-twitch fibers in zebrafish [136]. Furthermore, mammalian DYRK2 has been shown to negatively regulate Hh pathway activity by phosphorylating and degrading GLI2 [33] (Figure 2). It is therefore reasonable to speculate that DYRK2, through its negative influence on Hh signaling, might impact on the slow/fast-twitch fiber differentiation during muscle development.

A similarly antagonistic relationship between DYRK2 and Hh signaling might also play a role in *Drosophila*, which encodes three DYRKs: Minibrain/Dyrk1A, DmDyrk2, and DmDyrk3. Recent reports have shown that *DmDyrk2* is expressed in the developing third antennal segment, an anatomical structure responsible for smell, and in the morphogenetic furrow of the developing eye, where it contributes to the development of the visual system [137]. In addition, Hedgehog is a known regulator of morphogenetic furrow progression and ommatidial cell differentiation in the *Drosophila* eye disc [138,139]. If, in analogy to mammals, DmDYRK2 also regulates Hh signaling, it is intriguing to hypothesize that functional DYRK2-Hh cross-talk is involved in the specification of the *Drosophila* eye.

In comparison to class I DYRKs and DYRK2, the class II family members DYRK3 and DYRK4 show a very restricted expression profile with the strongest expression in erythroid progenitors and testes, respectively [140,141]. As can be assumed from this expression pattern, DYRK3 is involved in erythropoiesis. While *Dyrk3*$^{-/-}$ mice surprisingly present without a hematological phenotype, they develop increased numbers of red blood cells under conditions of anemia, suggesting that DYRK3 functions as a negative regulator of erythropoiesis [140]. Therefore, small-molecule DYRK3 inhibitors might be of interest to ameliorate anemic conditions. Although Hh ligands (mostly DHH and IHH) have also been shown to regulate erythropoiesis [142–144], it currently remains unclear whether cross-talk between DYRK3 and Hh signaling contributes to this process.

Significantly more work has been done on the function of DYRK3 on the cellular and molecular level, albeit a clear link to Hh signaling has so far not been established. Specifically, DYRK3 impinges on stress-associated mTOR signaling [145], as well as on endocytosis dynamics [146]. Endocytic sorting of SHH and PTCH1 in clathrin-coated vesicles is also critical for proper Hh signaling in signal producing and receiving cells [147–149]. However, whether DYRK3 is indeed involved in these steps awaits further experimentation.

Of all the DYRK kinases discussed so far, the least is known about DYRK4. The expression of this family member is strongly restricted to testicular tissue, with a strikingly selective peak of expression in step VIII spermatids [46,141], suggesting a role in male fertility. Surprisingly, however, analysis of *Dyrk4* null animals revealed no aberrant sperm phenotype or defects in male fertility [141]. DYRK4 was shown to be present in the duck ovary, and was more active or upregulated in the high egg production ovaries, which would suggest a hitherto unrecognized role in the female reproductive system, at least in some species [150]. Desert Hh (DHH) signaling also occurs in testes and, at least in certain species, also in ovaries, but the involvement of DYRK4 in DHH-mediated processes is unclear at the moment [151–153].

6. Conclusions

DYRK kinases are highly conserved during evolution from yeast to humans. Due to the evolutionary diversification, DYRKs might represent the requirement of more critical and specialized functions in vertebrates or might have contributed to this diversification. Multifaceted roles of DYRK kinases have been discussed in this review and their importance in various developmental processes has been stated. As of now, three of five mammalian DYRK kinases have been functionally linked to Hh signaling (DYRK1A, DYRK1B, DYRK2), arguing for a close regulatory connectivity to the developmentally important Hh system. Hence, although DYRK kinases are not absolutely required for Hh signaling, they function as modulators and it is therefore reasonable to hypothesize that they contribute to many Hh-driven steps during embryonic development. Unfortunately, however, more work is needed to provide a clear picture of the exact and tissue-specific cross-talk between DYRKs and Hh, particularly in in vivo settings. As certain DYRKs have a preferred expression in specific tissues (e.g., DYRK1A in neuronal and DYRK1B in muscle tissue), it is reasonable to speculate that the impact on the tissue-selective Hh pathway activity is specified by the respective DYRK enzyme. In other tissues or cell types, where several DYRK kinases are expressed together at comparable levels, a certain degree of functional redundancy might exist, particularly for the class I DYRKs. These questions are important to address in the future in light of the development of small-molecule inhibitors which might lack the necessary specificity and target several DYRKs simultaneously. Complicating the developmental interpretation is the fact that DYRK kinases also modulate other, non-Hh signaling systems, such as, e.g., the NFAT (nuclear factor of activated T-cells) [154] pathway or HIF (hypoxia-inducible factor) signaling [155]. Future work will reveal whether Hedgehog or any of the other signaling systems is particularly important for the physiological impact of DYRK kinases.

Most evidence for developmental cross-talk between DYRKs and Hh stems from studies on DYRK1A, neuronal development, and the Down syndrome. In addition, available data encourage speculations on DYRK1B and DYRK2 modulating Hh signaling in muscle development and on the involvement of DYRK1B in adipocyte differentiation. In light of recent reports strengthening the concept of Hh-pathway modulation by DYRKs in pathological conditions such as metabolic syndrome or cancer, it will be interesting to see whether future research unveils more cross-talk between this group of kinases and the Hh system in physiological processes. In this review, we have tried to outline the currently available knowledge on the DYRK family of kinases engaging in developmental biology, physiology, and pathology, focusing on its impact on Hh signaling.

Acknowledgments: We apologize to all colleagues who have contributed important work to this research field, but whose publications we could not cite due to space constraints. This work was supported by the University Medical Center Giessen and Marburg (UKGM) and the German Research Foundation (DFG-KFO325).

Conflicts of Interest: The authors declare no conflict of interest.

References

1. Nusslein-Volhard, C.; Wieschaus, E. Mutations affecting segment number and polarity in Drosophila. *Nature* **1980**, *287*, 795–801. [CrossRef] [PubMed]
2. Ingham, P.W.; McMahon, A.P. Hedgehog signaling in animal development: Paradigms and principles. *Genes. Dev.* **2001**, *15*, 3059–3087. [CrossRef] [PubMed]
3. Briscoe, J.; Therond, P.P. The mechanisms of Hedgehog signalling and its roles in development and disease. *Nat. Rev. Mol. Cell Biol.* **2013**, *14*, 416–429. [CrossRef] [PubMed]
4. Lee, R.T.; Zhao, Z.; Ingham, P.W. Hedgehog signalling. *Development* **2016**, *143*, 367–372. [CrossRef] [PubMed]
5. Di Magno, L.; Coni, S.; Di Marcotullio, L.; Canettieri, G. Digging a hole under Hedgehog: Downstream inhibition as an emerging anticancer strategy. *Biochim. Biophys. Acta* **2015**, *1856*, 62–72. [CrossRef] [PubMed]
6. Riddle, R.D.; Johnson, R.L.; Laufer, E.; Tabin, C. Sonic hedgehog mediates the polarizing activity of the ZPA. *Cell* **1993**, *75*, 1401–1416. [CrossRef]

7. Echelard, Y.; Epstein, D.J.; St-Jacques, B.; Shen, L.; Mohler, J.; McMahon, J.A.; McMahon, A.P. Sonic hedgehog, a member of a family of putative signaling molecules, is implicated in the regulation of CNS polarity. *Cell* **1993**, *75*, 1417–1430. [CrossRef]
8. Roelink, H.; Augsburger, A.; Heemskerk, J.; Korzh, V.; Norlin, S.; Ruiz i Altaba, A.; Tanabe, Y.; Placzek, M.; Edlund, T.; Jessell, T.M.; et al. Floor plate and motor neuron induction by vhh-1, a vertebrate homolog of hedgehog expressed by the notochord. *Cell* **1994**, *76*, 761–775. [CrossRef]
9. Cohn, M.J.; Tickle, C. Limbs: A model for pattern formation within the vertebrate body plan. *Trends Genet.* **1996**, *12*, 253–257. [CrossRef]
10. Jessell, T.M. Neuronal specification in the spinal cord: Inductive signals and transcriptional codes. *Nat. Rev. Genet.* **2000**, *1*, 20–29. [CrossRef] [PubMed]
11. McMahon, A.P.; Ingham, P.W.; Tabin, C.J. Developmental roles and clinical significance of hedgehog signaling. *Curr. Top. Dev. Biol.* **2003**, *53*, 1–114. [PubMed]
12. St-Jacques, B.; Hammerschmidt, M.; McMahon, A.P. Indian hedgehog signaling regulates proliferation and differentiation of chondrocytes and is essential for bone formation. *Genes Dev.* **1999**, *13*, 2072–2086. [CrossRef] [PubMed]
13. Park, S.Y.; Tong, M.; Jameson, J.L. Distinct roles for steroidogenic factor 1 and desert hedgehog pathways in fetal and adult Leydig cell development. *Endocrinology* **2007**, *148*, 3704–3710. [CrossRef] [PubMed]
14. Pandit, T.; Ogden, S. Contributions of Noncanonical Smoothened Signaling During Embryonic Development. *J. Dev. Biol.* **2017**, *5*, 11. [CrossRef]
15. Belgacem, Y.H.; Hamilton, A.M.; Shim, S.; Spencer, K.A.; Borodinsky, L.N. The Many Hats of Sonic Hedgehog Signaling in Nervous System Development and Disease. *J. Dev. Biol.* **2016**, *4*, 35. [CrossRef]
16. Ramsbottom, S.A.; Pownall, M.E. Regulation of Hedgehog Signalling Inside and Outside the Cell. *J. Dev. Biol.* **2016**, *4*, 23. [CrossRef] [PubMed]
17. Fernandes-Silva, H.; Correia-Pinto, J.; Moura, R. Canonical Sonic Hedgehog Signaling in Early Lung Development. *J. Dev. Biol.* **2017**, *5*, 3. [CrossRef]
18. Nakano, Y.; Guerrero, I.; Hidalgo, A.; Taylor, A.; Whittle, J.R.S.; Ingham, P.W. A protein with several possible membrane-spanning domains encoded by the Drosophila segment polarity gene patched. *Nature* **1989**, *341*, 508–513. [CrossRef] [PubMed]
19. Hooper, J.E.; Scott, M.P. The Drosophila patched gene encodes a putative membrane protein required for segmental patterning. *Cell* **1989**, *59*, 751–765. [CrossRef]
20. Taipale, J.; Cooper, M.K.; Maiti, T.; Beachy, P.A. Patched acts catalytically to suppress the activity of Smoothened. *Nature* **2002**, *418*, 892–897. [CrossRef] [PubMed]
21. Roberts, B.; Casillas, C.; Alfaro, A.C.; Jagers, C.; Roelink, H. Patched1 and Patched2 inhibit Smoothened non-cell autonomously. *Elife* **2016**, *5*, e17634. [CrossRef] [PubMed]
22. Hui, C.C.; Angers, S. Gli proteins in development and disease. *Annu. Rev. Cell Dev. Biol.* **2011**, *27*, 513–537. [CrossRef] [PubMed]
23. Riobo, N.A.; Lu, K.; Ai, X.; Haines, G.M.; Emerson, C.P., Jr. Phosphoinositide 3-kinase and Akt are essential for Sonic Hedgehog signaling. *Proc. Natl. Acad. Sci. USA* **2006**, *103*, 4505–4510. [CrossRef] [PubMed]
24. Lauth, M.; Bergstrom, A.; Toftgard, R. Phorbol esters inhibit the Hedgehog signalling pathway downstream of Suppressor of Fused, but upstream of Gli. *Oncogene* **2007**, *26*, 5163–5168. [CrossRef] [PubMed]
25. Wang, Y.; Ding, Q.; Yen, C.J.; Xia, W.; Izzo, J.G.; Lang, J.Y.; Li, C.W.; Hsu, J.L.; Miller, S.A.; Wang, X.; et al. The crosstalk of mTOR/S6K1 and Hedgehog pathways. *Cancer Cell* **2012**, *21*, 374–387. [CrossRef] [PubMed]
26. Wang, B.; Fallon, J.F.; Beachy, P.A. Hedgehog-regulated processing of Gli3 produces an anterior/posterior repressor gradient in the developing vertebrate limb. *Cell* **2000**, *100*, 423–434. [CrossRef]
27. Chen, Y.; Yue, S.; Xie, L.; Pu, X.H.; Jin, T.; Cheng, S.Y. Dual Phosphorylation of suppressor of fused (Sufu) by PKA and GSK3beta regulates its stability and localization in the primary cilium. *J. Biol. Chem.* **2011**, *286*, 13502–13511. [CrossRef] [PubMed]
28. Zhao, Z.; Lee, R.T.; Pusapati, G.V.; Iyu, A.; Rohatgi, R.; Ingham, P.W. An essential role for Grk2 in Hedgehog signalling downstream of Smoothened. *EMBO Rep.* **2016**, *17*, 739–752. [CrossRef] [PubMed]
29. Singh, R.; Dhanyamraju, P.K.; Lauth, M. DYRK1B blocks canonical and promotes non-canonical Hedgehog signaling through activation of the mTOR/AKT pathway. *Oncotarget* **2017**, *8*, 833–845. [CrossRef] [PubMed]

30. Lauth, M.; Bergstrom, A.; Shimokawa, T.; Tostar, U.; Jin, Q.; Fendrich, V.; Guerra, C.; Barbacid, M.; Toftgard, R. DYRK1B-dependent autocrine-to-paracrine shift of Hedgehog signaling by mutant RAS. *Nat. Struct. Mol. Biol.* **2010**, *17*, 718–725. [CrossRef] [PubMed]
31. Shimokawa, T.; Tostar, U.; Lauth, M.; Palaniswamy, R.; Kasper, M.; Toftgard, R.; Zaphiropoulos, P.G. Novel human glioma-associated oncogene 1 (GLI1) splice variants reveal distinct mechanisms in the terminal transduction of the hedgehog signal. *J. Biol. Chem.* **2008**, *283*, 14345–14354. [CrossRef] [PubMed]
32. Mao, J.; Maye, P.; Kogerman, P.; Tejedor, F.J.; Toftgard, R.; Xie, W.; Wu, G.; Wu, D. Regulation of Gli1 transcriptional activity in the nucleus by Dyrk1. *J. Biol. Chem.* **2002**, *277*, 35156–35161. [CrossRef] [PubMed]
33. Varjosalo, M.; Bjorklund, M.; Cheng, F.; Syvanen, H.; Kivioja, T.; Kilpinen, S.; Sun, Z.; Kallioniemi, O.; Stunnenberg, H.G.; He, W.W.; et al. Application of active and kinase-deficient kinome collection for identification of kinases regulating hedgehog signaling. *Cell* **2008**, *133*, 537–548. [CrossRef] [PubMed]
34. Gruber, W.; Hutzinger, M.; Elmer, D.P.; Parigger, T.; Sternberg, C.; Cegielkowski, L.; Zaja, M.; Leban, J.; Michel, S.; Hamm, S.; et al. DYRK1B as therapeutic target in Hedgehog/GLI-dependent cancer cells with Smoothened inhibitor resistance. *Oncotarget* **2016**, *7*, 7134–7148. [CrossRef] [PubMed]
35. Keramati, A.R.; Fathzadeh, M.; Go, G.W.; Singh, R.; Choi, M.; Faramarzi, S.; Mane, S.; Kasaei, M.; Sarajzadeh-Fard, K.; Hwa, J.; et al. A form of the metabolic syndrome associated with mutations in DYRK1B. *N. Engl. J. Med.* **2014**, *370*, 1909–1919. [CrossRef] [PubMed]
36. Schneider, P.; Bayo-Fina, J.M.; Singh, R.; Kumar Dhanyamraju, P.; Holz, P.; Baier, A.; Fendrich, V.; Ramaswamy, A.; Baumeister, S.; Martinez, E.D.; et al. Identification of a novel actin-dependent signal transducing module allows for the targeted degradation of GLI1. *Nat. Commun.* **2015**, *6*, 8023. [CrossRef] [PubMed]
37. Futreal, P.A.; Coin, L.; Marshall, M.; Down, T.; Hubbard, T.; Wooster, R.; Rahman, N.; Stratton, M.R. A census of human cancer genes. *Nat. Rev. Cancer* **2004**, *4*, 177–183. [CrossRef] [PubMed]
38. Lahiry, P.; Torkamani, A.; Schork, N.J.; Hegele, R.A. Kinase mutations in human disease: Interpreting genotype-phenotype relationships. *Nat. Rev. Genet.* **2010**, *11*, 60–74. [CrossRef] [PubMed]
39. Hanks, S.K.; Hunter, T. Protein kinases 6. The eukaryotic protein kinase superfamily: Kinase (catalytic) domain structure and classification. *FASEB J.* **1995**, *9*, 576–596. [PubMed]
40. Manning, G.; Whyte, D.B.; Martinez, R.; Hunter, T.; Sudarsanam, S. The protein kinase complement of the human genome. *Science* **2002**, *298*, 1912–1934. [CrossRef] [PubMed]
41. Miranda-Saavedra, D.; Barton, G.J. Classification and functional annotation of eukaryotic protein kinases. *Proteins Struct. Funct. Bioinform.* **2007**, *68*, 893–914. [CrossRef] [PubMed]
42. Manning, G.; Plowman, G.D.; Hunter, T.; Sudarsanam, S. Evolution of protein kinase signaling from yeast to man. *Trends Biochem. Sci.* **2002**, *27*, 514–520. [CrossRef]
43. Varjosalo, M.; Keskitalo, S.; Van Drogen, A.; Nurkkala, H.; Vichalkovski, A.; Aebersold, R.; Gstaiger, M. The protein interaction landscape of the human CMGC kinase grou. *Cell Rep.* **2013**, *3*, 1306–1320. [CrossRef] [PubMed]
44. Garrett, S.; Broach, J. Loss of Ras activity in Saccharomyces cerevisiae is suppressed by disruptions of a new kinase gene, YAKI, whose product may act downstream of the cAMP-dependent protein kinase. *Genes Dev.* **1989**, *3*, 1336–1348. [CrossRef] [PubMed]
45. Soppa, U.; Becker, W. DYRK protein kinases. *Curr. Biol.* **2015**, *25*, R488–R489. [CrossRef] [PubMed]
46. Becker, W.; Weber, Y.; Wetzel, K.; Eirmbter, K.; Tejedor, F.J.; Joost, H.G. Sequence characteristics, subcellular localization, and substrate specificity of DYRK-related kinases, a novel family of dual specificity protein kinases. *J. Biol. Chem.* **1998**, *273*, 25893–25902. [CrossRef] [PubMed]
47. Aranda, S.; Laguna, A.; de la Luna, S. DYRK family of protein kinases: Evolutionary relationships, biochemical properties, and functional roles. *FASEB J.* **2011**, *25*, 449–462. [CrossRef] [PubMed]
48. Galceran, J.; de Graaf, K.; Tejedor, F.J.; Becker, W. The MNB/DYRK1A protein kinase: Genetic and biochemical properties. *J. Neural. Transm. Suppl.* **2003**, *67*, 139–148.
49. Liu, Q.; Tang, Y.; Chen, L.; Liu, N.; Lang, F.; Liu, H.; Wang, P.; Sun, X. E3 Ligase SCFbetaTrCP-induced DYRK1A Protein Degradation Is Essential for Cell Cycle Progression in HEK293 Cells. *J. Biol. Chem.* **2016**, *291*, 26399–26409. [CrossRef] [PubMed]
50. Park, J.; Song, W.J.; Chung, K.C. Function and regulation of Dyrk1A: Towards understanding Down syndrome. *Cell Mol. Life Sci.* **2009**, *66*, 3235–3240. [CrossRef] [PubMed]

51. Alvarez, M.; Estivill, X.; de la Luna, S. DYRK1A accumulates in splicing speckles through a novel targeting signal and induces speckle disassembly. *J. Cell Sci.* **2003**, *116*, 3099–3107. [CrossRef] [PubMed]
52. Salichs, E.; Ledda, A.; Mularoni, L.; Alba, M.M.; de la Luna, S. Genome-wide analysis of histidine repeats reveals their role in the localization of human proteins to the nuclear speckles compartment. *PLoS Genet.* **2009**, *5*, e1000397. [CrossRef] [PubMed]
53. Lochhead, P.A.; Sibbet, G.; Morrice, N.; Cleghon, V. Activation-loop autophosphorylation is mediated by a novel transitional intermediate form of DYRKs. *Cell* **2005**, *121*, 925–936. [CrossRef] [PubMed]
54. Kinstrie, R.; Luebbering, N.; Miranda-Saavedra, D.; Sibbet, G.; Han, J.; Lochhead, P.A.; Cleghon, V. Characterization of a domain that transiently converts class 2 DYRKs into intramolecular tyrosine kinases. *Sci. Signal* **2010**, *3*, ra16. [CrossRef] [PubMed]
55. Walte, A.; Ruben, K.; Birner-Gruenberger, R.; Preisinger, C.; Bamberg-Lemper, S.; Hilz, N.; Bracher, F.; Becker, W. Mechanism of dual specificity kinase activity of DYRK1A. *FEBS. J.* **2013**, *280*, 4495–4511. [CrossRef] [PubMed]
56. Soundararajan, M.; Roos, A.K.; Savitsky, P.; Filippakopoulos, P.; Kettenbach, A.N.; Olsen, J.V.; Gerber, S.A.; Eswaran, J.; Knapp, S.; Elkins, J.M. Structures of Down syndrome kinases, DYRKs, reveal mechanisms of kinase activation and substrate recognition. *Structure* **2013**, *21*, 986–996. [CrossRef] [PubMed]
57. Alvarez, M.; Altafaj, X.; Aranda, S.; de la Luna, S. DYRK1A Autophosphorylation on Serine Residue 520 Modulates Its Kinase Activity via 14–3-3 Binding. *Mol. Biol. Cell* **2007**, *18*, 1167–1178. [CrossRef] [PubMed]
58. Gardiner, K.; Herault, Y.; Lott, I.T.; Antonarakis, S.E.; Reeves, R.H.; Dierssen, M. Down syndrome: From understanding the neurobiology to therapy. *J. Neurosci.* **2010**, *30*, 14943–14945. [CrossRef] [PubMed]
59. Presson, A.P.; Partyka, G.; Jensen, K.M.; Devine, O.J.; Rasmussen, S.A.; McCabe, L.L.; McCabe, E.R.B. Current Estimate of Down Syndrome Population Prevalence in the United States. *J. Pediatr.* **2013**, *163*, 1163–1168. [CrossRef] [PubMed]
60. Epstein, C.J. Down's syndrome: Critical genes in a critical region. *Nature* **2006**, *441*, 582–583. [CrossRef] [PubMed]
61. Montoya, J.C.; Fajardo, D.; Pena, A.; Sanchez, A.; Dominguez, M.C.; Satizabal, J.M.; Garcia-Vallejo, F. Global differential expression of genes located in the Down Syndrome Critical Region in normal human brain. *Colomb. Med. (Cali)* **2014**, *45*, 154–161. [PubMed]
62. Korenberg, J.R.; Kawashima, H.; Pulst, S.M.; Ikeuchi, T.; Ogasawara, N.; Yamamoto, K.; Schonberg, S.A.; West, R.; Allen, L.; Magenis, E.; et al. Molecular definition of a region of chromosome 21 that causes features of the Down syndrome phenotype. *Am. J. Hum. Genet.* **1990**, *47*, 236–246. [PubMed]
63. Tschop, K.; Conery, A.R.; Litovchick, L.; Decaprio, J.A.; Settleman, J.; Harlow, E.; Dyson, N. A kinase shRNA screen links LATS2 and the pRB tumor suppressor. *Genes Dev.* **2011**, *25*, 814–830. [CrossRef] [PubMed]
64. Litovchick, L.; Florens, L.A.; Swanson, S.K.; Washburn, M.P.; DeCaprio, J.A. DYRK1A protein kinase promotes quiescence and senescence through DREAM complex assembly. *Genes Dev.* **2011**, *25*, 801–813. [CrossRef] [PubMed]
65. Tejedor, F.; Zhu, X.R.; Kaltenbach, E.; Ackermann, A.; Baumann, A.; Canal, I.; Heisenberg, M.; Fischbach, K.F.; Pongs, O. Minibrain: A new protein kinase family involved in postembryonic neurogenesis in Drosophila. *Neuron* **1995**, *14*, 287–301. [CrossRef]
66. Contestabile, A.; Fila, T.; Ceccarelli, C.; Bonasoni, P.; Bonapace, L.; Santini, D.; Bartesaghi, R.; Ciani, E. Cell cycle alteration and decreased cell proliferation in the hippocampal dentate gyrus and in the neocortical germinal matrix of fetuses with Down syndrome and in Ts65Dn mice. *Hippocampus* **2007**, *17*, 665–678. [CrossRef] [PubMed]
67. Thomazeau, A.; Lassalle, O.; Iafrati, J.; Souchet, B.; Guedj, F.; Janel, N.; Chavis, P.; Delabar, J.; Manzoni, O.J. Prefrontal Deficits in a Murine Model Overexpressing the Down Syndrome Candidate Gene *Dyrk1a*. *J. Neurosci.* **2014**, *34*, 1138–1147. [CrossRef] [PubMed]
68. Dierssen, M. Down syndrome: The brain in trisomic mode. *Nat. Rev. Neurosci.* **2012**, *13*, 844–858. [CrossRef] [PubMed]
69. Martinez de Lagran, M.; Altafaj, X.; Gallego, X.; Marti, E.; Estivill, X.; Sahun, I.; Fillat, C.; Dierssen, M. Motor phenotypic alterations in TgDyrk1a transgenic mice implicate DYRK1A in Down syndrome motor dysfunction. *Neurobiol. Dis.* **2004**, *15*, 132–142. [CrossRef] [PubMed]

70. Van Bon, B.W.M.; Coe, B.P.; Bernier, R.; Green, C.; Gerdts, J.; Witherspoon, K.; Kleefstra, T.; Willemsen, M.H.; Kumar, R.; Bosco, P.; et al. Disruptive de novo mutations of DYRK1A lead to a syndromic form of autism and ID. *Mol. Psychiatry* **2016**, *21*, 126–132. [CrossRef] [PubMed]
71. Tejedor, F.J.; Hammerle, B. MNB/DYRK1A as a multiple regulator of neuronal development. *FEBS J.* **2010**, *278*, 223–235. [CrossRef] [PubMed]
72. Moller, R.S.; Kubart, S.; Hoeltzenbein, M.; Heye, B.; Vogel, I.; Hansen, C.P.; Menzel, C.; Ullmann, R.; Tommerup, N.; Ropers, H.H.; et al. Truncation of the Down syndrome candidate gene DYRK1A in two unrelated patients with microcephaly. *Am. J. Hum. Genet.* **2008**, *82*, 1165–1170. [CrossRef] [PubMed]
73. Dang, T.; Duan, W.Y.; Yu, B.; Tong, D.L.; Cheng, C.; Zhang, Y.F.; Wu, W.; Ye, K.; Zhang, W.X.; Wu, M.; et al. Autism-associated Dyrk1a truncation mutants impair neuronal dendritic and spine growth and interfere with postnatal cortical development. *Mol. Psychiatry* **2017**. [CrossRef] [PubMed]
74. Martinez de Lagran, M.; Benavides-Piccione, R.; Ballesteros-Yanez, I.; Calvo, M.; Morales, M.; Fillat, C.; Defelipe, J.; Ramakers, G.J.; Dierssen, M. Dyrk1A influences neuronal morphogenesis through regulation of cytoskeletal dynamics in mammalian cortical neurons. *Cereb. Cortex* **2012**, *22*, 2867–2877. [CrossRef] [PubMed]
75. Park, J.; Sung, J.Y.; Park, J.; Song, W.J.; Chang, S.; Chung, K.C. Dyrk1A negatively regulates the actin cytoskeleton through threonine phosphorylation of N-WASP. *J. Cell Sci.* **2012**, *125*, 67–80. [CrossRef] [PubMed]
76. Liu, T.; Sims, D.; Baum, B. Parallel RNAi screens across different cell lines identify generic and cell type-specific regulators of actin organization and cell morphology. *Genome Biol.* **2009**, *10*, R26. [CrossRef] [PubMed]
77. Gockler, N.; Jofre, G.; Papadopoulos, C.; Soppa, U.; Tejedor, F.J.; Becker, W. Harmine specifically inhibits protein kinase DYRK1A and interferes with neurite formation. *FEBS J.* **2009**, *276*, 6324–6337. [CrossRef] [PubMed]
78. Kim, Y.; Park, J.; Song, W.J.; Chang, S. Overexpression of Dyrk1A causes the defects in synaptic vesicle endocytosis. *Neurosignals* **2010**, *18*, 164–172. [CrossRef] [PubMed]
79. Van Bon, B.W.M.; Hoischen, A.; Hehir-Kwa, J.; de Brouwer, A.P.M.; Ruivenkamp, C.; Gijsbers, A.C.J.; Marcelis, C.L.; de Leeuw, N.; Veltman, J.A.; Brunner, H.G.; et al. Intragenic deletion in DYRK1A leads to mental retardation and primary microcephaly. *Clin. Genet.* **2011**, *79*, 296–299. [CrossRef] [PubMed]
80. Hogart, A.; Leung, K.N.; Wang, N.J.; Wu, D.J.; Driscoll, J.; Vallero, R.O.; Schanen, N.C.; LaSalle, J.M. Chromosome 15q11–13 duplication syndrome brain reveals epigenetic alterations in gene expression not predicted from copy number. *J. Med. Genet.* **2009**, *46*, 86–93. [CrossRef] [PubMed]
81. Strong, E.; Butcher, D.T.; Singhania, R.; Mervis, C.B.; Morris, C.A.; De Carvalho, D.; Weksberg, R.; Osborne, L.R. Symmetrical Dose-Dependent DNA-Methylation Profiles in Children with Deletion or Duplication of 7q11.23. *Am. J. Hum. Genet.* **2015**, *97*, 216–227. [CrossRef] [PubMed]
82. F Ferent, J.; Traiffort, E. Hedgehog: Multiple Paths for Multiple Roles in Shaping the Brain and Spinal Cord. *Neuroscientist* **2015**, *21*, 356–371. [CrossRef] [PubMed]
83. Stecca, B.; Ruiz i Altaba, A. Brain as a paradigm of organ growth: Hedgehog-Gli signaling in neural stem cells and brain tumors. *J. Neurobiol.* **2005**, *64*, 476–490. [CrossRef] [PubMed]
84. Chiang, C.; Litingtung, Y.; Lee, E.; Young, K.E.; Corden, J.L.; Westphal, H.; Beachy, P.A. Cyclopia and defective axial patterning in mice lacking Sonic hedgehog gene function. *Nature* **1996**, *383*, 407–413. [CrossRef] [PubMed]
85. Roessler, E.; Belloni, E.; Gaudenz, K.; Jay, P.; Berta, P.; Scherer, S.W.; Tsui, L.C.; Muenke, M. Mutations in the human Sonic Hedgehog gene cause holoprosencephaly. *Nat. Genet.* **1996**, *14*, 357–360. [CrossRef] [PubMed]
86. Traiffort, E.; Dubourg, C.; Faure, H.; Rognan, D.; Odent, S.; Durou, M.R.; David, V.; Ruat, M. Functional characterization of sonic hedgehog mutations associated with holoprosencephaly. *J. Biol. Chem.* **2004**, *279*, 42889–42897. [CrossRef] [PubMed]
87. Belloni, E.; Muenke, M.; Roessler, E.; Traverso, G.; Siegel-Bartelt, J.; Frumkin, A.; Mitchell, H.F.; Donis-Keller, H.; Helms, C.; Hing, A.V.; et al. Identification of Sonic hedgehog as a candidate gene responsible for holoprosencephaly. *Nat. Genet.* **1996**, *14*, 353–356. [CrossRef] [PubMed]
88. Shiohama, T.; Fujii, K.; Miyashita, T.; Mizuochi, H.; Uchikawa, H.; Shimojo, N. Brain morphology in children with nevoid basal cell carcinoma syndrome. *Am. J. Med. Genet. A* **2017**, *173*, 946–952. [CrossRef] [PubMed]

89. Roper, R.J.; Baxter, L.L.; Saran, N.G.; Klinedinst, D.K.; Beachy, P.A.; Reeves, R.H. Defective cerebellar response to mitogenic Hedgehog signaling in Down [corrected] syndrome mice. *Proc. Natl. Acad. Sci. USA* **2006**, *103*, 1452–1456. [CrossRef] [PubMed]
90. Das, I.; Park, J.M.; Shin, J.H.; Jeon, S.K.; Lorenzi, H.; Linden, D.J.; Worley, P.F.; Reeves, R.H. Hedgehog agonist therapy corrects structural and cognitive deficits in a Down syndrome mouse model. *Sci. Transl. Med.* **2013**, *5*, 201ra120. [CrossRef] [PubMed]
91. Dutka, T.; Hallberg, D.; Reeves, R.H. Chronic up-regulation of the SHH pathway normalizes some developmental effects of trisomy in Ts65Dn mice. *Mech. Dev.* **2015**, *135*, 68–80. [CrossRef] [PubMed]
92. Ehe, B.K.; Lamson, D.R.; Tarpley, M.; Onyenwoke, R.U.; Graves, L.M.; Williams, K.P. Identification of a DYRK1A-mediated phosphorylation site within the nuclear localization sequence of the hedgehog transcription factor GLI1. *Biochem. Biophys. Res. Commun.* **2017**, *491*, 767–772. [CrossRef] [PubMed]
93. Morrison, S.J.; Kimble, J. Asymmetric and symmetric stem-cell divisions in development and cancer. *Nature* **2006**, *441*, 1068–1074. [CrossRef] [PubMed]
94. Saade, M.; Gonzalez-Gobartt, E.; Escalona, R.; Usieto, S.; Marti, E. Shh-mediated centrosomal recruitment of PKA promotes symmetric proliferative neuroepithelial cell division. *Nat. Cell Biol.* **2017**, *19*, 493–503. [CrossRef] [PubMed]
95. Yang, R.; Wang, M.; Wang, J.; Huang, X.; Yang, R.; Gao, W.Q. Cell Division Mode Change Mediates the Regulation of Cerebellar Granule Neurogenesis Controlled by the Sonic Hedgehog Signaling. *Stem. Cell Rep.* **2015**, *5*, 816–828. [CrossRef] [PubMed]
96. Ferron, S.R.; Pozo, N.; Laguna, A.; Aranda, S.; Porlan, E.; Moreno, M.; Fillat, C.; de la Luna, S.; Sanchez, P.; Arbones, M.L.; et al. Regulated segregation of kinase Dyrk1A during asymmetric neural stem cell division is critical for EGFR-mediated biased signaling. *Cell Stem Cell* **2010**, *7*, 367–379. [CrossRef] [PubMed]
97. Hammerle, B.; Vera-Samper, E.; Speicher, S.; Arencibia, R.; Martinez, S.; Tejedor, F.J. Mnb/Dyrk1A is transiently expressed and asymmetrically segregated in neural progenitor cells at the transition to neurogenic divisions. *Dev. Biol.* **2002**, *246*, 259–273. [CrossRef] [PubMed]
98. Jin, N.; Yin, X.; Gu, J.; Zhang, X.; Shi, J.; Qian, W.; Ji, Y.; Cao, M.; Gu, X.; Ding, F.; et al. Truncation and Activation of Dual Specificity Tyrosine Phosphorylation-regulated Kinase 1A by Calpain I: A molecular mechanism linked to tau pathology in alzheimer disease. *J. Biol. Chem.* **2015**, *290*, 15219–15237. [CrossRef] [PubMed]
99. Mann, K.M.; Ward, J.M.; Yew, C.C.; Kovochich, A.; Dawson, D.W.; Black, M.A.; Brett, B.T.; Sheetz, T.E.; Dupuy, A.J.; Chang, D.K.; et al. Sleeping Beauty mutagenesis reveals cooperating mutations and pathways in pancreatic adenocarcinoma. *Proc. Natl. Acad. Sci. USA* **2012**, *109*, 5934–5941. [CrossRef] [PubMed]
100. Nizetic, D.; Groet, J. Tumorigenesis in Down's syndrome: Big lessons from a small chromosome. *Nat. Rev. Cancer* **2012**, *12*, 721–732. [CrossRef] [PubMed]
101. Sussan, T.E.; Yang, A.; Li, F.; Ostrowski, M.C.; Reeves, R.H. Trisomy represses Apc(Min)-mediated tumours in mouse models of Down's syndrome. *Nature* **2008**, *451*, 73–75. [CrossRef] [PubMed]
102. De Wit, N.J.; Burtscher, H.J.; Weidle, U.H.; Ruiter, D.J.; van Muijen, G.N. Differentially expressed genes identified in human melanoma cell lines with different metastatic behaviour using high density oligonucleotide arrays. *Melanoma Res.* **2002**, *12*, 57–69. [CrossRef] [PubMed]
103. Genovesi, L.A.; Ng, C.G.; Davis, M.J.; Remke, M.; Taylor, M.D.; Adams, D.J.; Rust, A.G.; Ward, J.M.; Ban, K.H.; Jenkins, N.A.; et al. Sleeping Beauty mutagenesis in a mouse medulloblastoma model defines networks that discriminate between human molecular subgroups. *Proc. Natl. Acad. Sci. USA* **2013**, *110*, E4325–E4334. [CrossRef] [PubMed]
104. Baek, K.H.; Zaslavsky, A.; Lynch, R.C.; Britt, C.; Okada, Y.; Siarey, R.J.; Lensch, M.W.; Park, I.H.; Yoon, S.S.; Minami, T.; et al. Down's syndrome suppression of tumour growth and the role of the calcineurin inhibitor DSCR1. *Nature* **2009**, *459*, 1126–1130. [CrossRef] [PubMed]
105. Reynolds, L.E.; Watson, A.R.; Baker, M.; Jones, T.A.; D'Amico, G.; Robinson, S.D.; Joffre, C.; Garrido-Urbani, S.; Rodriguez-Manzaneque, J.C.; Martino-Echarri, E.; et al. Tumour angiogenesis is reduced in the Tc1 mouse model of Down's syndrome. *Nature* **2010**, *465*, 813–817. [CrossRef] [PubMed]
106. Morita, K.; Lo Celso, C.; Spencer-Dene, B.; Zouboulis, C.C.; Watt, F.M. HAN11 binds mDia1 and controls GLI1 transcriptional activity. *J. Dermatol. Sci.* **2006**, *44*, 11–20. [CrossRef] [PubMed]
107. Jin, K.; Park, S.; Ewton, D.Z.; Friedman, E. The survival kinase Mirk/Dyrk1B is a downstream effector of

oncogenic K-ras in pancreatic cancer. *Cancer Res.* **2007**, *67*, 7247–7255. [CrossRef] [PubMed]

108. Davis, S.J.; Sheppard, K.E.; Pearson, R.B.; Campbell, I.G.; Gorringe, K.L.; Simpson, K.J. Functional analysis of genes in regions commonly amplified in high-grade serous and endometrioid ovarian cancer. *Clin. Cancer Res.* **2013**, *19*, 1411–1421. [CrossRef] [PubMed]
109. Lee, K.; Deng, X.; Friedman, E. Mirk protein kinase is a mitogen-activated protein kinase substrate that mediates survival of colon cancer cells. *Cancer Res.* **2000**, *60*, 3631–3637. [PubMed]
110. Becker, W.; Joost, H.G. Structural and functional characteristics of Dyrk, a novel subfamily of protein kinases with dual specificity. *Prog. Nucleic Acid. Res. Mol. Biol.* **1999**, *62*, 1–17. [PubMed]
111. Kentrup, H.; Becker, W.; Heukelbach, J.; Wilmes, A.; Schürmann, A.; Huppertz, C.; Kainulainen, H.; Joost, H.G. Dyrk, a dual specificity protein kinase with unique structural features whose activity is dependent on tyrosine residues between subdomains VII and VIII. *J. Biol. Chem.* **1996**, *271*, 3488–3495. [CrossRef] [PubMed]
112. Leder, S.; Weber, Y.; Altafaj, X.; Estivill, X.; Joost, H.G.; Becker, W. Cloning and characterization of DYRK1B, a novel member of the DYRK family of protein kinases. *Biochem. Biophys. Res. Commun.* **1999**, *254*, 474–479. [CrossRef] [PubMed]
113. Li, Q.Z.; Wang, C.Y.; Shi, J.D.; Ruan, Q.G.; Eckenrode, S.; Davoodi-Semiromi, A.; Kukar, T.; Gu, Y.; Lian, W.; Wu, D.; et al. Molecular cloning and characterization of the mouse and human TUSP gene, a novel member of the tubby superfamily. *Gene* **2001**, *273*, 275–284. [CrossRef]
114. Gao, J.; Zheng, Z.; Rawal, B.; Schell, M.J.; Bepler, G.; Haura, E.B. Mirk/Dyrk1B, a novel therapeutic target, mediates cell survival in non-small cell lung cancer cells. *Cancer Biol. Ther.* **2009**, *8*, 1671–1679. [CrossRef] [PubMed]
115. Mercer, S.E.; Friedman, E. Mirk/Dyrk1B: A multifunctional dual-specificity kinase involved in growth arrest, differentiation, and cell survival. *Cell Biochem. Biophys.* **2006**, *45*, 303–315. [CrossRef]
116. Deng, X.; Ewton, D.Z.; Pawlikowski, B.; Maimone, M.; Friedman, E. Mirk/dyrk1B is a Rho-induced kinase active in skeletal muscle differentiation. *J. Biol. Chem.* **2003**, *278*, 41347–41354. [CrossRef] [PubMed]
117. Mercer, S.E.; Ewton, D.Z.; Deng, X.; Lim, S.; Mazur, T.R.; Friedman, E. Mirk/Dyrk1B mediates survival during the differentiation of C2C12 myoblasts. *J. Biol. Chem.* **2005**, *280*, 25788–25801. [CrossRef] [PubMed]
118. Mercer, S.E.; Ewton, D.Z.; Shah, S.; Naqvi, A.; Friedman, E. Mirk/Dyrk1b mediates cell survival in rhabdomyosarcomas. *Cancer Res.* **2006**, *66*, 5143–5150. [CrossRef] [PubMed]
119. Koleva, M.; Kappler, R.; Vogler, M.; Herwig, A.; Fulda, S.; Hahn, H. Pleiotropic effects of sonic hedgehog on muscle satellite cells. *Cell Mol. Life Sci.* **2005**, *62*, 1863–1870. [CrossRef] [PubMed]
120. Kern, D.; Regl, G.; Hofbauer, S.W.; Altenhofer, P.; Achatz, G.; Dlugosz, A.; Schnidar, H.; Greil, R.; Hartmann, T.N.; Aberger, F. Hedgehog/GLI and PI3K signaling in the initiation and maintenance of chronic lymphocytic leukemia. *Oncogene* **2015**, *34*, 5341–5351. [CrossRef] [PubMed]
121. Johnson, J.S.; Meliton, V.; Kim, W.K.; Lee, K.B.; Wang, J.C.; Nguyen, K.; Yoo, D.; Jung, M.E.; Atti, E.; Tetradis, S.; et al. Novel oxysterols have pro-osteogenic and anti-adipogenic effects in vitro and induce spinal fusion in vivo. *J. Cell Biochem.* **2011**, *112*, 1673–1684. [CrossRef] [PubMed]
122. Kha, H.T.; Basseri, B.; Shouhed, D.; Richardson, J.; Tetradis, S.; Hahn, T.J.; Parhami, F. Oxysterols regulate differentiation of mesenchymal stem cells: Pro-bone and anti-fat. *J. Bone Miner. Res.* **2004**, *19*, 830–840. [CrossRef] [PubMed]
123. Nosavanh, L.; Yu, D.H.; Jaehnig, E.J.; Tong, Q.; Shen, L.; Chen, M.H. Cell-autonomous activation of Hedgehog signaling inhibits brown adipose tissue development. *Proc. Natl. Acad. Sci. USA* **2015**, *112*, 5069–5074. [CrossRef] [PubMed]
124. Abu Jhaisha, S.; Widowati, E.W.; Kii, I.; Sonamoto, R.; Knapp, S.; Papadopoulos, C.; Becker, W. DYRK1B mutations associated with metabolic syndrome impair the chaperone-dependent maturation of the kinase domain. *Sci. Rep.* **2017**, *7*, 6420. [CrossRef] [PubMed]
125. Lord, K.A.; Creasy, C.L.; King, A.G.; King, C.; Burns, B.M.; Lee, J.C.; Dillon, S.B. REDK, a novel human regulatory erythroid kinase. *Blood* **2000**, *95*, 2838. [PubMed]
126. Taira, N.; Nihira, K.; Yamaguchi, T.; Miki, Y.; Yoshida, K. DYRK2 is targeted to the nucleus and controls p53 via Ser46 phosphorylation in the apoptotic response to DNA damage. *Mol. Cell* **2007**, *25*, 725–738. [CrossRef] [PubMed]
127. Taira, N.; Yamamoto, H.; Yamaguchi, T.; Miki, Y.; Yoshida, K. ATM augments nuclear stabilization of

DYRK2 by inhibiting MDM2 in the apoptotic response to DNA damage. *J. Biol. Chem.* **2010**, *285*, 4909–4919. [CrossRef] [PubMed]

128. Taira, N.; Mimoto, R.; Kurata, M.; Yamaguchi, T.; Kitagawa, M.; Miki, Y.; Yoshida, K. DYRK2 priming phosphorylation of c-Jun and c-Myc modulates cell cycle progression in human cancer cells. *J. Clin. Investig.* **2012**, *122*, 859–872. [CrossRef] [PubMed]
129. Liu, H.; Wang, K.; Chen, S.; Sun, Q.; Zhang, Y.; Chen, L.; Sun, X. NFATc1 phosphorylation by DYRK1A increases its protein stability. *PLoS ONE* **2017**, *12*, e0172985. [CrossRef] [PubMed]
130. Kurabayashi, N.; Hirota, T.; Sakai, M.; Sanada, K.; Fukada, Y. DYRK1A and glycogen synthase kinase 3beta, a dual-kinase mechanism directing proteasomal degradation of CRY2 for circadian timekeeping. *Mol. Cell Biol.* **2010**, *30*, 1757–1768. [CrossRef] [PubMed]
131. Zou, Y.; Ewton, D.Z.; Deng, X.; Mercer, S.E.; Friedman, E. Mirk/dyrk1B kinase destabilizes cyclin D1 by phosphorylation at threonine 288. *J. Biol. Chem.* **2004**, *279*, 27790–27798. [CrossRef] [PubMed]
132. Deng, X.; Mercer, S.E.; Shah, S.; Ewton, D.Z.; Friedman, E. The cyclin-dependent kinase inhibitor p27Kip1 is stabilized in G by Mirk/dyrk1B kinase. *J. Biol. Chem.* **2004**, *279*, 22498–22504. [CrossRef] [PubMed]
133. Yamaguchi, N.; Mimoto, R.; Yanaihara, N.; Imawari, Y.; Hirooka, S.; Okamoto, A.; Yoshida, K. DYRK2 regulates epithelial-mesenchymal-transition and chemosensitivity through Snail degradation in ovarian serous adenocarcinoma. *Tumour Biol.* **2015**. [CrossRef] [PubMed]
134. Maddika, S.; Chen, J. Protein kinase DYRK2 is a scaffold that facilitates assembly of an E3 ligase. *Nat. Cell Biol.* **2009**, *11*, 409–419. [CrossRef] [PubMed]
135. Sun, W.; Jiao, S.; Tan, X.; Zhang, P.; You, F. DYRK2 displays muscle fiber type specific function during zebrafish early somitogenesis. *Int. J. Dev. Biol.* **2017**, *61*, 459–463. [CrossRef] [PubMed]
136. Jackson, H.E.; Ingham, P.W. Control of muscle fibre-type diversity during embryonic development: The zebrafish paradigm. *Mech. Dev.* **2013**. [CrossRef] [PubMed]
137. Luebbering, N.; Charlton-Perkins, M.; Kumar, J.P.; Lochead, P.A.; Rollmann, S.M.; Cook, T.; Cleghon, V. Drosophila Dyrk2 plays a role in the development of the visual system. *PLoS ONE* **2013**, *8*, e76775. [CrossRef] [PubMed]
138. Strutt, D.I.; Mlodzik, M. Hedgehog is an indirect regulator of morphogenetic furrow progression in the Drosophila eye disc. *Development* **1997**, *124*, 3233–3240. [PubMed]
139. Baker, N.E.; Bhattacharya, A.; Firth, L.C. Regulation of Hh signal transduction as Drosophila eye differentiation progresses. *Dev. Biol.* **2009**, *335*, 356–366. [CrossRef] [PubMed]
140. Bogacheva, O.; Bogachev, O.; Menon, M.; Dev, A.; Houde, E.; Valoret, E.I.; Prosser, H.M.; Creasy, C.L.; Pickering, S.J.; Grau, E.; et al. DYRK3 Dual-specificity Kinase Attenuates Erythropoiesis during Anemia. *J. Biol. Chem.* **2008**, *283*, 36665–36675. [CrossRef] [PubMed]
141. Sacher, F.; Moller, C.; Bone, W.; Gottwald, U.; Fritsch, M. The expression of the testis-specific Dyrk4 kinase is highly restricted to step 8 spermatids but is not required for male fertility in mice. *Mol. Cell Endocrinol.* **2007**, *267*, 80–88. [CrossRef] [PubMed]
142. Cridland, S.O.; Keys, J.R.; Papathanasiou, P.; Perkins, A.C. Indian hedgehog supports definitive erythropoiesis. *Blood Cells Mol. Dis.* **2009**, *43*, 149–155. [CrossRef] [PubMed]
143. Lau, C.I.; Outram, S.V.; Saldana, J.I.; Furmanski, A.L.; Dessens, J.T.; Crompton, T. Regulation of murine normal and stress-induced erythropoiesis by Desert Hedgehog. *Blood* **2012**, *119*, 4741–4751. [CrossRef] [PubMed]
144. Paulson, R.F.; Shi, L.; Wu, D.C. Stress erythropoiesis: New signals and new stress progenitor cells. *Curr. Opin. Hematol.* **2011**, *18*, 139–145. [CrossRef] [PubMed]
145. Wippich, F.; Bodenmiller, B.; Trajkovska, M.G.; Wanka, S.; Aebersold, R.; Pelkmans, L. Dual specificity kinase DYRK3 couples stress granule condensation/dissolution to mTORC1 signaling. *Cell* **2013**, *152*, 791–805. [CrossRef] [PubMed]
146. Pelkmans, L.; Fava, E.; Grabner, H.; Hannus, M.; Habermann, B.; Krausz, E.; Zerial, M. Genome-wide analysis of human kinases in clathrin- and caveolae/raft-mediated endocytosis. *Nature* **2005**, *436*, 78–86. [CrossRef] [PubMed]
147. Yue, S.; Tang, L.Y.; Tang, Y.; Tang, Y.; Shen, Q.H.; Ding, J.; Chen, Y.; Zhang, Z.; Yu, T.T.; Zhang, Y.E.; et al. Requirement of Smurf-mediated endocytosis of Patched1 in sonic hedgehog signal reception. *Elife* **2014**, *3*, e02555. [CrossRef] [PubMed]

148. Karpen, H.E.; Bukowski, J.T.; Hughes, T.; Gratton, J.P.; Sessa, W.C.; Gailani, M.R. The sonic hedgehog receptor patched associates with caveolin-1 in cholesterol-rich microdomains of the plasma membrane. *J. Biol. Chem.* **2001**, *276*, 19503–19511. [CrossRef] [PubMed]
149. Mao, H.; Diehl, A.M.; Li, Y.X. Sonic hedgehog ligand partners with caveolin-1 for intracellular transport. *Lab. Investig.* **2009**, *89*, 290–300. [CrossRef] [PubMed]
150. Tao, Z.; Song, W.; Zhu, C.; Xu, W.; Liu, H.; Zhang, S.; Huifang, L. Comparative transcriptomic analysis of high and low egg-producing duck ovaries. *Poult. Sci.* **2017**. [CrossRef] [PubMed]
151. Sahin, Z.; Szczepny, A.; McLaughlin, E.A.; Meistrich, M.L.; Zhou, W.; Ustunel, I.; Loveland, K.L. Dynamic Hedgehog signalling pathway activity in germline stem cells. *Andrology* **2014**, *2*, 267–274. [CrossRef] [PubMed]
152. O'Hara, W.A.; Azar, W.J.; Behringer, R.R.; Renfree, M.B.; Pask, A.J. Desert hedgehog is a mammal-specific gene expressed during testicular and ovarian development in a marsupial. *BMC Dev. Biol.* **2011**, *11*, 72. [CrossRef] [PubMed]
153. Yao, H.H.; Whoriskey, W.; Capel, B. Desert Hedgehog/Patched 1 signaling specifies fetal Leydig cell fate in testis organogenesis. *Genes Dev.* **2002**, *16*, 1433–1440. [CrossRef] [PubMed]
154. Gwack, Y.; Sharma, S.; Nardone, J.; Tanasa, B.; Iuga, A.; Srikanth, S.; Okamura, H.; Bolton, D.; Feske, S.; Hogan, P.G.; et al. A genome-wide Drosophila RNAi screen identifies DYRK-family kinases as regulators of NFAT. *Nature* **2006**, *441*, 646–650. [CrossRef] [PubMed]
155. Lee, S.B.; Frattini, V.; Bansal, M.; Castano, A.M.; Sherman, D.; Hutchinson, K.; Bruce, J.N.; Califano, A.; Liu, G.; Cardozo, T.; et al. An ID2-dependent mechanism for VHL inactivation in cancer. *Nature* **2016**, *529*, 172–177. [CrossRef] [PubMed]

4

Drosophila as a Model to Study the Link between Metabolism and Cancer

Héctor Herranz * and Stephen M. Cohen

Department of Cellular and Molecular Medicine, University of Copenhagen, Blegdamsvej 3, 2200 N Copenhagen, Denmark; scohen@sund.ku.dk
* Correspondence: hherranz@sund.ku.dk

Abstract: Cellular metabolism has recently been recognized as a hallmark of cancer. Investigating the origin and effects of the reprogrammed metabolism of tumor cells, and identifying its genetic mediators, will improve our understanding of how these changes contribute to disease progression and may suggest new approaches to therapy. *Drosophila melanogaster* is emerging as a valuable model to study multiple aspects of tumor formation and malignant transformation. In this review, we discuss the use of *Drosophila* as model to study how changes in cellular metabolism, as well as metabolic disease, contribute to cancer.

Keywords: *Drosophila*; animal model; metabolism; cancer; Warburg effect; obesity; type 2 diabetes

---

1. Introduction

Cancer is a genetic disease that involves the accumulation of mutations that affect multiple biological processes. Cancer-promoting changes include sustaining proliferative signalling, evading growth suppressors, resisting cell death, enabling replicative immortality, inducing angiogenesis, and activating invasion and metastasis, which can result in the death of the host [1]. Tumorigenesis is also dependent on changes in cellular metabolism. Nearly a century ago, Otto Warburg observed a metabolic shift in cancer cells toward increased use of glucose in a process called aerobic glycolysis. However, tumor metabolism has only recently been recognized as a hallmark of cancer [1,2]. Interest in cancer metabolism has grown rapidly in recent years, with an emphasis on understanding how cancer metabolism is regulated, how it influences disease progression, and how these factors could be exploited to improve cancer therapy [3].

Drosophila melanogaster is emerging as a valuable model for studying neurological disorders, cardiovascular diseases, and cancer. Flies have many experimental advantages, including sophisticated genetic tools, a short life cycle, and the ease with which large numbers of individuals can be examined in well-targeted genetic screens. Genetic conservation between flies and humans is substantial, with 75% of known human disease genes having related sequences in *Drosophila* [4]. Consequently, hypotheses generated using flies often prove to be relevant to biomedicine. *Drosophila* has served as a platform to develop models that recapitulate various aspects of human cancer, and its use has contributed to revealing different aspects of human malignant tumors [5]. This review summarizes the contribution of *Drosophila* to study how metabolism affects tumorigenesis and malignant transformation.

2. Metabolic Control of Tissue Growth

Nutritional status plays a pivotal role regulating metabolism and tissue growth during development. Nutrient-sensing systems coordinate growth by connecting nutrient availability and metabolism with the growth machinery. Insulin/insulin-like growth factor (IGF) signalling is a universal, nutrient-sensing system. It is conserved from mammals to *Drosophila* and plays a central role regulating growth in response to nutrient availability. insulin/IGF binds receptor tyrosine kinases, and this leads to the activation of the Ras/MAPK and PI-3K/Akt/TOR pathways. By activating those signalling pathways, insulin/IGF coordinates multiple processes frequently deregulated in cancer, including cellular growth, proliferation, metabolism, and cell survival. Key elements of these signalling cascades, including PI-3K, Akt, TOR, and PTEN, are recognized oncogenes and tumor suppressors [6]. Research in *Drosophila* has served to determine how components of the insulin/IGF system, including Rheb, Tsc1, and Tsc2, are organized in the insulin/Akt pathway, and has contributed to defining how insulin/Akt controls systemic and cellular growth. Those studies (reviewed in depth elsewhere [7–9]) have expanded our knowledge of the normal physiological roles that these oncogenes and tumor suppressors play, as well as how their dysregulation promotes disease.

3. Warburg Effect in *Drosophila*

Nearly a century ago Otto Warburg observed that tumor cells had altered glucose metabolism. In the presence of oxygen, normal cells convert glucose to pyruvate through glycolysis. Pyruvate dehydrogenase (PDH) mediates the conversion of Pyruvate to Acetyl-CoA, which feeds into the Tricarboxylic acid (TCA) cycle and will eventually be metabolized to CO_2 in the mitochondria via oxidative phosphorylation. Under anaerobic conditions, pyruvate is converted to lactate by lactate dehydrogenase (LDH). Warburg observed that tumor cells use glycolysis rather than the TCA cycle and oxidative phosphorylation, even in the presence of oxygen [10,11]. This, ultimately leads to the generation of lactate by the tumor cells and is known as aerobic glycolysis, and the ability of cancer cells to obtain this metabolic state was termed Warburg effect. Recent studies have noted that the Warburg effect contributes to cancer progression [2].

LDH is a key enzyme in this process. Human LDH enzymes are encoded by four separate genes, LDHA-D, which catalyse the conversion of pyruvate to lactate. The LDHA isoform is predominantly expressed in cancer cells and has higher affinity for pyruvate, favouring the conversion of pyruvate to lactate [12,13]. Tumor cells frequently grow in an oxygen-poor environment and hypoxia has been proposed as the primary signal that induces metabolic reprogramming in tumor cells [14]. In an oxygen-poor environment, the Hypoxia inducible factor-1α (Hif-1α in *Drosophila*) is stabilized. At normal oxygen levels, Hif-1α is rapidly degraded. Hif-1α has been shown to regulate LDHA expression in cancer cell lines [15]. Even though this supports the metabolic switch observed in hypoxia, it does not explain how aerobic glycolysis is maintained in normoxic conditions. LDH is encoded by a single gene (*ImpL3*) in *Drosophila*. Biochemical studies demonstrated that fly LDH, like mammalian LDHA, favors the conversion of pyruvate to lactate [16]. The presence of a single LDH gene in flies eliminates potential genetic redundancy of the LDH family members and facilitates the analysis of its function. This, together with the sophisticated toolkit available in *Drosophila*, has allowed the use of *Drosophila* to study the Warburg effect.

3.1. Warburg Effect in Drosophila Tumors

The *Drosophila* imaginal discs are epithelial structures that proliferate extensively during larval development and, after metamorphosis, give rise to the external structures of the adult fly. The imaginal discs have been used to model different aspects of tumorigenesis [17]. Activation of the PDGF/VEGF-receptor (Pvr) in the imaginal discs leads to the formation of epithelial tumors that up-regulate LDH and show aerobic glycolysis [18,19]. Pvr activation in the fly imaginal epithelium has therefore been used as a model to dissect how aerobic glycolysis is induced and maintained in epithelial tumor cells [19]. Pvr activation causes metabolic reprogramming by affecting glycolysis and mitochondrial respiration. On the one hand, cells activating Pvr upregulate LDH and several glycolytic enzymes in a Hif-1α dependent manner. On the other hand, Pvr attenuates mitochondrial respiration by downregulating genes encoding mitochondria proteins, and by inhibiting PDH activity. One of the primary consequences of mitochondrial dysfunction is the accumulation of reactive oxygen species (ROS) [20,21]. ROS accumulation has two important outcomes: (1) it stabilizes Hif-1α; and (2) ROS activates the c-Jun amino (N)-terminal kinase pathway (JNK) pathway, which has been shown to inhibit PDH activity [22,23]. Therefore, experiments performed in *Drosophila* have demonstrated that the accumulation of ROS in tumor cells is used as a feedback mechanism to consolidate the high-LDH/low-PDH state, and thereby to stabilize the aerobic glycolysis state.

The Notch signalling pathway regulates cell proliferation and differentiation, and has been shown to function as both an oncogene or as a tumor suppressor [24,25]. Notch behaves as an oncogene in the *Drosophila* imaginal discs, in which Notch activation leads to tissue hyperplasia and tumor formation [26]. Notch activation also induces a switch towards increased glycolysis and reduced TCA cycle and mitochondrial respiration in the imaginal epithelia. Interestingly, several key metabolic regulators like the glucose transporter Glut-1, the glycolytic enzyme hexokinase A, and LDH have been shown to be direct target genes of the Notch pathway [27].

These studies have shed light on the molecular mechanisms by which the Pvr and Notch oncogenes promote a shift toward Warburg effect metabolism in epithelial tumors [19,27].

3.2. Warburg Effect in Normal Development

Warburg effect metabolism is thought to promote cancer cell growth by diverting glucose metabolism toward production of amino acids at the expense of efficiency in ATP production via the TCA cycle [28]. Although apparently inefficient, this shift favors increased biomass by the converting nutrients into cellular building blocks to support rapid growth during normal larval development in the fly. This depends on the activity of the *Drosophila* estrogen-related receptor (dERR) [29], a nuclear receptor that regulates the expression of genes involved in glucose metabolism. dERR mutants show lower expression of almost all glycolytic genes and die in the larval stage, presenting high levels of circulating glucose and reduced levels of lactate and triacylglycerides. The metabolic state induced by dERR aims to support larval growth by generating biomass from dietary carbohydrates, and it resembles the metabolic switch observed in Warburg effect. Interestingly, multiple studies have implicated human ERRs in cancer [30–33]. The results obtained in *Drosophila* raise the possibility that ERRs might control tumor development, at least in part, by modulating cellular metabolism in a manner that promotes a Warburg-like phenotype.

Recent work has used the *Drosophila* intestinal stem cell model to analyse how changes in the glucose metabolism affect intestinal homeostasis. The *Drosophila* gut is a simple system in which intestinal stem cells divide to self-renew and to give rise to a class of progenitors called enteroblasts, which differentiate (without cell division) into two different cell lineages: absorptive intestinal epithelial cells (enterocytes) and endocrine cells [34–36]. The mitochondrial pyruvate carrier (MPC) mediates pyruvate uptake by the mitochondria and is crucially linked to glycolysis and mitochondrial oxidation. In the *Drosophila* intestine, MPC shows high levels of expression in the more differentiated endocrine cells, and its expression is lower in the proliferative stem cells. Interestingly, MPC mutant intestinal cells proliferate faster than their wild type counterparts, and guts mutant for MPC develop tumorous growths with epithelial multi-layering, characteristics of cancer. Reducing LDH levels in the pool of intestinal stem cells has the opposite effect and leads to a reduction in the proliferation rate [37]. These results demonstrate that pyruvate oxidative metabolism plays a central role in regulating stem cell proliferation, differentiation, and tissue homeostasis, as well as in providing additional evidence supporting a central role of metabolism in development and tumorigenesis.

3.3. Mitochondrial Dysfunction and Tumor Progression

Pyruvate produced in glycolysis is used in the mitochondria to generate ATP and metabolites. Mitochondrial dysfunction is common in human cancers. However, whether mitochondrial dysfunction is a consequence of the increase in the glycolytic flux, or whether it is a driving force in tumorigenesis, is an ongoing subject of debate [38–41]. The Ras^{V12} oncogenic mutation induces cell proliferation and benign tumors in *Drosophila* epithelial cells [42] and has been used as a sensitized genetic background to search for cooperating factors that promote formation of neoplastic tumors [43–45]. One screen identified multiple mitochondrial genes that, when mutated, cooperate with Ras^{V12} in the formation of malignant tumors and metastasis [44]. Interestingly, while the genetically modified cells induce p53 and undergo cell cycle arrest and Ras-induced cellular senescence [46], they act in a non-autonomous manner to drive the formation of tumors in the surrounding normal cells [44]. These surprising findings may make sense in light of recent studies showing that senescent cells secrete cytokines and growth factors known as the senescence-associated secretory phenotype (SASP) [47]. Senescence, therefore, has concurrent and opposing roles: it has an autonomous suppressive role aimed at eliminating tumor cells, and a non-autonomous signalling effect mediated by the SASP phenotype that can promote overproliferation, leading to tumor formation in nearby cells.

Drosophila genetics has been used to explore the molecular connection between mitochondrial dysfunction and the SASP phenotype. *Drosophila* epithelial cells defective in mitochondrial function and expressing Ras^{V12} produce ROS, which activates the JNK stress-response pathway. This represses the Hippo tumor-suppressor pathway and leads to upregulation of the secreted proteins Unpaired/interleukin-6 and Wingless/WNT. Unpaired and Wingless induce cell proliferation and benign hyperplasia of neighbouring tissue. Interestingly, when Ras^{V12} is active in the surrounding cells, activation of the Unpaired and Wingless pathways promotes malignant tumors and metastasis [44]. These observations suggest that mitochondrial defects, rather than being a consequence of the increase in the glycolytic flux observed in cancer cells, can play an instructive role in cancer progression. These studies may provide a context to understand the effects of mutations in respiratory-chain complex genes that have been observed in pancreatic cancer [40], which, in most cases, also have activating mutations in Ras [48].

4. The Link between Metabolic Diseases and Cancer

In addition to changes in cellular metabolism, metabolic dysfunctions can influence cancer. Obesity is a complex metabolic disorder that results in excessive accumulation of dietary lipids. Changes in lifestyle, linked to increased consumption of energy rich foods and more sedentary behaviour, increase the incidence of obesity-associated metabolic disease, including type 2 diabetes. There is a strong clinical association between these metabolic disorders and the risk of specific types of cancer, and patients with metabolic dysfunction have higher cancer-related mortality [49–53]. However, the mechanisms that link metabolic morbidities and cancer are not well understood. We discuss here how *Drosophila* has been used to investigate the obesity-cancer interaction.

4.1. Factors Controlling Obesity

In addition to the impact of changes in life style, human obesity has a strong genetic component [54]. Genome-wide screens using *Drosophila* have been used to identify novel genetic factors that can contribute to obesity [55]. An RNAi-based in vivo screen identified ~500 genes that led to changes in fat levels in adult flies. 60% of these genes have human orthologues. Depletion of those candidates, specifically in the fly adipose tissue, created a role for the conserved hedgehog (hh) signalling pathway. Downregulation of the hedgehog pathway in the fat body, via the depletion of *hh*, *iHog*, and *smoothened*, or the transcription pathway *cubitus interruptus*, led to a marked increase in triglyceride levels. In a mouse model, fat-specific hedgehog-activation led to lean mice with a strong reduction in fat tissue, revealing conservation between flies and mammals in adipocyte tissue physiology [55].

Genome-wide transcriptome profiling comparing starved versus fed flies implicated the lipase gene *brummer* (*bmm*) in obesity [56]. The expression of *bmm* is controlled by the nutritional status of the animal. *bmm* downregulation is sufficient to cause obesity and, reciprocally, flies overexpressing *bmm* showed a depletion in organismal fat stores, resembling nutrient-deprived animals. *bmm* encodes the lipid-storage, droplet-associated triacylglycerol lipase, which is a homolog of the human adipocyte triglyceride lipase. The obesity-related phenotype observed in flies depleted of *bmm*, together with this evolutionary conservation, suggest that regulation of this family of genes could be involved in human obesity [56].

In addition to genetics and the environment, studies in human populations have demonstrated that the parental nutritional state can influence the metabolism of the offspring for multiple generations [57–60]. Even though different molecular mechanisms for imprinting, including DNA methylation, histone modifications, and non-coding RNA have been implicated in this process [61], how intergenerational phenotypes are initiated, stabilized, and transmitted remains largely unknown. As in humans, the nutritional state of the parents affects the metabolism of the offspring in flies: an increase in paternal dietary sugar affects metabolism in the offspring. The analysis of flies grown in higher sugar diet (HSD) detected changes in the chromatin state of mature sperm and in the embryos they produce that led to obesity in the offspring. Interestingly, many genes involved in cytosolic and mitochondrial metabolism are found in those chromatin regions, which might account for the metabolic changes observed in the offspring. The use of a *Drosophila* has allowed the identification of gene networks involved in parental-induced intergenerational metabolic reprogramming. This gene signature is conserved from flies to mice and humans [62], suggesting that similar gene networks may regulate obesity susceptibility in humans and *Drosophila*.

4.2. Connecting Obesity and Cancer

There is a clinical association between obesity and several types of cancer. A primary consequence of obesity is the development of insulin-resistant type 2 diabetes. When the body becomes resistant to insulin, it compensates by producing more insulin [63]. In addition to its glucose regulatory function, insulin is a mitogen that promotes tissue growth [64], and, thus, chronic elevation of systemic insulin levels is a candidate for promoting tumorigenesis in obese and diabetic patients.

The insulin pathway is conserved from flies to humans, and feeding flies with HSD induces hyperinsulinemia and insulin resistance [65], which resembles the metabolic defects observed in obese humans. Activation of the oncogenes Ras and Src is common in multiple human cancers [66], and this genetic background has been used in *Drosophila* to study epithelial tumor formation [67]. Clones of cells activating Ras and Src in the imaginal discs of *Drosophila* do not develop into tumors when raised on a control-normal diet; however, they form malignant tumors with aggressive metastatic-like behaviour when the flies are fed on a HSD [68]. While most tissues in flies fed in HSD are insulin-resistant, cells activating the oncogenes Ras-Src express high levels of the insulin receptor. This makes these tumors sensitive to the elevated levels of systemic insulin produced by these nutritional conditions. High insulin signalling in the Ras-Src-expressing cells upregulates the signalling molecule Wingless (WNT), which, in turn, causes a further increase in Insulin receptor levels, creating a positive feedback loop that reinforces this program. The increase of insulin receptor levels in Ras-Src tumors makes them sensitive to insulin, allowing them to use circulating sugar to fuel their growth [68].

How do tumors in animals fed on high sugar trigger the initial upregulation of the insulin receptors to benefit from high systemic insulin? Further studies by the Cagan group identified the salt-inducible kinases (SIKs) as proteins induced by activation of Ras and Src in flies fed in HSD [69]. SIKs are serine/threonine kinases of the AMP kinase family of proteins that control metabolic homeostasis in mammals and flies [70–75]. SIKs can induce tissue growth by repressing the Hippo tumor suppressor pathway [76]. HSD induces SIKs in cells activating the oncogenes Ras and Src. This inhibits the Hippo kinase cascade that results in the activation of the Hippo transcriptional activator, the oncogene Yorkie, which induces the expression of its target gene Wingless [69]. SIKs act as sugar sensors that link the nutritional status to Hippo activity. By using this mechanism, the Ras-Src cells respond to changes in the diet to promote tumor growth in conditions of HSD. The finding that diet influences tumor development in *Drosophila* opens the possibility to use genetic tools to further analyse the molecular links between obesity and the risk of cancer.

4.3. Drosophila Fat Body to Identify Drugs that Modulate Lipid Metabolism

Drosophila has also been used as a platform to search for chemical compounds regulating signalling pathways involved in cancer and metabolism [77]. WNT signalling, in addition to its implications in specific types of cancer, has been shown to control adipocyte development [78,79]. Adipocytes play central roles in controlling fat metabolism and storage in mammals. The adipocytes in the *Drosophila* fat body contain numerous lipid droplets and store most of the triglycerides of the fly, having a similar metabolic role to mammalian adipose tissues. Axin is a negative regulator of the Wingless cascade in *Drosophila*, and *Axin* depletion results in an increase in Wingless pathway activity. Interestingly, *Axin* mutants show reduced triglycerides and a smaller amount of fat body tissue, and the adipocytes in *Axin* mutants are reduced in size and contain fewer oil droplets. The *Axin* mutant background was used to screen for compounds that regulate the Wingless pathway and lipid metabolism. That screening resulted in the identification of peptide boronic acids (a family of proteasome inhibitors) as chemical compounds that reduce Wingless signalling and rescue the fat body defects observed in *Axin* mutants [77]. This work exemplifies how fly models can be used to identify candidate drugs.

4.4. TGF-βs as Regulators of Obesity and Insulin Resistance

A high-fat diet can also induce type 2 diabetes in humans. Adipose tissue, in addition to functioning as the site of fat storage, behaves as an endocrine organ that secretes signalling molecules that control metabolic homeostasis [80,81]. The signalling molecule TGF-β is secreted by the adipose tissue and has been identified as a critical mediator of diet-induced insulin resistance in mice, although the mechanism by which TGF-β controlled insulin resistance was not elucidated [82]. In *Drosophila*, a high fat diet (HFD) leads to elevated circulating triglyceride and glucose levels, and insulin resistance, akin to obese mammals. This has allowed the use of *Drosophila* genetics to explore diet-induced insulin resistance [83]. The *Drosophila* gene *glass bottom boat* (*gbb*) is a TGF-β family member with functions in growth, morphogenesis, and energy homeostasis [84–87]. Consistent with the result obtained in mice, the protein levels of the TGF-β Gbb are increased in HFD conditions, and overexpression of *gbb* in the *Drosophila* fat body is sufficient to trigger obesity and insulin resistance, recapitulating the HFD phenotypes. Conversely, depletion of *gbb* results in a rescue of the HFD-induced metabolic phenotypes. Gbb signalling promotes insulin resistance by increasing the expression of the gene *tribbles* [88]. TRB3, a mammalian homolog of *Drosophila tribbles*, functions as a negative modulator of Insulin pathway in mice by regulating the serine-threonine kinase Akt, which is a central mediator of insulin signalling. TRB3 physically interacts with Akt and represses its kinase activity, causing insulin resistance [89]. In agreement with those observations, *Drosophila* Tribbles represses Akt in the fat body. HFD-fed mice upregulate TGF-β and the orthologous protein TRB3, which inhibited insulin signalling. The combination of studies in mice and flies shed light on the mechanisms by which TGF-β controls obesity and diabetic phenotypes [88,89].

Defects in mitochondrial activity have been associated with obesity, and both mitochondrial perturbations and obesity influence cancer progression. Accumulating evidence shows that mitochondrial alterations in one organ can have systemic effects through hormonal signalling [90,91]. *Drosophila* has been used to explore how mitochondrial defects can modulate fat accumulation in distant organs. This analysis found that mitochondrial perturbations in the muscle can influence mitochondrial function and metabolism in the fat body, leading to triglyceride accumulation, characteristic of obesity [83]. This work identified the TGF-β ligand, Activin, as the key element mediating mitochondrial synchrony between the muscle and the fat body. *Activin* is expressed at low levels in wild-type muscles, but its expression is increased in muscles with defects in mitochondrial activity. Activin secreted by the muscles acts systemically to impair mitochondrial activity and increase lipid accumulation in the fat body. This work identified a molecular mechanism whereby Activin behaved as a hormone to regulate "mitochondrial synchrony" between organs. The results obtained in *Drosophila* are in good agreement with previous observations showing that Activin represses mitochondrial function and lipid accumulation in mammals [92], and they suggest that this mechanism is conserved.

5. *Drosophila* as a Model to Study Cachexia

More than 80% of advanced human cancer patients show signs of loss of body mass caused by degeneration of skeletal muscle and fat storage wasting, a disorder known as cachexia [93]. Cachexia remains as a major hurdle in cancer treatment, in part because the molecular mechanisms that drive wasting remain elusive. As in humans, the presence of malignant tumors in *Drosophila* impairs muscle function and causes systemic wasting [94,95]. The induction of neoplastic tumors in flies leads to metabolic reprograming in the muscle that results in a strong downregulation of genes involved in energy metabolism, including glycolysis, TCA cycle, and Oxidative phosphorylation. This leads to mitochondrial degeneration and a reduction in ATP levels in muscle cells. *Drosophila* tumors produce *ImpL2*, a homolog of the secreted insulin growth factor-binding protein. Elevated ImpL2 release can reduce systemic insulin signalling in flies with malignant tumors. Depletion of insulin activity is sufficient to induce cachexia-like phenotype. Notably, suppression of ImpL2 increases insulin signalling and alleviates some wasting phenotypes associated with these tumors [94,95], suggesting a potential therapeutic approach to cachexia.

6. Metabolism and Angiogenesis

Angiogenesis is the physiological process through which new blood vasculature is formed. Tumors often grow in nutrient- and oxygen-poor environments, and angiogenesis is one of the strategies used to obtain nutrients and oxygen. Consequently, angiogenesis is a key target in cancer treatment [1]. In *Drosophila*, the tracheal system has an analogous role to the mammalian circulatory system, providing internal organs with oxygen [96,97]. Similar to vertebrate capillaries, the tracheal system responds to low oxygen levels by generating new branches via Hif-1α and Fibroblast Growth Factor signalling [98,99]. Angiogenesis also operates in neoplastic *Drosophila* tumors [100]. Angiogenesis is thought to be driven by signals secreted by oxygen-deprived target cells. However, a recent study in *Drosophila* has identified a new mechanism by which nutritional status regulates angiogenesis in the *Drosophila* larval gut [101]. Tracheal tubes in the gut are in close proximity to axons that emanate from insulin-producing neurons in the *Drosophila* brain. Those neurons guide changes in tracheal architecture through expression of Insulin-like peptide 7 (Ilp7), as well as pigment dispersing factor (Pdf), a neuropeptide that shares similarities to the vertebrate Vasoactive Intestinal Polypeptide. Ilp7 and Pdf control the growth of the tracheal cell terminal branches in response to nutritional cues received by those neurons. This work illustrates a new mechanism that couples the metabolic state with the process of angiogenesis. It will be of interest to see whether future studies show similar links between systemic metabolism and angiogenesis in human cancer progression.

7. Metabolism and Cell Competition

Tumors have several strategies, beyond angiogenesis, for obtaining nutrients. Cancer cells can use amino acids produced from the degradation of extracellular proteins internalized via macropicnocytosis, from entosis of living cells, from phagocytosis of apoptotic bodies, and from lipids from the extracellular phospholipids [2]. For example, cells activating the oncogene K-Ras kill and engulf surrounding cells by a process known as cell competition [102], and accumulating evidence suggests that cell competition contributes to cancer [103–106]. Cell competition was first described in *Drosophila* in the 1970s [107] and serves as a fitness selection mechanism by which "more fit" cells (winners) recognize and eliminate their "less fit" neighbours (losers) [105].

Metabolic pathways active in tumor cells are controlled by oncogenes [108,109], and oncogene activation can trigger cell competition [110]. Oncogenic Ras promotes cell growth and proliferation in *Drosophila* by activating the PI-3K pathway and by increasing Myc protein levels [111]. *Drosophila* cells expressing high levels of the *Myc* oncogene acquire a "supercompetitor" status and induce apoptosis in normal surrounding cells that, when surrounded by those supercompetitor cells, become losers [112,113]. Myc controls cellular metabolism in different ways. In the *Drosophila* fat tissue, Myc regulates glucose and lipid metabolism to favor the storage of nutrients lipids and sugars [114]. In a context of cell competition, *Myc*-expressing cells reprogram their metabolism towards enhancing glycolysis in a p53-dependent manner. In absence of p53, cells upregulating Myc lose their supercompetitor status and cannot induce apoptosis in surrounding normal cells, so that the population of *Myc*-expressing cells cannot increase in size [115]. This suggests that metabolic reprogramming might play a crucial role in early phases of disease, whereby emerging cancer cells might expand by boosting glycolysis.

Cell competition involves the induction of apoptosis in the loser cells and, in some cases, this is accompanied by the engulfment of the losers by the winners [102,104,116,117]. A recent study identified the conserved miRNA, *miR-8*, as a cooperative partner of the *Epidermal Growth Factor Receptor* (*EGFR*) oncogene [104]. Coexpression of *EGFR* and *miR-8* in the imaginal wing epithelium results in the formation of metastatic tumors. An interesting characteristic of those tumors is that they contain giant polyploid cells, which induce apoptosis and engulf their neighbouring cells. Interestingly, inhibition of apoptosis or engulfment completely suppress the formation of the polyploid cells and, consequently, tumors and metastasis are not formed. These observations suggest that cancer polyploid cells may need to engulf surrounding cells in order to obtain nutrients that allow them to grow and become malignant [104].

Downregulation of the gene *calderón* (*cald*) also induces cell competition in *Drosophila* epithelia. *cald* encodes a protein with homology on the major facilitator super-family of organic cation transporters, and these transporters are believed to function mainly in the uptake of sugars and other organic metabolites. The *cald* gene is regulated by the insulin pathway and is required for insulin-regulated cell growth and proliferation. Interestingly, cells mutant for *cald* are viable when surrounded by other *cald* mutant cells, but they are eliminated by cell competition when their neighbours are normal cells [118]. This report suggests that the metabolic state of the cell can impact in the "winner" vs. "loser" status and hence induce cell competition and tumor progression.

8. Perspectives

In this review, we have discussed how the use of *Drosophila melanogaster* has contributed to the study of mechanistic links between metabolism and cancer. Metabolic changes are emerging as important drivers of malignancy, and cancer-specific metabolic alterations are becoming an important research theme. *Drosophila* has been broadly used to study how metabolism is orchestrated to support biological processes, including growth, proliferation, and apoptosis in normal development. The fact that *Drosophila* shares a great degree of gene conservation with humans has made of this model organism a valuable tool for the identification and analysis of genes involved in cancer metabolism. To date, most studies have been focused on alterations in glucose metabolism. However, cancer cells also use nutrients like glutamine, amino acids, essential fatty acids, etc. Given the underlying similarity in metabolism, future studies using fly models should be used to explore links between nutrient utilization and disease progression. A variety of obesity-related models have been developed in *Drosophila*. The combination of these with available fly tumor models should be a fruitful path to investigate the molecular connections between diet, obesity, diabetes, and cancer. Finally, the impact of metabolites produced by gut microbiota on tumor progression is only starting to be appreciated. *Drosophila* has been extensively used to investigate the contribution of microbiota on insect behaviour, immunity, and metabolism. This will allow the fly to be used to study the influence that metabolic changes mediated by the activity of the microbiota have on disease progression.

Acknowledgments: This work was supported by the Neye Foundation, and by the NovoNordisk Foundation (grant NNF12OC0000552).

Author Contributions: H.H. and S.M.C. wrote the manuscript.

Conflicts of Interest: The authors declare no conflict of interest.

References

1. Hanahan, D.; Weinberg, R.A. Hallmarks of cancer: The next generation. *Cell* **2011**, *144*, 646–674. [CrossRef] [PubMed]
2. Pavlova, N.N.; Thompson, C.B. The emerging hallmarks of cancer metabolism. *Cell Metab.* **2016**, *23*, 27–47. [CrossRef] [PubMed]
3. Hay, N. Reprogramming glucose metabolism in cancer: Can it be exploited for cancer therapy? *Nat. Rev. Cancer* **2016**, *16*, 635–649. [CrossRef] [PubMed]
4. Reiter, L.T.; Potocki, L.; Chien, S.; Gribskov, M.; Bier, E. A systematic analysis of human disease-associated gene sequences in drosophila melanogaster. *Genome Res.* **2001**, *11*, 1114–1125. [CrossRef] [PubMed]
5. Gonzalez, C. Drosophila melanogaster: A model and a tool to investigate malignancy and identify new therapeutics. *Nat. Rev. Cancer* **2013**, *13*, 172–183. [CrossRef] [PubMed]
6. Cully, M.; You, H.; Levine, A.J.; Mak, T.W. Beyond pten mutations: The pi3k pathway as an integrator of multiple inputs during tumorigenesis. *Nat. Rev. Cancer* **2006**, *6*, 184–192. [CrossRef] [PubMed]
7. Hietakangas, V.; Cohen, S.M. Regulation of tissue growth through nutrient sensing. *Annu. Rev. Genet.* **2009**, *43*, 389–410. [CrossRef] [PubMed]
8. Oldham, S.; Hafen, E. Insulin/IGF and target of rapamycin signaling: A tor de force in growth control. *Trends Cell Biol.* **2003**, *13*, 79–85. [CrossRef]
9. Teleman, A.A. Molecular mechanisms of metabolic regulation by insulin in Drosophila. *Biochem. J.* **2009**, *425*, 13–26. [CrossRef] [PubMed]
10. Warburg, O. On respiratory impairment in cancer cells. *Science* **1956**, *124*, 269–270. [PubMed]
11. Warburg, O. On the origin of cancer cells. *Science* **1956**, *123*, 309–314. [CrossRef] [PubMed]
12. Doherty, J.R.; Cleveland, J.L. Targeting lactate metabolism for cancer therapeutics. *J. Clin. Investig.* **2013**, *123*, 3685–3692. [CrossRef] [PubMed]
13. Valvona, C.J.; Fillmore, H.L.; Nunn, P.B.; Pilkington, G.J. The regulation and function of lactate dehydrogenase a: Therapeutic potential in brain tumor. *Brain Pathol.* **2016**, *26*, 3–17. [CrossRef] [PubMed]
14. Gatenby, R.A.; Gillies, R.J. Why do cancers have high aerobic glycolysis? *Nat. Rev. Cancer* **2004**, *4*, 891–899. [CrossRef] [PubMed]
15. Dang, C.V.; Semenza, G.L. Oncogenic alterations of metabolism. *Trends Biochem. Sci.* **1999**, *24*, 68–72. [CrossRef]
16. Rechsteiner, M.C. Drosophila lactate dehydrogenase and alpha-glycerolphosphate dehydrogenase: Distribution and change in activity during development. *J. Insect Physiol.* **1970**, *16*, 1179–1192. [CrossRef]
17. Herranz, H.; Eichenlaub, T.; Cohen, S.M. Cancer in drosophila: Imaginal discs as a model for epithelial tumor formation. *Curr. Top. Dev. Biol.* **2016**, *116*, 181–199. [PubMed]
18. Rosin, D.; Schejter, E.; Volk, T.; Shilo, B.Z. Apical accumulation of the drosophila pdgf/vegf receptor ligands provides a mechanism for triggering localized actin polymerization. *Development* **2004**, *131*, 1939–1948. [CrossRef] [PubMed]
19. Wang, C.W.; Purkayastha, A.; Jones, K.T.; Thaker, S.K.; Banerjee, U. In vivo genetic dissection of tumor growth and the warburg effect. *eLife* **2016**, *5*, e18126. [CrossRef] [PubMed]
20. Glushakova, L.G.; Judge, S.; Cruz, A.; Pourang, D.; Mathews, C.E.; Stacpoole, P.W. Increased superoxide accumulation in pyruvate dehydrogenase complex deficient fibroblasts. *Mol. Genet. Metab.* **2011**, *104*, 255–260. [CrossRef] [PubMed]
21. Sun, W.; Zhou, S.; Chang, S.S.; McFate, T.; Verma, A.; Califano, J.A. Mitochondrial mutations contribute to hif1alpha accumulation via increased reactive oxygen species and up-regulated pyruvate dehydrogenease kinase 2 in head and neck squamous cell carcinoma. *Clin. Cancer Res.* **2009**, *15*, 476–484. [CrossRef] [PubMed]
22. Zhou, Q.; Lam, P.Y.; Han, D.; Cadenas, E. C-jun N-terminal kinase regulates mitochondrial bioenergetics by modulating pyruvate dehydrogenase activity in primary cortical neurons. *J. Neurochem.* **2008**, *104*, 325–335. [CrossRef] [PubMed]
23. Zhou, Q.; Lam, P.Y.; Han, D.; Cadenas, E. Activation of c-jun-n-terminal kinase and decline of mitochondrial pyruvate dehydrogenase activity during brain aging. *FEBS Lett.* **2009**, *583*, 1132–1140. [CrossRef] [PubMed]

24. Estella, C.; Baonza, A. Cell proliferation control by notch signalling during imaginal discs development in drosophila. *AIMS Genet.* **2015**, *2*, 70–96. [CrossRef]
25. Ntziachristos, P.; Lim, J.S.; Sage, J.; Aifantis, I. From fly wings to targeted cancer therapies: A centennial for notch signaling. *Cancer Cell* **2014**, *25*, 318–334. [CrossRef] [PubMed]
26. Go, M.J.; Eastman, D.S.; Artavanis-Tsakonas, S. Cell proliferation control by notch signaling in drosophila development. *Development* **1998**, *125*, 2031–2040. [PubMed]
27. Slaninova, V.; Krafcikova, M.; Perez-Gomez, R.; Steffal, P.; Trantirek, L.; Bray, S.J.; Krejci, A. Notch stimulates growth by direct regulation of genes involved in the control of glycolysis and the tricarboxylic acid cycle. *Open Biol.* **2016**, *6*, 150155. [CrossRef] [PubMed]
28. Vander Heiden, M.G.; Cantley, L.C.; Thompson, C.B. Understanding the warburg effect: The metabolic requirements of cell proliferation. *Science* **2009**, *324*, 1029–1033. [CrossRef] [PubMed]
29. Tennessen, J.M.; Baker, K.D.; Lam, G.; Evans, J.; Thummel, C.S. The drosophila estrogen-related receptor directs a metabolic switch that supports developmental growth. *Cell Metab.* **2011**, *13*, 139–148. [CrossRef] [PubMed]
30. Ariazi, E.A.; Clark, G.M.; Mertz, J.E. Estrogen-related receptor alpha and estrogen-related receptor gamma associate with unfavorable and favorable biomarkers, respectively, in human breast cancer. *Cancer Res.* **2002**, *62*, 6510–6518. [PubMed]
31. Deblois, G.; Giguere, V. Oestrogen-related receptors in breast cancer: Control of cellular metabolism and beyond. *Nat. Rev. Cancer* **2013**, *13*, 27–36. [CrossRef] [PubMed]
32. Stein, R.A.; Chang, C.Y.; Kazmin, D.A.; Way, J.; Schroeder, T.; Wergin, M.; Dewhirst, M.W.; McDonnell, D.P. Estrogen-related receptor alpha is critical for the growth of estrogen receptor-negative breast cancer. *Cancer Res.* **2008**, *68*, 8805–8812. [CrossRef] [PubMed]
33. Yu, S.; Wong, Y.C.; Wang, X.H.; Ling, M.T.; Ng, C.F.; Chen, S.; Chan, F.L. Orphan nuclear receptor estrogen-related receptor-beta suppresses in vitro and in vivo growth of prostate cancer cells via p21(waf1/cip1) induction and as a potential therapeutic target in prostate cancer. *Oncogene* **2008**, *27*, 3313–3328. [CrossRef] [PubMed]
34. Jiang, H.; Edgar, B.A. Intestinal stem cell function in drosophila and mice. *Curr. Opin. Genet. Dev.* **2012**, *22*, 354–360. [CrossRef] [PubMed]
35. Lemaitre, B.; Miguel-Aliaga, I. The digestive tract of Drosophila melanogaster. *Annu. Rev. Genet.* **2013**, *47*, 377–404. [CrossRef] [PubMed]
36. Li, H.; Jasper, H. Gastrointestinal stem cells in health and disease: From flies to humans. *Dis. Model. Mech.* **2016**, *9*, 487–499. [CrossRef] [PubMed]
37. Schell, J.C.; Wisidagama, D.R.; Bensard, C.; Zhao, H.; Wei, P.; Tanner, J.; Flores, A.; Mohlman, J.; Sorensen, L.K.; Earl, C.S.; et al. Control of intestinal stem cell function and proliferation by mitochondrial pyruvate metabolism. *Nat. Cell Biol.* **2017**, *19*, 1027–1036. [CrossRef] [PubMed]
38. Brandon, M.; Baldi, P.; Wallace, D.C. Mitochondrial mutations in cancer. *Oncogene* **2006**, *25*, 4647–4662. [CrossRef] [PubMed]
39. Carew, J.S.; Huang, P. Mitochondrial defects in cancer. *Mol. Cancer* **2002**, *1*, 9. [CrossRef] [PubMed]
40. Modica-Napolitano, J.S.; Kulawiec, M.; Singh, K.K. Mitochondria and human cancer. *Curr. Mol. Med.* **2007**, *7*, 121–131. [CrossRef] [PubMed]
41. Pedersen, P.L. Tumor mitochondria and the bioenergetics of cancer cells. *Prog. Exp. Tumor Res.* **1978**, *22*, 190–274. [PubMed]
42. Karim, F.D.; Rubin, G.M. Ectopic expression of activated ras1 induces hyperplastic growth and increased cell death in drosophila imaginal tissues. *Development* **1998**, *125*, 1–9. [PubMed]
43. Brumby, A.M.; Richardson, H.E. Scribble mutants cooperate with oncogenic ras or notch to cause neoplastic overgrowth in drosophila. *EMBO J.* **2003**, *22*, 5769–5779. [CrossRef] [PubMed]
44. Ohsawa, S.; Sato, Y.; Enomoto, M.; Nakamura, M.; Betsumiya, A.; Igaki, T. Mitochondrial defect drives non-autonomous tumour progression through hippo signalling in drosophila. *Nature* **2012**, *490*, 547–551. [CrossRef] [PubMed]
45. Pagliarini, R.A.; Xu, T. A genetic screen in drosophila for metastatic behavior. *Science* **2003**, *302*, 1227–1231. [CrossRef] [PubMed]
46. Nakamura, M.; Ohsawa, S.; Igaki, T. Mitochondrial defects trigger proliferation of neighbouring cells via a senescence-associated secretory phenotype in drosophila. *Nat. Commun.* **2014**, *5*, 5264. [CrossRef] [PubMed]

47. Perez-Mancera, P.A.; Young, A.R.; Narita, M. Inside and out: The activities of senescence in cancer. *Nat. Rev. Cancer* **2014**, *14*, 547–558. [CrossRef] [PubMed]
48. Downward, J. Targeting ras signalling pathways in cancer therapy. *Nat. Rev. Cancer* **2003**, *3*, 11–22. [CrossRef] [PubMed]
49. Calle, E.E.; Rodriguez, C.; Walker-Thurmond, K.; Thun, M.J. Overweight, obesity, and mortality from cancer in a prospectively studied cohort of u.S. Adults. *N. Engl. J. Med.* **2003**, *348*, 1625–1638. [CrossRef] [PubMed]
50. Giovannucci, E.; Harlan, D.M.; Archer, M.C.; Bergenstal, R.M.; Gapstur, S.M.; Habel, L.A.; Pollak, M.; Regensteiner, J.G.; Yee, D. Diabetes and cancer: A consensus report. *Diabetes Care* **2010**, *33*, 1674–1685. [CrossRef] [PubMed]
51. Harding, J.L.; Shaw, J.E.; Peeters, A.; Cartensen, B.; Magliano, D.J. Cancer risk among people with type 1 and type 2 diabetes: Disentangling true associations, detection bias, and reverse causation. *Diabetes Care* **2015**, *38*, 264–270. [CrossRef] [PubMed]
52. Incio, J.; Liu, H.; Suboj, P.; Chin, S.M.; Chen, I.X.; Pinter, M.; Ng, M.R.; Nia, H.T.; Grahovac, J.; Kao, S.; et al. Obesity-induced inflammation and desmoplasia promote pancreatic cancer progression and resistance to chemotherapy. *Cancer Discov.* **2016**, *6*, 852–869. [CrossRef] [PubMed]
53. Lauby-Secretan, B.; Scoccianti, C.; Loomis, D.; Grosse, Y.; Bianchini, F.; Straif, K.; International Agency for Research on Cancer Handbook Working Group. Body fatness and cancer—Viewpoint of the iarc working group. *N. Engl. J. Med.* **2016**, *375*, 794–798. [CrossRef] [PubMed]
54. Farooqi, I.S.; O'Rahilly, S. Genetic factors in human obesity. *Obes. Rev.* **2007**, *8* (Suppl. S1), 37–40. [CrossRef] [PubMed]
55. Pospisilik, J.A.; Schramek, D.; Schnidar, H.; Cronin, S.J.; Nehme, N.T.; Zhang, X.; Knauf, C.; Cani, P.D.; Aumayr, K.; Todoric, J.; et al. Drosophila genome-wide obesity screen reveals hedgehog as a determinant of brown versus white adipose cell fate. *Cell* **2010**, *140*, 148–160. [CrossRef] [PubMed]
56. Gronke, S.; Mildner, A.; Fellert, S.; Tennagels, N.; Petry, S.; Muller, G.; Jackle, H.; Kuhnlein, R.P. Brummer lipase is an evolutionary conserved fat storage regulator in drosophila. *Cell Metab.* **2005**, *1*, 323–330. [CrossRef] [PubMed]
57. De Rooij, S.R.; Painter, R.C.; Phillips, D.I.; Osmond, C.; Michels, R.P.; Godsland, I.F.; Bossuyt, P.M.; Bleker, O.P.; Roseboom, T.J. Impaired insulin secretion after prenatal exposure to the dutch famine. *Diabetes Care* **2006**, *29*, 1897–1901. [CrossRef] [PubMed]
58. Li, Y.; Jaddoe, V.W.; Qi, L.; He, Y.; Wang, D.; Lai, J.; Zhang, J.; Fu, P.; Yang, X.; Hu, F.B. Exposure to the chinese famine in early life and the risk of metabolic syndrome in adulthood. *Diabetes Care* **2011**, *34*, 1014–1018. [CrossRef] [PubMed]
59. Lumey, L.H.; Stein, A.D.; Kahn, H.S.; Romijn, J.A. Lipid profiles in middle-aged men and women after famine exposure during gestation: The dutch hunger winter families study. *Am. J. Clin. Nutr.* **2009**, *89*, 1737–1743. [CrossRef] [PubMed]
60. Ravelli, A.C.; van Der Meulen, J.H.; Osmond, C.; Barker, D.J.; Bleker, O.P. Obesity at the age of 50 y in men and women exposed to famine prenatally. *Am. J. Clin. Nutr.* **1999**, *70*, 811–816. [PubMed]
61. Rando, O.J. Daddy issues: Paternal effects on phenotype. *Cell* **2012**, *151*, 702–708. [CrossRef] [PubMed]
62. Ost, A.; Lempradl, A.; Casas, E.; Weigert, M.; Tiko, T.; Deniz, M.; Pantano, L.; Boenisch, U.; Itskov, P.M.; Stoeckius, M.; et al. Paternal diet defines offspring chromatin state and intergenerational obesity. *Cell* **2014**, *159*, 1352–1364. [CrossRef] [PubMed]
63. Muoio, D.M.; Newgard, C.B. Mechanisms of disease:Molecular and metabolic mechanisms of insulin resistance and beta-cell failure in type 2 diabetes. *Nat. Rev. Mol. Cell Biol.* **2008**, *9*, 193–205. [CrossRef] [PubMed]
64. Ish-Shalom, D.; Christoffersen, C.T.; Vorwerk, P.; Sacerdoti-Sierra, N.; Shymko, R.M.; Naor, D.; De Meyts, P. Mitogenic properties of insulin and insulin analogues mediated by the insulin receptor. *Diabetologia* **1997**, *40* (Suppl. S2), S25–S31. [CrossRef] [PubMed]
65. Musselman, L.P.; Fink, J.L.; Narzinski, K.; Ramachandran, P.V.; Hathiramani, S.S.; Cagan, R.L.; Baranski, T.J. A high-sugar diet produces obesity and insulin resistance in wild-type drosophila. *Dis. Model. Mech.* **2011**, *4*, 842–849. [CrossRef] [PubMed]
66. Ishizawar, R.; Parsons, S.J. C-src and cooperating partners in human cancer. *Cancer Cell* **2004**, *6*, 209–214. [CrossRef] [PubMed]

67. Vidal, M.; Warner, S.; Read, R.; Cagan, R.L. Differing src signaling levels have distinct outcomes in drosophila. *Cancer Res.* **2007**, *67*, 10278–10285. [CrossRef] [PubMed]
68. Hirabayashi, S.; Baranski, T.J.; Cagan, R.L. Transformed drosophila cells evade diet-mediated insulin resistance through wingless signaling. *Cell* **2013**, *154*, 664–675. [CrossRef] [PubMed]
69. Hirabayashi, S.; Cagan, R.L. Salt-inducible kinases mediate nutrient-sensing to link dietary sugar and tumorigenesis in drosophila. *eLife* **2015**, *4*, e08501. [CrossRef] [PubMed]
70. Choi, S.; Kim, W.; Chung, J. Drosophila salt-inducible kinase (sik) regulates starvation resistance through camp-response element-binding protein (creb)-regulated transcription coactivator (crtc). *J. Biol. Chem.* **2011**, *286*, 2658–2664. [CrossRef] [PubMed]
71. Choi, S.; Lim, D.S.; Chung, J. Feeding and fasting signals converge on the lkb1-sik3 pathway to regulate lipid metabolism in drosophila. *PLoS Genet.* **2015**, *11*, e1005263. [CrossRef] [PubMed]
72. Dentin, R.; Liu, Y.; Koo, S.H.; Hedrick, S.; Vargas, T.; Heredia, J.; Yates, J., 3rd; Montminy, M. Insulin modulates gluconeogenesis by inhibition of the coactivator torc2. *Nature* **2007**, *449*, 366–369. [CrossRef] [PubMed]
73. Patel, K.; Foretz, M.; Marion, A.; Campbell, D.G.; Gourlay, R.; Boudaba, N.; Tournier, E.; Titchenell, P.; Peggie, M.; Deak, M.; et al. The lkb1-salt-inducible kinase pathway functions as a key gluconeogenic suppressor in the liver. *Nat. Commun.* **2014**, *5*, 4535. [CrossRef] [PubMed]
74. Wang, B.; Goode, J.; Best, J.; Meltzer, J.; Schilman, P.E.; Chen, J.; Garza, D.; Thomas, J.B.; Montminy, M. The insulin-regulated creb coactivator torc promotes stress resistance in drosophila. *Cell Metab.* **2008**, *7*, 434–444. [CrossRef] [PubMed]
75. Wang, B.; Moya, N.; Niessen, S.; Hoover, H.; Mihaylova, M.M.; Shaw, R.J.; Yates, J.R., 3rd; Fischer, W.H.; Thomas, J.B.; Montminy, M. A hormone-dependent module regulating energy balance. *Cell* **2011**, *145*, 596–606. [CrossRef] [PubMed]
76. Wehr, M.C.; Holder, M.V.; Gailite, I.; Saunders, R.E.; Maile, T.M.; Ciirdaeva, E.; Instrell, R.; Jiang, M.; Howell, M.; Rossner, M.J.; et al. Salt-inducible kinases regulate growth through the hippo signalling pathway in drosophila. *Nat. Cell Biol.* **2013**, *15*, 61–71. [CrossRef] [PubMed]
77. Zhang, T.; Hsu, F.N.; Xie, X.J.; Li, X.; Liu, M.; Gao, X.; Pei, X.; Liao, Y.; Du, W.; Ji, J.Y. Reversal of hyperactive wnt signaling-dependent adipocyte defects by peptide boronic acids. *Proc. Natl. Acad. Sci. USA* **2017**, *114*, E7469–E7478. [CrossRef] [PubMed]
78. Christodoulides, C.; Lagathu, C.; Sethi, J.K.; Vidal-Puig, A. Adipogenesis and wnt signalling. *Trends Endocrinol. Metab.* **2009**, *20*, 16–24. [CrossRef] [PubMed]
79. Zeve, D.; Seo, J.; Suh, J.M.; Stenesen, D.; Tang, W.; Berglund, E.D.; Wan, Y.; Williams, L.J.; Lim, A.; Martinez, M.J.; et al. Wnt signaling activation in adipose progenitors promotes insulin-independent muscle glucose uptake. *Cell Metab.* **2012**, *15*, 492–504. [CrossRef] [PubMed]
80. Havel, P.J. Update on adipocyte hormones: Regulation of energy balance and carbohydrate/lipid metabolism. *Diabetes* **2004**, *53* (Suppl. S1), S143–S151. [CrossRef] [PubMed]
81. Rosen, E.D.; Spiegelman, B.M. Adipocytes as regulators of energy balance and glucose homeostasis. *Nature* **2006**, *444*, 847–853. [CrossRef] [PubMed]
82. Yadav, H.; Quijano, C.; Kamaraju, A.K.; Gavrilova, O.; Malek, R.; Chen, W.; Zerfas, P.; Zhigang, D.; Wright, E.C.; Stuelten, C.; et al. Protection from obesity and diabetes by blockade of tgf-beta/smad3 signaling. *Cell Metab.* **2011**, *14*, 67–79. [CrossRef] [PubMed]
83. Song, W.; Owusu-Ansah, E.; Hu, Y.; Cheng, D.; Ni, X.; Zirin, J.; Perrimon, N. Activin signaling mediates muscle-to-adipose communication in a mitochondria dysfunction-associated obesity model. *Proc. Natl. Acad. Sci. USA* **2017**. [CrossRef] [PubMed]
84. Ballard, S.L.; Jarolimova, J.; Wharton, K.A. Gbb/bmp signaling is required to maintain energy homeostasis in drosophila. *Dev. Biol.* **2010**, *337*, 375–385. [CrossRef] [PubMed]
85. McCabe, B.D.; Marques, G.; Haghighi, A.P.; Fetter, R.D.; Crotty, M.L.; Haerry, T.E.; Goodman, C.S.; O'Connor, M.B. The bmp homolog gbb provides a retrograde signal that regulates synaptic growth at the drosophila neuromuscular junction. *Neuron* **2003**, *39*, 241–254. [CrossRef]
86. O'Connor, M.B.; Umulis, D.; Othmer, H.G.; Blair, S.S. Shaping bmp morphogen gradients in the drosophila embryo and pupal wing. *Development* **2006**, *133*, 183–193. [CrossRef] [PubMed]
87. Parker, L.; Stathakis, D.G.; Arora, K. Regulation of bmp and activin signaling in drosophila. *Prog. Mol. Subcell. Biol.* **2004**, *34*, 73–101. [PubMed]

88. Hong, S.H.; Kang, M.; Lee, K.S.; Yu, K. High fat diet-induced tgf-beta/gbb signaling provokes insulin resistance through the tribbles expression. *Sci. Rep.* **2016**, *6*, 30265. [CrossRef] [PubMed]
89. Du, K.; Herzig, S.; Kulkarni, R.N.; Montminy, M. Trb3: A tribbles homolog that inhibits akt/pkb activation by insulin in liver. *Science* **2003**, *300*, 1574–1577. [CrossRef] [PubMed]
90. Bostrom, P.; Wu, J.; Jedrychowski, M.P.; Korde, A.; Ye, L.; Lo, J.C.; Rasbach, K.A.; Bostrom, E.A.; Choi, J.H.; Long, J.Z.; et al. A pgc1-alpha-dependent myokine that drives brown-fat-like development of white fat and thermogenesis. *Nature* **2012**, *481*, 463–468. [CrossRef] [PubMed]
91. Durieux, J.; Wolff, S.; Dillin, A. The cell-non-autonomous nature of electron transport chain-mediated longevity. *Cell* **2011**, *144*, 79–91. [CrossRef] [PubMed]
92. Casalena, G.; Daehn, I.; Bottinger, E. Transforming growth factor-beta, bioenergetics, and mitochondria in renal disease. *Semin. Nephrol.* **2012**, *32*, 295–303. [CrossRef] [PubMed]
93. Lok, C. Cachexia: The last illness. *Nature* **2015**, *528*, 182–183. [CrossRef] [PubMed]
94. Figueroa-Clarevega, A.; Bilder, D. Malignant drosophila tumors interrupt insulin signaling to induce cachexia-like wasting. *Dev. Cell* **2015**, *33*, 47–55. [CrossRef] [PubMed]
95. Kwon, Y.; Song, W.; Droujinine, I.A.; Hu, Y.; Asara, J.M.; Perrimon, N. Systemic organ wasting induced by localized expression of the secreted insulin/IGF antagonist impl2. *Dev. Cell* **2015**, *33*, 36–46. [CrossRef] [PubMed]
96. Fraisl, P.; Mazzone, M.; Schmidt, T.; Carmeliet, P. Regulation of angiogenesis by oxygen and metabolism. *Dev. Cell* **2009**, *16*, 167–179. [CrossRef] [PubMed]
97. Uv, A.; Cantera, R.; Samakovlis, C. Drosophila tracheal morphogenesis: Intricate cellular solutions to basic plumbing problems. *Trends Cell Biol.* **2003**, *13*, 301–309. [CrossRef]
98. Centanin, L.; Dekanty, A.; Romero, N.; Irisarri, M.; Gorr, T.A.; Wappner, P. Cell autonomy of hif effects in drosophila: Tracheal cells sense hypoxia and induce terminal branch sprouting. *Dev. Cell* **2008**, *14*, 547–558. [CrossRef] [PubMed]
99. Ghabrial, A.; Luschnig, S.; Metzstein, M.M.; Krasnow, M.A. Branching morphogenesis of the drosophila tracheal system. *Annu. Rev. Cell Dev. Biol.* **2003**, *19*, 623–647. [CrossRef] [PubMed]
100. Grifoni, D.; Sollazzo, M.; Fontana, E.; Froldi, F.; Pession, A. Multiple strategies of oxygen supply in drosophila malignancies identify tracheogenesis as a novel cancer hallmark. *Sci. Rep.* **2015**, *5*, 9061. [CrossRef] [PubMed]
101. Linneweber, G.A.; Jacobson, J.; Busch, K.E.; Hudry, B.; Christov, C.P.; Dormann, D.; Yuan, M.; Otani, T.; Knust, E.; de Bono, M.; et al. Neuronal control of metabolism through nutrient-dependent modulation of tracheal branching. *Cell* **2014**, *156*, 69–83. [CrossRef] [PubMed]
102. Sun, Q.; Luo, T.; Ren, Y.; Florey, O.; Shirasawa, S.; Sasazuki, T.; Robinson, D.N.; Overholtzer, M. Competition between human cells by entosis. *Cell Res.* **2014**, *24*, 1299–1310. [CrossRef] [PubMed]
103. Di Giacomo, S.; Sollazzo, M.; de Biase, D.; Ragazzi, M.; Bellosta, P.; Pession, A.; Grifoni, D. Human cancer cells signal their competitive fitness through myc activity. *Sci. Rep.* **2017**, *7*, 12568. [CrossRef] [PubMed]
104. Eichenlaub, T.; Cohen, S.M.; Herranz, H. Cell competition drives the formation of metastatic tumors in a drosophila model of epithelial tumor formation. *Curr. Biol.* **2016**, *26*, 419–427. [CrossRef] [PubMed]
105. Merino, M.M.; Levayer, R.; Moreno, E. Survival of the fittest: Essential roles of cell competition in development, aging, and cancer. *Trends Cell Biol.* **2016**, *26*, 776–788. [CrossRef] [PubMed]
106. Suijkerbuijk, S.J.; Kolahgar, G.; Kucinski, I.; Piddini, E. Cell competition drives the growth of intestinal adenomas in drosophila. *Curr. Biol.* **2016**, *26*, 428–438. [CrossRef] [PubMed]
107. Morata, G.; Ripoll, P. Minutes: Mutants of drosophila autonomously affecting cell division rate. *Dev. Biol.* **1975**, *42*, 211–221. [CrossRef]
108. DeBerardinis, R.J.; Lum, J.J.; Hatzivassiliou, G.; Thompson, C.B. The biology of cancer: Metabolic reprogramming fuels cell growth and proliferation. *Cell Metab.* **2008**, *7*, 11–20. [CrossRef] [PubMed]
109. Hsu, P.P.; Sabatini, D.M. Cancer cell metabolism: Warburg and beyond. *Cell* **2008**, *134*, 703–707. [CrossRef] [PubMed]
110. Moreno, E. Is cell competition relevant to cancer? *Nat. Rev. Cancer* **2008**, *8*, 141–147. [CrossRef] [PubMed]
111. Prober, D.A.; Edgar, B.A. Interactions between ras1, dmyc, and dpi3k signaling in the developing drosophila wing. *Genes Dev.* **2002**, *16*, 2286–2299. [CrossRef] [PubMed]
112. De la Cova, C.; Abril, M.; Bellosta, P.; Gallant, P.; Johnston, L.A. Drosophila myc regulates organ size by inducing cell competition. *Cell* **2004**, *117*, 107–116. [CrossRef]
113. Moreno, E.; Basler, K. Dmyc transforms cells into super-competitors. *Cell* **2004**, *117*, 117–129. [CrossRef]

114. Parisi, F.; Riccardo, S.; Zola, S.; Lora, C.; Grifoni, D.; Brown, L.M.; Bellosta, P. Dmyc expression in the fat body affects dilp2 release and increases the expression of the fat desaturase desat1 resulting in organismal growth. *Dev. Biol.* **2013**, *379*, 64–75. [CrossRef] [PubMed]
115. De la Cova, C.; Senoo-Matsuda, N.; Ziosi, M.; Wu, D.C.; Bellosta, P.; Quinzii, C.M.; Johnston, L.A. Supercompetitor status of drosophila myc cells requires p53 as a fitness sensor to reprogram metabolism and promote viability. *Cell Metab.* **2014**, *19*, 470–483. [CrossRef] [PubMed]
116. Li, W.; Baker, N.E. Engulfment is required for cell competition. *Cell* **2007**, *129*, 1215–1225. [CrossRef] [PubMed]
117. Ohsawa, S.; Sugimura, K.; Takino, K.; Xu, T.; Miyawaki, A.; Igaki, T. Elimination of oncogenic neighbors by jnk-mediated engulfment in drosophila. *Dev. Cell* **2011**, *20*, 315–328. [CrossRef] [PubMed]
118. Herranz, H.; Morata, G.; Milan, M. Calderon encodes an organic cation transporter of the major facilitator superfamily required for cell growth and proliferation of drosophila tissues. *Development* **2006**, *133*, 2617–2625. [CrossRef] [PubMed]

5

Modeling Alzheimer's and Other Age Related Human Diseases in Embryonic Systems

Chu Hsien Lim [1] **and Ajay S. Mathuru** [1,2,*]

[1] Yale NUS College, 16 College Avenue West, Singapore 138610, Singapore; limchuhsien@yale-nus.edu.sg
[2] Institute of Molecular and Cell Biology, 61 Biopolis Drive, Singapore 137673, Singapore
* Correspondence: ajay.mathuru@yale-nus.edu.sg

Abstract: Modeling human disease in animals is an important strategy to discover potential methods of intervention. We suggest that there is much to be gained by employing a multi-model approach that takes advantage of different animal systems used in the laboratory simultaneously. We use the example of modeling Alzheimer's disease in *Drosophila melanogaster*, *Caenorhabditis elegans*, and *Danio rerio* to illustrate how such an approach can be employed to investigate the pathophysiology of the disease.

Keywords: Alzheimer's disease; CRY2; *Drosophila melanogaster*; *Caenorhabditis elegans*; *Danio rerio*; optogenetics

---

1. Introduction

Animal models are indispensable tools for understanding biology and for designing strategies to mitigate human disease. An important question to address then is the choice of the animal species when modeling a disorder or a disease. After all, animals may not suffer from the same ailments as humans do. A popular idea in the field of neuroethology is to study a 'champion' species or a specialist in the particular motor or sensory function of interest [1–3]. An echo-locating bat may reveal the workings of auditory processing or a cephalopod may help us identify a neural mechanism for camouflage. Indeed, unusual insights into cancer pathology have come to us via naked mole rats that are highly resistant to cancers in spite of a life spanning around three decades [4,5]. However, this is not entirely suitable to tease apart the molecular and cellular processes underlying disease progression in many cases that often require extensive experimentation and elaborate screens.

Historically, rodents and non-human primates have received the most attention in medicine as mammalian models, particularly in the final phases of drug or vaccine development prior to clinical trials in humans. However, the costs and infrastructure associated with running such experiments at scale are fast becoming prohibitive. Instead *Drosophila melanogaster*, *Caenorhabditis elegans*, and *Danio rerio* (colloquially, fruit flies, nematode worms, and zebrafish) are increasingly used to study various aspects of diseases for exactly the same reasons that made them popular for studying fundamental biology—the ease of rearing them in the lab, performing genetic manipulations, molecular and biochemical analyses, and behavioral studies. For instance, fruit fly, worm, and zebrafish models have been critical to elucidate the RAS signaling pathway in cancer, polyglutamine-expansion disorders, and melanoma respectively among other diseases [6–10]. We suggest that using all three in combination may circumvent some of the challenges posed in the search of a 'champion' animal model that satisfactorily recapitulates a disease and allows studies of genetics, physiology, and behavior. In addition, the tools available for studying development in these systems become highly advantageous for dissociating incidental age-related changes from changes causally linked in an age dependent disease.

2. Alzheimer's Disease: An Unresolved Problem

In this commentary, we turn our attention to Alzheimer's disease (AD). AD is the most common cause of dementia, and a debilitating age-associated neurodegenerative disease currently lacking effective treatment [11,12]. According to a non-profit, voluntary organization specializing in Alzheimer's care, AD is the sixth leading cause of death in America, and there are now 5.5 million Americans living with it [13]. Globally, there are 46.8 million people living with dementia, including AD and this number is expected to increase to reach 131.5 million people by 2050 [14]. Global burden of disease studies suggest that the cost of AD and other dementia-related illness is in the range of USD 250 billion in America, and over USD 600 billion globally [13,15]. Such statistics highlight the prevalence of AD and other dementia-related illnesses and the grave public health challenge that future generations face.

Even after several decades of research, the exact cause of AD is yet to be clearly established. However, two classic neuropathological signs are associated with the disease. Namely, neurofibrillary tangles and extracellular Amyloid-β (Aβ) plaques in the human brain [16,17]. Tangles are thought to be caused by misfolding and hyperphosphorylation of tau proteins normally involved in microtubule stabilization [17]. Plaques, on the other hand, are aggregates of the beta-amyloid peptide—a cleavage product of a protein that is normally expressed throughout the ubiquitously called the amyloid precursor protein (APP) [18].

While both these processes are associated with AD, here, for the purpose of illustration, we only consider the amyloid hypothesis. According to this hypothesis, the accumulation of Aβ in various forms, as plaques or as soluble oligomers cause neuronal damage and the consequent inflammatory response that eventually leads to the deleterious symptoms associated with AD [19]. Evidence underpinning this association comes from familial cases of early-onset AD, which is linked to germline mutations in APP [20] and the presenilin genes involved in cleaving Aβ from APP [21]. The amyloid hypothesis proponents use pathological similarities between familial and sporadic AD as an argument, as well as observations of other age-related diseases such as the myositis-like phenotype caused by Aβ accumulation in transgenic mice overexpressing Aβ in muscle tissue [22].

Despite such evidence, the pathophysiology of Aβ plaques and their relation to AD is not a settled issue. Critics of the amyloid hypothesis point out findings of Aβ plaques in brains of individuals who did not show any symptoms of AD at the time of death. Other unresolved issues including poor correlation between plaque load and cognitive scores and heterogeneity in pathology have called into question the causal association ascribed to Aβ plaques [23]. There has been a shift of focus from plaques to the other potential malfunctions including to role of soluble forms of Aβ oligomerisation that further triggers a downstream cascade [24]. A recent paper also demonstrated that blood-derived Aβ resulted in pathogenic symptoms of AD in mice models, suggesting that Aβ plaques need not originate in the brain to evoke pathological symptoms [25]. Given the current status, whether extracellular Aβ plaques have a detrimental role in AD, or if they have a protective role before other malfunctions accumulate continues to be a central question [26]. Further, even though extracellular plaques are considered a hallmark of AD, consensus on whether problems arise due to Aβ aggregates formed intracellularly in neurons, or from secreted peptides extracellulary (or either), is also lacking [27–29].

3. Combining Forces

To unravel these questions, one approach could be to first clearly evaluate if Aβ plaque formation has any impact on neuronal health and cognitive performance. One approach is to use a multi-model study design that capitalizes on the unique advantages of flies, nematodes, and zebrafish.

Research using *Drosophila* has been vital to dissect genetic pathways critical for the formation, migration, and activity of Aβ plaques [6]. Using the galactose-inducible yeast system, commonly known as the Gal4/UAS system [30], transgenic flies can be easily generated to analyze several genetic pathways in a short period. The genetic pathways identified from such analyses can be further investigated in *C. elegans* biochemically. *C. elegans* has been widely employed to investigate the effects of aging and various neurodegenerative diseases, including AD [7]. With a short life cycle of three days from egg to adult, they are highly amenable to evaluate the "druggability" of targets among the biochemical pathways [31]. Feeding assays can be designed to implement large-scale drug screens or screens based on RNA interference (RNAi) knockdown [7]. Another factor when modeling human pathologies is to recapitulate the biological outcomes described in humans. Compared to the two invertebrate animal models, the vertebrate nervous system of zebrafish is more appropriate to study the consequences of Aβ plaques in the brain. Due to their transparent nature, zebrafish are also an excellent tool to investigate neural activity in vivo, thereby making them particularly useful when studying neurophysiology. Using whole brain imaging and cell-type specific labeling techniques, unique neurons in any region of the brain can be studied with ease [32]. In addition, several behavioral paradigms have already been developed that can be employed to evaluate the cognitive impact if neurodegeneration is occurring [33].

Advancements in imaging and genetic techniques further enhance the utility of this multi-model approach. One such technique is optogenetics—an extremely versatile tool using light to examine physiological processes [34]. Compared to the constraints posed by electrical recordings and pharmacological interventions in the past, optogenetics can be used to examine specific cellular activities with extremely high precision and speed in vivo. Recently, cryptochrome2 (CRY2) protein system from the *Arabidopsis* was modified as an optogenetic tool to study the control of cortical actin dynamics in cell culture and cell contractility during tissue morphogenesis in *Drosophila* embryos [35,36]. Activation of the CRY2 oligomerization at 488 nm and inhibition in the dark allows researchers to manipulate protein–protein interactions without the need for other external agents [37].

It is not hard to imagine that the same CRY2 optogenetic system can be employed with Aβ to create an in vivo, light-dependent, oligomerization switch for the formation or dissolution of Aβ aggregates (Figure 1). By generating transgenics in all three organisms in the neurons, or cells of interest, the spatiotemporal kinetics of Aβ plaque formation can be controlled at will to address the question of their relationship and AD pathogenesis. Light-inducible aggregates in worms will allow fast screening for drugs that are effective in dissolution when administered before, or after, or when plaques form repeatedly. Transparency of zebrafish allows imaging the kinetics of plaque formation. We envisage that the use of such a technique in fish will also shed light on the link between intracellular Aβ peptide aggregates and extracellular plaques. In addition, promising drugs screened in the worm studies can be rapidly evaluated in the context of a vertebrate brain. A battery of behavioral assays reporting on learning, memory, anxiety, and depression can then be employed to assess the impact [33]. Finally, flies that express Aβ CRY2 in specific organs, at specific times, will be highly informative to understand the spatiotemporal dynamics of plaque location for any pathogenic effect to be visible, as well as to identify genetic interactors at different stages in case pathogenicity is observed [25]. Together, these will likely provide firm answers to the questions raised by the amyloid hypothesis of AD, especially on the role of soluble vs. Aβ plaques as plaque formation can be engineered externally.

The potential of using such a combinatory approach of various model organisms and optogenetics is therefore vast, can be applied to other diseases and will only continue to expand as further improvements in technology become available.

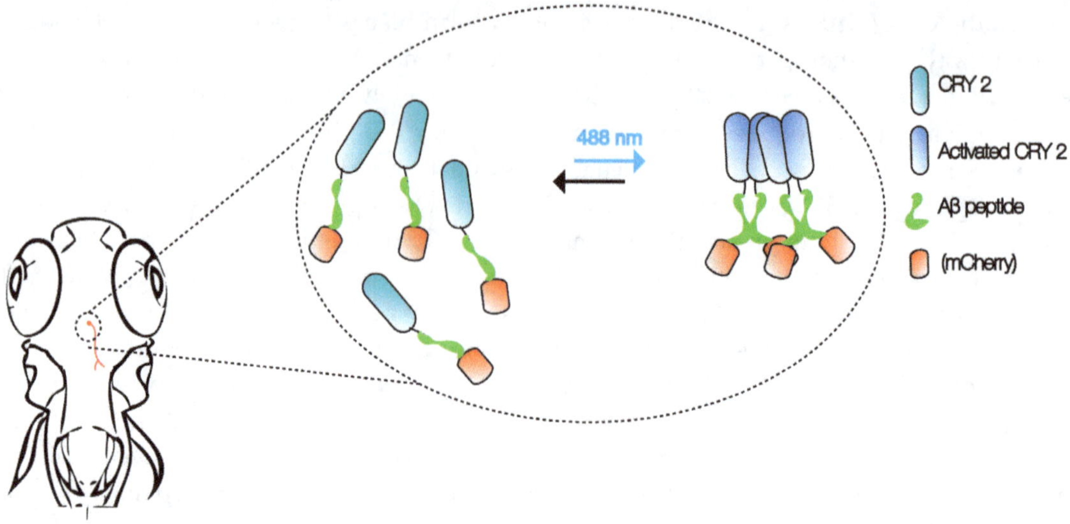

Figure 1. Aβ CRY2 optogenetic system. The schematic shows an example of in vivo, light-dependent, oligomerization switch for the formation or dissolution of Aβ aggregates in region of interest (neurons) in a transgenic zebrafish.

Acknowledgments: We thank the two anonymous reviewers and Nicholas Tolwinski for comments. This work was supported by Yale-NUS College through grants R-607-265-225-121 and IG16-LR003; and by Agency for Science, Technology and Research (A*STAR).

Author Contributions: C.H.L and A.S.M wrote the manuscript.

Conflicts of Interest: The authors declare no conflict of interest.

References

1. Bullock, T.H. Comparative neuroscience holds promise for quiet revolutions. *Science* **1984**, *225*, 473–478. [CrossRef] [PubMed]
2. Bullock, T.H. "Simple" model systems need comparative studies: Differences are as important as commonalities. *Trends Neurosci.* **1986**, *9*, 470–472. [CrossRef]
3. Heiligenberg, W. The neural basis of behavior: A neuroethological view. *Annu. Rev. Neurosci.* **1991**, *14*, 247–267. [CrossRef] [PubMed]
4. Tian, X.; Azpurua, J.; Hine, C.; Vaidya, A.; Myakishev-Rempel, M.; Ablaeva, J.; Mao, Z.; Nevo, E.; Gorbunova, V.; Seluanov, A. High-molecular-mass hyaluronan mediates the cancer resistance of the naked mole rat. *Nature* **2013**, *499*, 346. [CrossRef] [PubMed]
5. Delaney, M.A.; Ward, J.M.; Walsh, T.F.; Chinnadurai, S.K.; Kerns, K.; Kinsel, M.J.; Treuting, P.M. Initial case reports of cancer in naked mole-rats (heterocephalus glaber). *Vet. Pathol.* **2016**, *53*, 691–696. [CrossRef] [PubMed]
6. Pandey, U.B.; Nichols, C.D. Human disease models in drosophila melanogaster and the role of the fly in therapeutic drug discovery. *Pharmacol. Rev.* **2011**, *63*, 411–436. [CrossRef] [PubMed]
7. Markaki, M.; Tavernarakis, N. Modeling human diseases in caenorhabditis elegans. *Biotechnol. J.* **2010**, *5*, 1261–1276. [CrossRef] [PubMed]
8. Santoriello, C.; Zon, L.I. Hooked! Modeling human disease in zebrafish. *J. Clin. Investig.* **2012**, *122*, 2337–2343. [CrossRef] [PubMed]
9. Moulton, M.J.; Letsou, A. Modeling congenital disease and inborn errors of development in drosophila melanogaster. *Dis. Models Mech.* **2016**, *9*, 253–269. [CrossRef] [PubMed]

10. Kessler, M.; Rottbauer, W.; Just, S. Recent progress in the use of zebrafish for novel cardiac drug discovery. *Expert Opin. Drug Discov.* **2015**, *10*, 1231–1241. [CrossRef] [PubMed]
11. Kumar, A.; Singh, A.; Ekavali. A review on alzheimer's disease pathophysiology and its management: An update. *Pharmacol. Rep.* **2015**, *67*, 195–203. [CrossRef] [PubMed]
12. De-Paula, V.J.; Radanovic, M.; Diniz, B.S.; Forlenza, O.V. Alzheimer's disease. *Sub-Cell. Biochem.* **2012**, *65*, 329–352.
13. Alzheimer's Association. 2017 Alzheimer's Disease Facts and Figures. Available online: https://www.alz.org/documents_custom/2017-facts-and-figures.pdf (accessed on 10 November 2017).
14. Alzheimer's Disease International. *World Alzheimer Report 2016: Improving Healthcare for People Living with Dementia*; Alzheimer's Disease International (ADI): London, UK, 2016.
15. Alzheimers.net. 2016 Alzheimer's Statistics. Available online: https://www.alzheimers.net/resources/alzheimers-statistics/ (accessed on 10 November 2017).
16. Alexander, A.G.; Marfil, V.; Li, C. Use of caenorhabditis elegans as a model to study alzheimer's disease and other neurodegenerative diseases. *Front. Genet.* **2014**, *5*, 279. [CrossRef] [PubMed]
17. Iqbal, K.; Alonso Adel, C.; Chen, S.; Chohan, M.O.; El-Akkad, E.; Gong, C.X.; Khatoon, S.; Li, B.; Liu, F.; Rahman, A.; et al. Tau pathology in alzheimer disease and other tauopathies. *Biochim. Biophys. Acta* **2005**, *1739*, 198–210. [CrossRef] [PubMed]
18. Kang, J.; Lemaire, H.G.; Unterbeck, A.; Salbaum, J.M.; Masters, C.L.; Grzeschik, K.H.; Multhaup, G.; Beyreuther, K.; Muller-Hill, B. The precursor of alzheimer's disease amyloid a4 protein resembles a cell-surface receptor. *Nature* **1987**, *325*, 733–736. [CrossRef] [PubMed]
19. Eskelinen, M.H.; Kivipelto, M. Caffeine as a protective factor in dementia and alzheimer's disease. *J. Alzheimer's Dis. JAD* **2010**, *20* (Suppl. 1), S167–S174. [CrossRef] [PubMed]
20. Chartier-Harlin, M.C.; Crawford, F.; Houlden, H.; Warren, A.; Hughes, D.; Fidani, L.; Goate, A.; Rossor, M.; Roques, P.; Hardy, J.; et al. Early-onset alzheimer's disease caused by mutations at codon 717 of the beta-amyloid precursor protein gene. *Nature* **1991**, *353*, 844–846. [CrossRef] [PubMed]
21. Cruts, M.; Hendriks, L.; Van Broeckhoven, C. The presenilin genes: A new gene family involved in alzheimer disease pathology. *Hum. Mol. Genet.* **1996**, *5*, 1449–1455. [CrossRef] [PubMed]
22. Sugarman, M.C.; Yamasaki, T.R.; Oddo, S.; Echegoyen, J.C.; Murphy, M.P.; Golde, T.E.; Jannatipour, M.; Leissring, M.A.; LaFerla, F.M. Inclusion body myositis-like phenotype induced by transgenic overexpression of beta app in skeletal muscle. *Proc. Natl. Acad. Sci. USA* **2002**, *99*, 6334–6339. [CrossRef] [PubMed]
23. Morris, G.P.; Clark, I.A.; Vissel, B. Inconsistencies and controversies surrounding the amyloid hypothesis of alzheimer's disease. *Acta Neuropathol. Commun.* **2014**, *2*, 135. [CrossRef] [PubMed]
24. Joshi, P.; Turola, E.; Ruiz, A.; Bergami, A.; Libera, D.D.; Benussi, L.; Giussani, P.; Magnani, G.; Comi, G.; Legname, G.; et al. Microglia convert aggregated amyloid-beta into neurotoxic forms through the shedding of microvesicles. *Cell Death Differ.* **2014**, *21*, 582–593. [CrossRef] [PubMed]
25. Bu, X.L.; Xiang, Y.; Jin, W.S.; Wang, J.; Shen, L.L.; Huang, Z.L.; Zhang, K.; Liu, Y.H.; Zeng, F.; Liu, J.H.; et al. Blood-derived amyloid-β protein induces alzheimer's disease pathologies. *Mol. Psychiatry* **2017**. [CrossRef] [PubMed]
26. Wang, S.; Mims, P.N.; Roman, R.J.; Fan, F. Is beta-amyloid accumulation a cause or consequence of alzheimer's disease? *J. Alzheimer's Parkinsonism Dement.* **2016**, *1*, 007.
27. Friedrich, R.P.; Tepper, K.; Rönicke, R.; Soom, M.; Westermann, M.; Reymann, K.; Kaether, C.; Fändrich, M. Mechanism of amyloid plaque formation suggests an intracellular basis of aβ pathogenicity. *Proc. Natl. Acad. Sci. USA* **2010**, *107*, 1942–1947. [CrossRef] [PubMed]
28. Gouras, G.K.; Tampellini, D.; Takahashi, R.H.; Capetillo-Zarate, E. Intraneuronal β-amyloid accumulation and synapse pathology in alzheimer's disease. *Acta Neuropathol.* **2010**, *119*, 523–541. [CrossRef] [PubMed]
29. Zhang, Y.-W.; Thompson, R.; Zhang, H.; Xu, H. App processing in alzheimer's disease. *Mol. Brain* **2011**, *4*, 3. [CrossRef] [PubMed]
30. Duffy, J.B. Gal4 system in drosophila: A fly geneticist's swiss army knife. *Genesis* **2002**, *34*, 1–15. [CrossRef] [PubMed]
31. Brenner, S. The genetics of caenorhabditis elegans. *Genetics* **1974**, *77*, 71–94. [PubMed]
32. Kawashima, T.; Zwart, M.F.; Yang, C.T.; Mensh, B.D.; Ahrens, M.B. The serotonergic system tracks the outcomes of actions to mediate short-term motor learning. *Cell* **2016**, *167*, 933–946. [CrossRef] [PubMed]

33. Edward, D.; Levin, D.T.C. *Methods of Behavior Analysis in Neuroscience*, 2 ed.; CRC Press/Taylor & Francis: Boca Raton, FL, USA, 2009.
34. Fenno, L.; Yizhar, O.; Deisseroth, K. The development and application of optogenetics. *Annu. Rev. Neurosci.* **2011**, *34*, 389–412. [CrossRef] [PubMed]
35. Guglielmi, G.; Barry, J.D.; Huber, W.; De Renzis, S. An optogenetic method to modulate cell contractility during tissue morphogenesis. *Dev. Cell* **2015**, *35*, 646–660. [CrossRef] [PubMed]
36. Idevall-Hagren, O.; Dickson, E.J.; Hille, B.; Toomre, D.K.; De Camilli, P. Optogenetic control of phosphoinositide metabolism. *Proc. Natl. Acad. Sci. USA* **2012**, *109*, E2316–E2323. [CrossRef] [PubMed]
37. Kennedy, M.J.; Hughes, R.M.; Peteya, L.A.; Schwartz, J.W.; Ehlers, M.D.; Tucker, C.L. Rapid blue-light-mediated induction of protein interactions in living cells. *Nat. Methods* **2010**, *7*, 973–975. [CrossRef] [PubMed]

6

Dynamic Tissue Rearrangements during Vertebrate Eye Morphogenesis: Insights from Fish Models

Florencia Cavodeassi

Institute of Medical and Biomedical Education, St George's, University of London, London SW17 0RE, UK; fcavodea@sgul.ac.uk

Abstract: Over the last thirty years, fish models, such as the zebrafish and medaka, have become essential to pursue developmental studies and model human disease. Community efforts have led to the generation of wide collections of mutants, a complete sequence of their genomes, and the development of sophisticated genetic tools, enabling the manipulation of gene activity and labelling and tracking of specific groups of cells during embryonic development. When combined with the accessibility and optical clarity of fish embryos, these approaches have made of them an unbeatable model to monitor developmental processes in vivo and in real time. Over the last few years, live-imaging studies in fish have provided fascinating insights into tissue morphogenesis and organogenesis. This review will illustrate the advantages of fish models to pursue morphogenetic studies by highlighting the findings that, in the last decade, have transformed our understanding of eye morphogenesis.

Keywords: live-imaging; eye; morphogenesis

---

1. Introduction

Organogenesis is a highly orchestrated process that involves the coordination of cell fate specification and the morphogenesis of differentiating tissues to give rise to the final architecture of mature organs. Tissue morphogenesis involves cell proliferation, changes in cell shape and cell polarity, and precisely coordinated cell movements. As they reorganise, cells in a tissue influence their surroundings, and by exerting forces and tension on neighbouring cells they contribute to the formation of a functional, mature organ. The recent development of new labelling and imaging techniques has allowed for starting to record, in real time and with cellular and subcellular resolution, the cell dynamics that accompany tissue morphogenesis (reviewed in [1,2]). Tools have been developed to measure forces in developing tissues, and to manipulate molecular and mechanical cell properties, and assess their impact on morphogenesis [3,4]. Tropical fish, such as zebrafish and medaka, are amongst the most popular model organisms to pursue this type of analysis. Indeed, the small, transparent fish embryos allow detailed live imaging of any morphogenetic process, even those involving cell rearrangements deep in the embryo. When combined with the availability of transgenic tools to label specific tissues, cellular and subcellular structures, and the extensive array of embryological and genetic manipulations that can be performed in fish, the last few years have witnessed an explosion of imaging studies that provide detailed analysis of many organogenesis events (see for example [5–8]). The power of live imaging to uncover cellular mechanisms driving tissue morphogenesis is best exemplified by the studies that, during the last 10 years, have transformed our understanding of eye morphogenesis. This review will focus on these studies and will provide a reflection of where these findings are taking us next.

2. Overview of Eye Morphogenesis

The eye primordium is specified in the anterior portion of the neural plate (ANP) as a single domain spanning the midline. As the neural plate folds to give rise to the neural tube, the eye field cells are displaced laterally and evaginate to give rise to the optic vesicles, the first morphological manifestation of the eyes in the embryo [9,10]. Optic vesicle evagination is followed by the asymmetric folding of the distal portion of the eye primordium to give rise to the optic cup, which is a bilayered structure where the external layer will become the retinal pigment epithelium (RPE) and the internal layer will become the neural retina (NR). The asymmetric folding of the cup leaves a gap along the ventral portion of the optic primordium known as the choroid (or optic) fissure, which at late stages of eye maturation, will close and fuse to give rise to continuous NR and RPE layers. Throughout this whole process the distal eye structures are physically connected to the brain by the optic stalk. The optic stalk will eventually remodel to give rise to the cell types that ensheathe the optic nerve [11]. Further maturation of the NR involves the proliferation and differentiation of retinal progenitors to give rise to all the neuronal and glial cell types that compose the retina [12,13]. RPE cells, instead, undergo marked changes in cell morphology and eventually differentiate to give rise to a squamous, pigmented epithelium that covers the whole apical surface of the retina and preserves photoreceptors function [14,15] (Figure 1). The extensive reshaping of the eye primordium as development proceeds is highly conserved across vertebrates and can be reproduced in vitro in organoid models [16,17], emphasising the highly self-regulative properties of this process. It is likely that changes in individual cells' shape and organisation will somehow drive the remodelling of the eye at a tissue level; however, to understand the impact of cellular and subcellular processes on eye morphogenesis, it is necessary to capture the dynamics of changes in cell shape and organisation in real time. Live imaging of eye development has been performed in detail in fish models, and has allowed documenting cell behaviours from the earliest dynamic reorganisations of eye field cells, to the detailed differentiation paths taken by retinal progenitors in the mature optic cup.

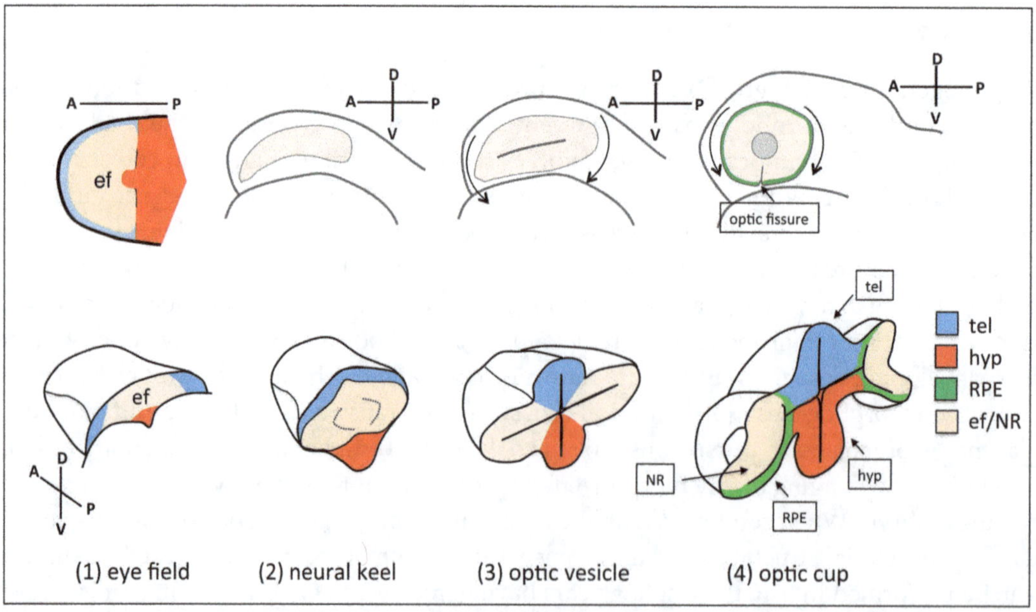

Figure 1. Schematic of eye morphogenesis in zebrafish. Eye field specification in the anterior portion of the neural plate (ANP) (**1**) is followed by the rearrangement of the tissue as the neural plate folds into a keel (**2**), and the evagination of the optic vesicles (**3**). Subsequent folding of the optic vesicle over itself leads to the formation of the optic cup (**4**). Top-left panel is a dorsal view with anterior to the left; the orientation of all the other panels is indicated in the figure. tel: telencephalon; hyp: hypothalamus; RPE: retinal pigment epithelium; ef: eye field; NR: neural retina.

3. Visualising Global Cellular Rearrangements and Tissue Remodelling

The specification of the eye precursors occurs concomitant to the subdivision of the ANP in several domains that will give rise to the different parts of the brain. Once specified, eye progenitors follow a distinctive morphogenetic program. By labelling and recording the organisation of ANP cells during the narrowing and initial folding of the neural plate, England and colleagues [18] showed that eye cells initially condense along the anterior-posterior (AP) axis, concomitant with the anterior movement of future hypothalamic cells. Subsequently, eye field cells at the ventral midline are displaced anteriorly and outwards by the anterior relocation of future dorsal diencephalic cells. As this occurs, more dorsally located eye field cells are also laterally displaced by the dorsal convergence and closure of the future telencephalon. The reorganisation of the ANP, and in particular of the eye field, was also documented by Rembold and colleagues [19], using the medaka fish as a model, and a transgenic line (*Tg{rx3::GFP}*) that specifically labels eye field cells. They further showed that the evagination of the eye field from the lateral walls of the neural keel to give rise to the two optic vesicles is a highly dynamic process reminiscent of active cell migration. These studies highlighted the active reorganisation that eye cells undergo during optic vesicle formation in relation to surrounding tissues, and how tightly coordinated eye and forebrain morphogenesis are. They also led to new hypotheses around the molecular mechanisms driving eye field cell rearrangements. For example, the Wnt molecule Wnt11 is required locally in the eye field to maintain the cohesion of this domain [20], but it also indirectly affects the ability of eye cells to move laterally prior to evagination, due to its role in driving neural plate narrowing and anterior migration of the prechordal plate [18]. The subsequent lateral expansion of the optic vesicles required the function of the eye-specification gene *rx3*, such that in the absence of Rx3 function eyeless embryos develop [19,21]). The analysis by Rembold and colleagues, extended later by Brown and colleagues [22] suggested that this is due to an inability of eye cells devoid of Rx3 to move away from the midline, behaving instead similarly to telencephalic cells.

As the optic vesicle matures and transforms into an optic cup, further remodelling of the tissue occurs. Global analysis of this whole process was performed by Kwan and colleagues [23], who generated over 12 hour movies spanning from the last stages of optic vesicle evagination to optic cup stages. By retrospectively tracking cells allocated in different optic cup regions at the end of the movie, they were able to determine the initial position of future RPE and NR cells in the optic vesicle. This study, together with others [23–25], also showed that as the optic vesicle folds to give rise to the optic cup, the cells in the primordium reorganise extensively; future RPE cells spread throughout the external layer of the optic cup, while a subset of NR cells initially positioned in the external layer of the primordium relocate to the internal layer to acquire their final position in the optic cup. These cell rearrangements may provide some of the driving force that is required for the efficient folding of the eye primordium.

All of these studies made use of conventional confocal microscopy and a whole repertoire of transgenic lines and other labelling approaches to document cell movements and tissue rearrangements, but early eye morphogenesis has also been documented by light-sheet microscopy, which is an advanced imaging technique that allows for the analysis of the whole cohort of cells in the embryo [26,27]. These studies, although not specifically directed to analysing eye morphogenesis, have provided movie sets with a high level of cellular resolution allowing the reconstruction of cellular paths over extended developmental times.

4. Following Changes in Cell Shape and Cytoskeletal Dynamics

The studies above analysed global cell dynamics, but did not address the structure of the tissue at a cellular level, nor the shape of the cells and the organisation of their cytoskeleton and junctional complexes. More recent studies revisited eye morphogenesis to provide a detailed understanding of the changes in cell shape and cell polarity that drive optic vesicle evagination and folding. Ivanovitch and colleagues [28] re-analysed the transition from eye field to optic vesicle, in order to uncover the organisation of eye cells within the tissue during these stages. By imaging eye morphogenesis from a transverse point of view and labelling cell contours and apical compartments, Ivanovitch et al. found that a subset of eye field cells at the margin of the neural keel are highly organised as a columnar epithelium from the onset of evagination. Following in detail individual cell shape and polarity revealed that this columnar epithelium thickens as neuroepithelial cells elongate, and becomes expanded mediolaterally by the gradual intercalation, elongation, and polarisation of eye cells at the core of the eye field domain (Figure 2). The net result of these cellular rearrangements is the lateral evagination and expansion of the eye field to give rise to the optic vesicles. Their study further showed that the initial trigger for the columnar organisation and polarisation of eye cells was the local assembly of a Laminin-rich basement membrane around the eye field; in the absence of Laminin function, eye cells fail to organise and elongate, and optic vesicle evagination is severely disrupted; conversely, an exogenous source of Laminin placed in the centre of the eye field instructs polarity around itself and distorts the organisation of eye field cells, thus generating ectopic lumens within the eye primordia.

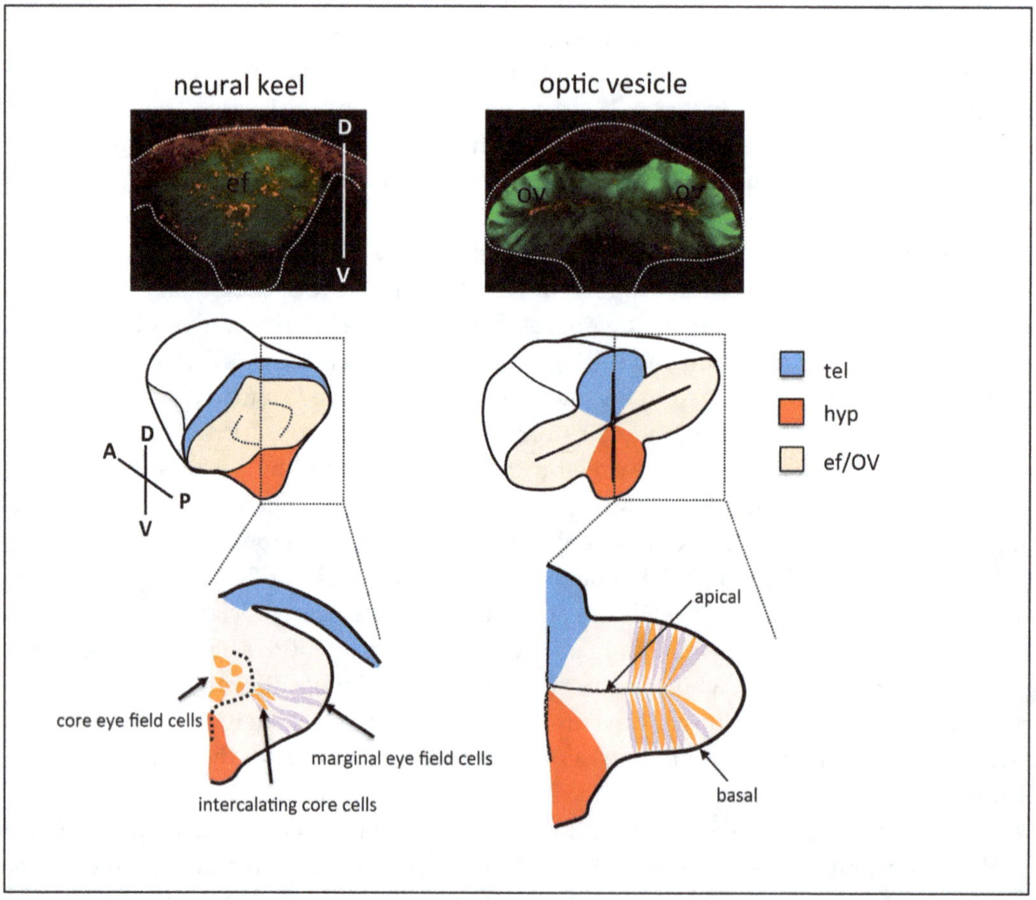

Figure 2. Top row: frontal views of neural keel (**left**) and optic vesicle (**right**) stage *Tg{rx3::GFP}* zebrafish embryos, immunostained to highlight the eye field/optic vesicles (GFP expression, green) and the apical domain of the tissue (ZO-1, red). The contour of the embryos is outlined. Middle and bottom row: schematic of cell rearrangements during optic vesicle evagination. Cells at the margin of the eye field (purple) organise as a columnar epithelium at the onset of optic vesicle evagination. Subsequent expansion of the optic vesicles is accompanied by the gradual intercalation and elongation of cells in the core of the eye field (orange) amongst marginal cells. tel: telencephalon; hyp: hypothalamus; ef: eye field; OV: optic vesicle.

Subcellular labelling of cytoskeletal components was also exploited to analyse the mechanisms driving subsequent stages of eye morphogenesis. As described above, the invagination of the optic vesicle to give rise to the bilayered optic cup is accompanied by a flow of cells around the rim of the primordium and their relocation from the external to the internal layer of the optic cup. Even though the driving force for this dynamic reorganisation is currently unclear, marked changes in cell shape have been documented during this process (Figure 3). A small subset of cells in the proximal region of the external layer of the optic vesicle (future RPE cells [23,29]) become shorter in the apico-basal (AB) axis and expand their surface, potentially displacing surrounding cells towards more distal regions as they do so. As prospective NR cells in the external layer approach the rim of the primordium, they elongate and undergo a marked apical constriction, while at their basal side, they display active lamellipodial activity. The net result is the displacement of externally located NR cells around the rim, and their relocation to the internal layer of the optic primordium [30]. Concomitant to these cell shape changes in the external layer of the optic cup, future NR cells in the internal layer of the invaginating primordium elongate in the AB axis and compact in the plane of the epithelium [24]. In addition, their basal surfaces constrict, while maintaining a wider apical surface, so that the global effect is the bending of the basal side of the NR [31–33].

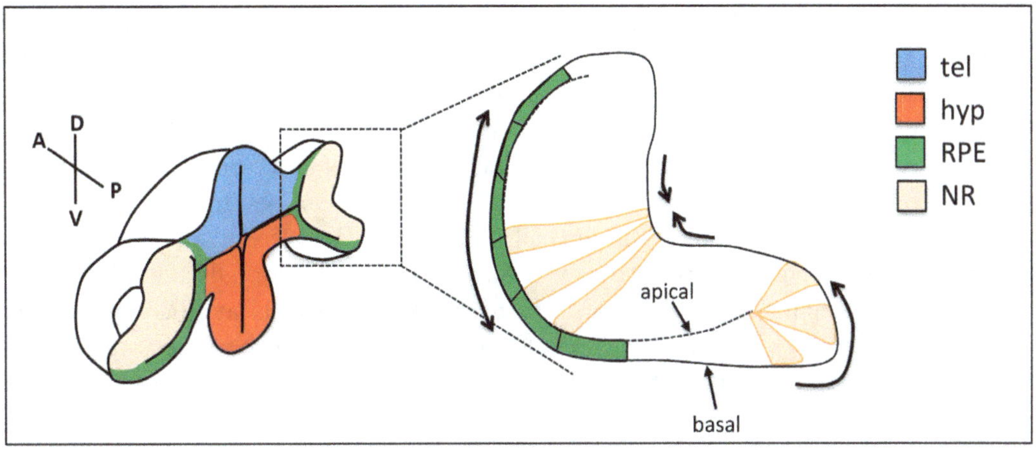

Figure 3. Changes in cell shape during optic cup folding. Prospective retinal pigment epithelium (RPE) cells (green) flatten, cells at the rim show apical constriction and basal lamellipodia, and neural retina (NR) cells show basal constriction. Arrows show the expected direction of tissue movement. tel: telencephalon; hyp: hypothalamus; RPE: retinal pigment epithelium; NR: neural retina.

Optic cup folding generates a fissure along the ventral portion of the primordium, and during late stages of optic cup formation, the lips of the optic fissure fuse to give rise to a continuous retina and RPE. This poorly understood process involves the fusion of two epithelia along their basal sides. For this fusion to occur, the basement membrane that covers them needs to be broken down, and cell rearrangements need to occur that align NR and RPE at both sides of the fissure (reviewed in [11]). Recent imaging studies have started to unravel the dynamics of these processes. Indeed, basement membrane breakdown precedes optic fissure fusion, and persistent basement membrane interferes with fusion [34]. Tethering of both sides of the optic fissure to each other, basement membrane breakdown, and eventually fissure closure requires the migration of a population of periocular mesenchymal (POM) cells into the fissure and the formation of the hyaloid vasculature [34–37], highlighting the dynamic interaction between the different tissues that are required for late stages of optic cup morphogenesis.

Ultimately, the changes in cell shape described during optic cup folding and choroid fissure closure are likely driven by the dynamic reorganisation of the cells' cytoskeleton and adhesion complexes. For example, NR cells basal constriction involves the stabilisation of basal localisation of focal adhesion complexes, which is a process that is controlled by ojoplano (opo), a conserved protein that localises to the basal end feet of NR cells, where it cooperates with the clathrin adaptors Numb and Numbl to regulate integrin endocytosis [31,32]. Analysis of the distribution of cytoskeletal components during optic cup folding revealed an enrichment of actomyosin at the basal surface of NR cells [30,33].

Recording actomyosin dynamics further revealed that this enriched cytoskeletal network promotes pulsatile behaviour of NR cell membranes, ultimately leading to the constriction of the basal surface of the NR [33]. Indeed, pulsatile behaviour and gradual constriction is associated to the episodic formation of foci of myosin accumulation, which seem to pull from the underlying ECM. Thus, the cell intrinsic gradual constriction of NR basal end feet is likely translated into tension at the tissue level by the linkage of the cells' cytoskeleton to the ECM through focal adhesions. Indeed, interfering with either the contraction of the cortical cytoskeleton, the assembly of focal adhesions or the assembly of the ECM eventually results in a release of tissue tension and failure of the NR to bend [31–33,38]. The changes in cell shape driving rim cell involution are also controlled by the cell cytoskeleton, which is remodelled at the basal surface leading to the extension of dynamic actin-rich protrusions in the direction of movement. Interference with rim cell involution eventually leads to defects in optic cup folding, and to the ectopic location of NR cells in the external layer of the optic cup [24,25,30].

The studies discussed above have identified a number of cellular behaviours (flattening of the prospective RPE cells, basal constriction of NR cells, active collective cell migration at the optic cup rim; active migration of the POM cells into the optic fissure) that are likely to contribute to the efficient folding of the optic cup and to the accurate positioning of NR and RPE cells into the eye primordium. However, it is still unclear from these studies whether there is a functional coupling between these different mechanisms, or whether they act independently and cooperatively to shape the optic cup. In the absence of Opo activity, the optic cup does not fold efficiently, and NR cells end up ectopically positioned in the external layer of the primordium, suggesting that interference with NR basal cells constriction may impact on rim cell involution [33]. Indeed, the tension generated by NR cell compaction and basal constriction may be enough to "pull" from cells at the rim, generating the force driving rim cell involution. Similarly, the flattening of RPE cells and their expansion in the surface of that tissue may "push" surrounding cells towards the rim promoting their relocation to the internal layer. However, determining the exact contribution of each tissue to the morphogenesis of the optic cup is not trivial, since the same molecular machinery seems to control all these cell behaviours. The development of tools to locally interfere with cytoskeleton dynamics in each of these optic cup regions will be crucial to resolve these issues.

5. Assessing the Coordination between Fate Specification and Morphogenesis

The shaping of the eye primordium occurs simultaneously to the gradual restriction in the fate of discrete groups of eye cells. Indeed, as the optic vesicle evaginates, it simultaneously becomes patterned to give rise to a proximal, future optic stalk domain, and distal RPE/NR domains (reviewed in [10,39]). In addition, axial patterning of the prospective NR starts to be established at the onset of optic vesicle evagination. Future nasal and temporal domains can already be visualised by the expression of specific nasal/temporal fate markers from the earliest stages of optic vesicle evagination. Studies in zebrafish have shown that signals from the Fibroblast-growth-factor (Fgf) and Hedgehog (Hh) families are required to establish these domains [24,40]. Optic vesicle evagination involves extensive cell rearrangements, and thus to have a full picture of how the architecture of the tissue is generated, one needs to be able to monitor cell fates as they are established in vivo and relate this to the morphogenetic rearrangement of the differentiating tissue. Crucial for this is the availability of tools that allow for us to visualise signalling pathway activation and fate establishment in vivo. In recent years many tools of this type have been developed [41–48], but they have not been applied to this purpose during eye morphogenesis.

The establishment of nasal/temporal patterning during optic vesicle evagination nicely exemplifies how this type of tools might help us understand the tight coordination of morphogenesis and fate acquisition. As already discussed, eye field cells at the margin of the neural keel are organised as a columnar epithelium prior to optic vesicle evagination. As optic vesicle evagination proceeds, this columnar epithelium thickens and becomes expanded mediolaterally by the subsequent intercalation, elongation and polarisation of the remaining eye cells, which were initially located at the core of the eye field domain [28]. This process of cell intercalation is apparently random, since intercalating cells populate all the regions of the evaginating optic vesicles. Dorsal (future nasal) and ventral (future temporal) domains of differential gene expression can already be detected from the onset of evagination, prefiguring the nasal/temporal pattern of the eye primordium [24]. Thus, intercalation of core eye cells occurs in an already patterned primordium, and as they integrate in either the dorsal or the ventral domain, they acquire nasal or temporal fates. How this process is coordinated is currently unclear. Do core cells intercalate randomly and subsequently acquire the corresponding cell fate, or instead, their intercalation into the dorsal or ventral domain is driven by their predetermination to one of these two fates? Recent studies showing that nasal and temporal fates are established by the coordinated action of Fgf and Shh signals [24,40] provide the necessary background knowledge to answer these questions. Indeed, the availability of nasal and temporal fate markers [24,40], and the existence of Shh and Fgf signalling pathway reporter lines [46,47], provides an entry point to assess how signals are received and interpreted by cells that are constantly changing their position. The design and application of similar tools to other stages of eye morphogenesis will allow us in the long term to gain a complete picture of the early events leading to the shaping and patterning of the eye primordium.

6. Perspective

The studies discussed above have provided us with a detailed understanding of the changes in cell shape and organisation that are involved in the shaping of the eye primordium in fish models. They have also allowed for us to start grasping the mechanical consequences of such changes at a cellular level, and to start formulating hypotheses on the mechanisms driving eye morphogenesis. Nevertheless, it is still unclear up to what extent these mechanisms are conserved in other organisms, particularly in mammals. One crucial limitation is the fact that a similar dynamic analysis of eye morphogenesis has not been conducted in mammals, and all of the available information comes from static analysis of fixed tissues. In addition, the structure and organisation of the embryonic tissues at the onset of eye morphogenesis in mammals is different to that described for fish embryogenesis. For example, while in the zebrafish the neural plate is not yet organised as a mature neuroepithelium at the onset of eye evagination, in mammals, this is indeed the case. However, many similarities do also exist. Indeed, a similar sequence of cell elongation at the onset of optic vesicle evagination has been suggested from the analysis of sectioned mouse tissue [49]. Moreover, recent experiments inducing in vitro optic cup formation in three-dimensional (3D) cultures of mouse and human embryonic stem cells have shown that the successful morphogenesis of eye organiods requires, as in zebrafish, the presence of a Laminin-rich extracellular matrix [16,17]. Subsequent folding of the optic cup progresses in similar ways in fish models and mammals, and the final bilayered structure of the optic cup is mostly identical in all vertebrates. In mammals, the RPE has been proposed to provide a rigid scaffold that promotes the folding of the optic cup [50,51], and marked changes in cell shape have been documented at the interface between the RPE and NR during optic cup folding in eye organiods [16,17]. Thus, the mechanistic differences that may exist early on likely reflects the fact that in fish models, early development occurs at a very fast pace, and the specification of tissues, their assembly and initial spatial reorganisation extensively overlap in time.

Not only global eye tissue rearrangements are largely comparable across species, but also the gene regulatory network (GRN) that is involved in triggering fate specification and regional patterning during eye development is mostly conserved [15,52–54]. Moreover, many of the genes and signalling pathways that are crucial for eye specification, morphogenesis, and patterning in fish are affected in human ocular malformations (reviewed in [55]; some examples in [56–59]). We are however still far from understanding how the GRN involved in controlling eye patterning ultimately controls cell behaviours and tissue morphogenesis, and how the emerging biomechanical properties of the differentiating optic cup domains impacts on organ shape. We can be certain that the zebrafish will make a crucial contribution to these open questions in the years to come.

Acknowledgments: The author is grateful to G. Gestri and K. Ivanovitch for critical reading of the manuscript, and to P. Ataliotis and F. Gibson for their support. This work has been supported by the Centre for Biomedical Education, IMBE, St. George's University of London.

Conflicts of Interest: The author declares no conflict of interest.

References

1. Pantazis, P.; Supatto, W. Advances in whole-embryo imaging: A quantitative transition is underway. *Nat. Rev. Mol. Cell Biol.* **2014**, *15*, 327–339. [CrossRef] [PubMed]
2. Vacaru, A.M.; Unlu, G.; Spitzner, M.; Mione, M.; Knapik, E.W.; Sadler, K.C. In vivo cell biology in zebrafish—Providing insights into vertebrate development and disease. *J. Cell Sci.* **2014**, *127*, 485–495. [CrossRef] [PubMed]
3. Grill, S.W. Growing up is stressful: Biophysical laws of morphogenesis. *Curr. Opin. Genet. Dev.* **2011**, *21*, 647–652. [CrossRef] [PubMed]
4. Colombelli, J.; Solon, J. Force communication in multicellular tissues addressed by laser nanosurgery. *Cell Tissue Res.* **2013**, *352*, 133–147. [CrossRef] [PubMed]
5. Behrndt, M.; Salbreux, G.; Campinho, P.; Hauschild, R.; Oswald, F.; Roensch, J.; Grill, S.W.; Heisenberg, C.P. Forces driving epithelial spreading in zebrafish gastrulation. *Science* **2012**, *338*, 257–260. [CrossRef] [PubMed]
6. Pestel, J.; Ramadass, R.; Gauvrit, S.; Helker, C.; Herzog, W.; Stainier, D.Y. Real-time 3d visualization of cellular rearrangements during cardiac valve formation. *Development* **2016**, *143*, 2217–2227. [CrossRef] [PubMed]
7. Blaser, H.; Reichman-Fried, M.; Castanon, I.; Dumstrei, K.; Marlow, F.L.; Kawakami, K.; Solnica-Krezel, L.; Heisenberg, C.P.; Raz, E. Migration of zebrafish primordial germ cells: A role for myosin contraction and cytoplasmic flow. *Dev. Cell* **2006**, *11*, 613–627. [CrossRef] [PubMed]
8. Durdu, S.; Iskar, M.; Revenu, C.; Schieber, N.; Kunze, A.; Bork, P.; Schwab, Y.; Gilmour, D. Luminal signalling links cell communication to tissue architecture during organogenesis. *Nature* **2014**, *515*, 120–124. [CrossRef] [PubMed]
9. Fuhrmann, S. Eye morphogenesis and patterning of the optic vesicle. *Curr. Top. Dev. Biol.* **2010**, *93*, 61–84. [PubMed]
10. Fuhrmann, S.; Zou, C.; Levine, E.M. Retinal pigment epithelium development, plasticity, and tissue homeostasis. *Exp. Eye Res.* **2014**, *123*, 141–150. [CrossRef] [PubMed]
11. Patel, A.; Sowden, J.C. Genes and pathways in optic fissure closure. *Semin. Cell Dev. Biol.* **2017**. [CrossRef] [PubMed]
12. Livesey, F.J.; Cepko, C.L. Vertebrate neural cell-fate determination: Lessons from the retina. *Nat. Rev. Neurosci.* **2001**, *2*, 109–118. [CrossRef] [PubMed]
13. Boije, H.; MacDonald, R.B.; Harris, W.A. Reconciling competence and transcriptional hierarchies with stochasticity in retinal lineages. *Curr. Opin. Neurobiol.* **2014**, *27*, 68–74. [CrossRef] [PubMed]
14. Strauss, O. The retinal pigment epithelium in visual function. *Physiol. Rev.* **2005**, *85*, 845–881. [CrossRef] [PubMed]
15. Amram, B.; Cohen-Tayar, Y.; David, A.; Ashery-Padan, R. The retinal pigmented epithelium—From basic developmental biology research to translational approaches. *Int. J. Dev. Biol.* **2017**, *61*, 225–234. [CrossRef] [PubMed]
16. Eiraku, M.; Takata, N.; Ishibashi, H.; Kawada, M.; Sakakura, E.; Okuda, S.; Sekiguchi, K.; Adachi, T.; Sasai, Y. Self-organizing optic-cup morphogenesis in three-dimensional culture. *Nature* **2011**, *472*, 51–56. [CrossRef] [PubMed]
17. Nakano, T.; Ando, S.; Takata, N.; Kawada, M.; Muguruma, K.; Sekiguchi, K.; Saito, K.; Yonemura, S.; Eiraku, M.; Sasai, Y. Self-formation of optic cups and storable stratified neural retina from human escs. *Cell Stem Cell* **2012**, *10*, 771–785. [CrossRef] [PubMed]
18. England, S.J.; Blanchard, G.B.; Mahadevan, L.; Adams, R.J. A dynamic fate map of the forebrain shows how vertebrate eyes form and explains two causes of cyclopia. *Development* **2006**, *133*, 4613–4617. [CrossRef] [PubMed]
19. Rembold, M.; Loosli, F.; Adams, R.J.; Wittbrodt, J. Individual cell migration serves as the driving force for optic vesicle evagination. *Science* **2006**, *313*, 1130–1134. [CrossRef] [PubMed]
20. Cavodeassi, F.; Carreira-Barbosa, F.; Young, R.M.; Concha, M.L.; Allende, M.L.; Houart, C.; Tada, M.;

Wilson, S.W. Early stages of zebrafish eye formation require the coordinated activity of wnt11, fz5, and the wnt/beta-catenin pathway. *Neuron* **2005**, *47*, 43–56. [CrossRef] [PubMed]

21. Loosli, F.; Staub, W.; Finger-Baier, K.C.; Ober, E.A.; Verkade, H.; Wittbrodt, J.; Baier, H. Loss of eyes in zebrafish caused by mutation of chokh/rx3. *EMBO Rep.* **2003**, *4*, 894–899. [CrossRef] [PubMed]
22. Brown, K.E.; Keller, P.J.; Ramialison, M.; Rembold, M.; Stelzer, E.H.; Loosli, F.; Wittbrodt, J. Nlcam modulates midline convergence during anterior neural plate morphogenesis. *Dev. Biol.* **2010**, *339*, 14–25. [CrossRef] [PubMed]
23. Kwan, K.M.; Otsuna, H.; Kidokoro, H.; Carney, K.R.; Saijoh, Y.; Chien, C.B. A complex choreography of cell movements shapes the vertebrate eye. *Development* **2012**, *139*, 359–372. [CrossRef] [PubMed]
24. Picker, A.; Cavodeassi, F.; Machate, A.; Bernauer, S.; Hans, S.; Abe, G.; Kawakami, K.; Wilson, S.W.; Brand, M. Dynamic coupling of pattern formation and morphogenesis in the developing vertebrate retina. *PLoS Biol.* **2009**, *7*, e1000214. [CrossRef] [PubMed]
25. Heermann, S.; Schutz, L.; Lemke, S.; Krieglstein, K.; Wittbrodt, J. Eye morphogenesis driven by epithelial flow into the optic cup facilitated by modulation of bone morphogenetic protein. *eLife* **2015**, *4*, e05216. [CrossRef] [PubMed]
26. Keller, P.J.; Schmidt, A.D.; Wittbrodt, J.; Stelzer, E.H. Reconstruction of zebrafish early embryonic development by scanned light sheet microscopy. *Science* **2008**, *322*, 1065–1069. [CrossRef] [PubMed]
27. Icha, J.; Schmied, C.; Sidhaye, J.; Tomancak, P.; Preibisch, S.; Norden, C. Using light sheet fluorescence microscopy to image zebrafish eye development. *J. Vis. Exp.* **2016**, *10*, e53966. [CrossRef] [PubMed]
28. Ivanovitch, K.; Cavodeassi, F.; Wilson, S.W. Precocious acquisition of neuroepithelial character in the eye field underlies the onset of eye morphogenesis. *Dev. Cell* **2013**, *27*, 293–305. [CrossRef] [PubMed]
29. Cechmanek, P.B.; McFarlane, S. Retinal pigment epithelium expansion around the neural retina occurs in two separate phases with distinct mechanisms. *Dev. Dyn.* **2017**, *246*, 598–609. [CrossRef] [PubMed]
30. Sidhaye, J.; Norden, C. Concerted action of neuroepithelial basal shrinkage and active epithelial migration ensures efficient optic cup morphogenesis. *eLife* **2017**, *6*, e22689. [CrossRef] [PubMed]
31. Martinez-Morales, J.R.; Rembold, M.; Greger, K.; Simpson, J.C.; Brown, K.E.; Quiring, R.; Pepperkok, R.; Martin-Bermudo, M.D.; Himmelbauer, H.; Wittbrodt, J. Ojoplano-mediated basal constriction is essential for optic cup morphogenesis. *Development* **2009**, *136*, 2165–2175. [CrossRef] [PubMed]
32. Bogdanovic, O.; Delfino-Machin, M.; Nicolas-Perez, M.; Gavilan, M.P.; Gago-Rodrigues, I.; Fernandez-Minan, A.; Lillo, C.; Rios, R.M.; Wittbrodt, J.; Martinez-Morales, J.R. Numb/numbl-opo antagonism controls retinal epithelium morphogenesis by regulating integrin endocytosis. *Dev. Cell* **2012**, *23*, 782–795. [CrossRef] [PubMed]
33. Nicolas-Perez, M.; Kuchling, F.; Letelier, J.; Polvillo, R.; Wittbrodt, J.; Martinez-Morales, J.R. Analysis of cellular behavior and cytoskeletal dynamics reveal a constriction mechanism driving optic cup morphogenesis. *eLife* **2016**, *5*, e15797. [CrossRef] [PubMed]
34. James, A.; Lee, C.; Williams, A.M.; Angileri, K.; Lathrop, K.L.; Gross, J.M. The hyaloid vasculature facilitates basement membrane breakdown during choroid fissure closure in the zebrafish eye. *Dev. Biol.* **2016**, *419*, 262–272. [CrossRef] [PubMed]
35. McMahon, C.; Gestri, G.; Wilson, S.W.; Link, B.A. Lmx1b is essential for survival of periocular mesenchymal cells and influences fgf-mediated retinal patterning in zebrafish. *Dev. Biol.* **2009**, *332*, 287–298. [CrossRef] [PubMed]
36. Lupo, G.; Gestri, G.; O'Brien, M.; Denton, R.M.; Chandraratna, R.A.; Ley, S.V.; Harris, W.A.; Wilson, S.W. Retinoic acid receptor signaling regulates choroid fissure closure through independent mechanisms in the ventral optic cup and periocular mesenchyme. *Proc. Natl. Acad. Sci. USA* **2011**, *108*, 8698–8703. [CrossRef] [PubMed]
37. Wen, W.; Pillai-Kastoori, L.; Wilson, S.G.; Morris, A.C. Sox4 regulates choroid fissure closure by limiting hedgehog signaling during ocular morphogenesis. *Dev. Biol.* **2015**, *399*, 139–153. [CrossRef] [PubMed]
38. Bryan, C.D.; Chien, C.B.; Kwan, K.M. Loss of laminin alpha 1 results in multiple structural defects and divergent effects on adhesion during vertebrate optic cup morphogenesis. *Dev. Biol.* **2016**, *416*, 324–337. [CrossRef] [PubMed]
39. Heavner, W.; Pevny, L. Eye development and retinogenesis. *Cold Spring Harb. Perspect. Biol.* **2012**, *4*, a008391. [CrossRef] [PubMed]
40. Hernandez-Bejarano, M.; Gestri, G.; Spawls, L.; Nieto-Lopez, F.; Picker, A.; Tada, M.; Brand, M.; Bovolenta, P.; Wilson, S.W.; Cavodeassi, F. Opposing shh and fgf signals initiate nasotemporal patterning of the zebrafish retina. *Development* **2015**, *142*, 3933–3942. [CrossRef] [PubMed]

41. Facchinello, N.; Schiavone, M.; Vettori, A.; Argenton, F.; Tiso, N. Monitoring wnt signaling in zebrafish using fluorescent biosensors. *Methods Mol. Biol.* **2016**, *1481*, 81–94. [PubMed]
42. Moro, E.; Ozhan-Kizil, G.; Mongera, A.; Beis, D.; Wierzbicki, C.; Young, R.M.; Bournele, D.; Domenichini, A.; Valdivia, L.E.; Lum, L.; et al. In vivo wnt signaling tracing through a transgenic biosensor fish reveals novel activity domains. *Dev. Biol.* **2012**, *366*, 327–340. [CrossRef] [PubMed]
43. Moro, E.; Vettori, A.; Porazzi, P.; Schiavone, M.; Rampazzo, E.; Casari, A.; Ek, O.; Facchinello, N.; Astone, M.; Zancan, I.; et al. Generation and application of signaling pathway reporter lines in zebrafish. *Mol. Genet. Genomics* **2013**, *288*, 231–242. [CrossRef] [PubMed]
44. Miesfeld, J.B.; Link, B.A. Establishment of transgenic lines to monitor and manipulate yap/taz-tead activity in zebrafish reveals both evolutionarily conserved and divergent functions of the hippo pathway. *Mech. Dev.* **2014**, *133*, 177–188. [CrossRef] [PubMed]
45. Collery, R.F.; Link, B.A. Dynamic smad-mediated bmp signaling revealed through transgenic zebrafish. *Dev. Dyn.* **2011**, *240*, 712–722. [CrossRef] [PubMed]
46. Schwend, T.; Loucks, E.J.; Ahlgren, S.C. Visualization of gli activity in craniofacial tissues of hedgehog-pathway reporter transgenic zebrafish. *PLoS ONE* **2010**, *5*, e14396. [CrossRef] [PubMed]
47. Molina, G.A.; Watkins, S.C.; Tsang, M. Generation of fgf reporter transgenic zebrafish and their utility in chemical screens. *BMC Dev. Biol.* **2007**, *7*, 62. [CrossRef] [PubMed]
48. Barolo, S. Transgenic wnt/tcf pathway reporters: All you need is lef? *Oncogene* **2006**, *25*, 7505–7511. [CrossRef] [PubMed]
49. Svoboda, K.K.; O'Shea, K.S. An analysis of cell shape and the neuroepithelial basal lamina during optic vesicle formation in the mouse embryo. *Development* **1987**, *100*, 185–200. [PubMed]
50. Eiraku, M.; Adachi, T.; Sasai, Y. Relaxation-expansion model for self-driven retinal morphogenesis: A hypothesis from the perspective of biosystems dynamics at the multi-cellular level. *Bioessays* **2012**, *34*, 17–25. [CrossRef] [PubMed]
51. Carpenter, A.C.; Smith, A.N.; Wagner, H.; Cohen-Tayar, Y.; Rao, S.; Wallace, V.; Ashery-Padan, R.; Lang, R.A. Wnt ligands from the embryonic surface ectoderm regulate 'bimetallic strip' optic cup morphogenesis in mouse. *Development* **2015**, *142*, 972–982. [CrossRef] [PubMed]
52. Beccari, L.; Marco-Ferreres, R.; Bovolenta, P. The logic of gene regulatory networks in early vertebrate forebrain patterning. *Mech. Dev.* **2013**, *130*, 95–111. [CrossRef] [PubMed]
53. Martinez-Morales, J. Vertebrate eye gene regulatory networks. In *Organogenetic Gene Networks*; Castelli-Gair Hombria, J., Bovolenta, P., Eds.; Springer International Publishing AG: Cham, Switzerland, 2016.
54. Cavodeassi, F.; Moreno-Marmol, T.; Hernandez-Bejarano, M.; Bovolenta, P. Principles of early vertebrate forebrain formation. In *Organogenetic Gene Networks*; Castelli-Gair Hombria, J., Bovolenta, P., Eds.; Springer International Publishing AG: Cham, Switzerland, 2016.
55. Williamson, K.A.; FitzPatrick, D.R. The genetic architecture of microphthalmia, anophthalmia and coloboma. *Eur. J. Med. Genet.* **2014**, *57*, 369–380. [CrossRef] [PubMed]
56. Voronina, V.A.; Kozhemyakina, E.A.; O'Kernick, C.M.; Kahn, N.D.; Wenger, S.L.; Linberg, J.V.; Schneider, A.S.; Mathers, P.H. Mutations in the human rax homeobox gene in a patient with anophthalmia and sclerocornea. *Hum. Mol. Genet.* **2004**, *13*, 315–322. [CrossRef] [PubMed]
57. Landsend, E.S.; Utheim, O.A.; Pedersen, H.R.; Lagali, N.; Baraas, R.C.; Utheim, T.P. The genetics of congenital aniridia—A guide for the ophthalmologist. *Surv. Ophthalmol.* **2018**, *63*, 105–113. [CrossRef] [PubMed]
58. Williamson, K.A.; Rainger, J.; Floyd, J.A.; Ansari, M.; Meynert, A.; Aldridge, K.V.; Rainger, J.K.; Anderson, C.A.; Moore, A.T.; Hurles, M.E.; et al. Heterozygous loss-of-function mutations in yap1 cause both isolated and syndromic optic fissure closure defects. *Am. J. Hum. Genet.* **2014**, *94*, 295–302. [CrossRef] [PubMed]
59. Liu, C.; Widen, S.A.; Williamson, K.A.; Ratnapriya, R.; Gerth-Kahlert, C.; Rainger, J.; Alur, R.P.; Strachan, E.; Manjunath, S.H.; Balakrishnan, A.; et al. A secreted wnt-ligand-binding domain of fzd5 generated by a frameshift mutation causes autosomal dominant coloboma. *Hum. Mol. Genet.* **2016**, *25*, 1382–1391. [CrossRef] [PubMed]

7

The Role of Apoptotic Signaling in Axon Guidance

Riley Kellermeyer †, Leah M. Heydman †, Grant S. Mastick and Thomas Kidd *

Department of Biology/MS 314, University of Nevada, 1664 N. Virginia St., Reno, NV 89557, USA; rileykellermeyer@gmail.com (R.K.); leah.heydman@gmail.com (L.M.H.); gmastick@unr.edu (G.S.M.)
* Correspondence: tkidd@unr.edu
† These authors contributed equally to this work.

Abstract: Navigating growth cones are exposed to multiple signals simultaneously and have to integrate competing cues into a coherent navigational response. Integration of guidance cues is traditionally thought to occur at the level of cytoskeletal dynamics. *Drosophila* studies indicate that cells exhibit a low level of continuous caspase protease activation, and that axon guidance cues can activate or suppress caspase activity. We base a model for axon guidance on these observations. By analogy with other systems in which caspase signaling has non-apoptotic functions, we propose that caspase signaling can either reinforce repulsion or negate attraction in response to external guidance cues by cleaving cytoskeletal proteins. Over the course of an entire trajectory, incorrectly navigating axons may pass the threshold for apoptosis and be eliminated, whereas axons making correct decisions will survive. These observations would also explain why neurotrophic factors can act as axon guidance cues and why axon guidance systems such as Slit/Robo signaling may act as tumor suppressors in cancer.

Keywords: axon guidance; growth cone; cytoskeleton; caspases; apoptosis; signal integration; basal level of caspase activity; death associated inhibitor of apoptosis; axon branching; Netrin; DCC; Frazzled; Slit; Robo; *Drosophila*

---

1. Introduction

The navigational center of growing axons is the growth cone, a highly dynamic expansion of the axon shaft that samples the environment and integrates multiple cues to generate directed extension, retraction, and turning [1,2]. Traditionally, axon attractants such as Netrins are thought to increase cytoskeletal outgrowth towards a cue, whereas axon repellents such as Slits inhibit cytoskeletal growth [3,4]. The net effect of attractive and repulsive cues on the cytoskeleton results in growth towards attractive cues and away from repulsive cues. Integration has been demonstrated to also occur through interactions between cell surface receptors, and through intracellular kinases [5,6]. More recent evidence suggests that the traditional view of growth cones navigating up or down gradients of guidance cues such as Netrin may not be accurate or even valid in vivo [7–11]. Our own work led us to identify a role for the apoptotic machinery in the growth cone that is likely functioning to integrate opposing guidance cues. This review examines models for how the cell death machinery could be involved in axon guidance with an emphasis on results from *Drosophila*.

Drosophila Netrin-B Is a Neurotrophic Factor That Blocks Cell Death

Netrins are diffusible axon guidance cues most famous for attracting axons to the CNS midline [12]. The fly has two Netrin genes, *NetA* and *NetB*, that are required for midline and motor neuron axon guidance [13,14]. The genes appear to be the product of tandem duplication and display a high degree of functional overlap. Netrins are expressed by the CNS midline and embryos lacking both

Netrins (*NetAB*) have axon guidance defects and increased cell death. Both phenotypes can be rescued by midline expression of either gene. Localized sources of Netrins therefore appear to provide a navigational cue. In contrast to midline expression, pan-neuronal expression of either Netrin in wild type embryos leads to axon phenotypes, either due to a lack of positional information specific to the midline or through attraction to non-midline areas [13,14]. It was therefore surprising to find that pan-neuronal expression of *NetB* alone can rescue axon guidance defects in *NetAB* embryos [15]. In contrast, pan-neural expression of *NetA* increases the severity of the *NetAB* mutant phenotype, establishing a clear difference between the proteins. NetB was subsequently identified as a neurotrophic factor when over-expressed, because blocking cell death using the baculovirus p35 caspase inhibitor in discrete subsets of neurons can rescue *NetAB* midline guidance defects [15]. These findings substantiate a model for a non-apoptotic role for caspases in the growth cone, as suggested by prior results in Xenopus retinal growth cones and other systems [16–18].

2. The Apoptotic Machinery and Guidance Receptors

Programmed cell death or apoptosis is a key part in the development of multicellular organisms and in the maintenance of the correct number of cells in mature animals [27]. The classic role for cell death in neural development is to eliminate unneeded connections that fail to compete for a limiting survival factor, the neurotrophic hypothesis [28]. For example, axons that are misguided and fail to reach their target tissue die [29]. Further research has shown that neurotrophic factors can also induce death in certain contexts, and this activity is functionally conserved in the fly [30–32]. Apoptosis operates through molecular cascades, offering several potential signaling nodes that could intersect with axon guidance signaling pathways. A crucial event in apoptosis is the activation of specialized cysteine-aspartic acid proteases called caspases (Figure 1A). Signals derived from the mitochondria, in response to external pro-apoptotic signals or the withdrawal of trophic support molecules, trigger activation of initiator caspases [33–35]. Activation relies on multimerization and/or conformational changes of initiator caspases, like Dronc, leading to proteolytic activation of effector caspases such as Drice and Dcp-1. This amplification of effector caspases is then responsible for the extensive proteolytic cleavage that occurs during apoptosis. Caspase activity is buffered by the death-associated inhibitor of apoptosis protein family (*Diap1/thread*), the viral p35 protein and other specialized inhibitors which oppose caspase activation [36–41]. Many caspase inhibitors, like Diap1, are continuously active to prevent apoptosis by marking initiator caspases for degradation [42,43]. To initiate apoptosis, Diap1 is cleaved by inhibitor of apoptosis antagonists, notably Hid, Grim and Reaper (RHG proteins), ultimately allowing caspase activation [34]. In addition to apoptosis, caspases have been implicated in actin dynamics of *Drosophila* spermatid individualization [44,45], sense organ precursor selection [42,46], and dendrite retraction [47,48] (reviewed in [35]). Together, this cascade of apoptotic regulators is present in all cells at low levels, but normally kept in check by specific and tightly regulated modulators.

The apoptotic machinery is present in the growth cone of extending axons and can be activated by external signals such as Netrin [16]. The principal vertebrate Netrin receptor is DCC (Deleted in Colorectal Cancer; Frazzled in the fly), which dimerizes upon ligand binding to stimulate signaling pathways that alter cytoskeletal dynamics [12]. Netrins are also capable of repelling axons through the Unc-5 receptor [49]. In vertebrates, failure of DCC to homodimerize triggers cell death via the initiator caspase-9 and the effector caspase-3 [50]. The dual function of DCC to transduce both migratory and apoptotic signals is known as the dependence receptor hypothesis, wherein cell survival is dependent on ligand occupancy of receptors. Dependence receptors such as DCC are characterized by having caspase cleavage sites in their cytoplasmic domains and that the absence of the ligand triggers caspase cleavage of the receptor. The DCC caspase cleavage site is required to trigger apoptosis, and mutation of the site in mice prevents tumor suppression [51,52]. However, the ability of Netrin to act as a survival factor, particularly in spinal cord development is controversial as conflicting survival phenotypes have been observed by different groups [53,54]. Additional recent evidence strongly suggests that the dependence receptor mechanism is not operating in the mouse spinal cord [55].

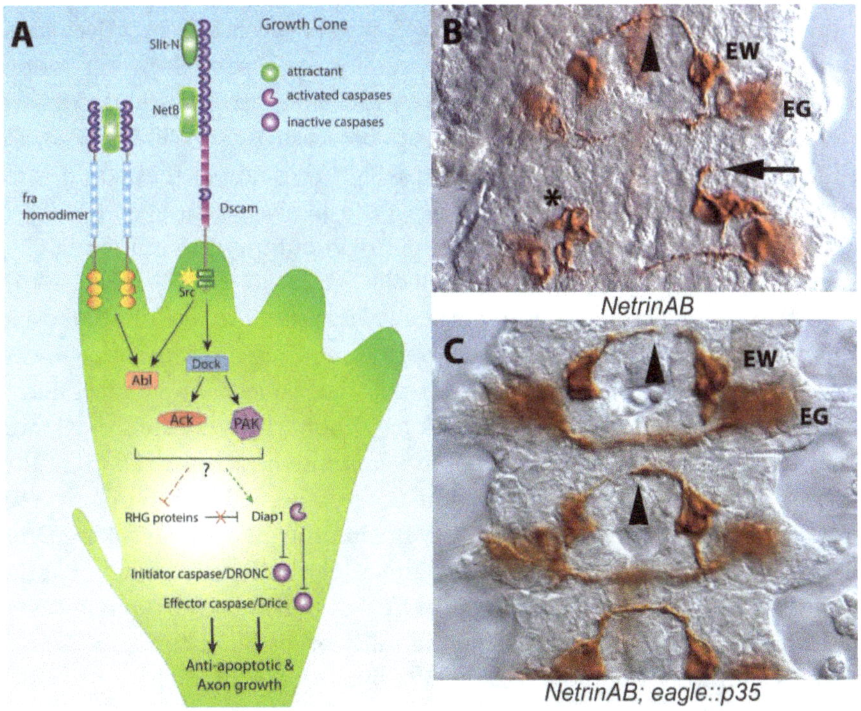

Figure 1. Apoptotic machinery and axons. (**A**) Schematic of axon guidance signaling pathways that potentially interact with the apoptotic machinery. Candidate downstream effectors are shown based on proteins that can promote survival and are known to interact with the receptors shown in *Drosophila* [19–24]. Cytoplasmic signaling components may act on caspase regulators, such as inactivating RHG proteins, promoting Diap1 function or could act directly on the caspases (uncertainty indicated by dotted lines). (**B**) A *Drosophila* embryo lacking the *Netrin-A* and *Netrin-B* axon guidance genes (*NetAB*) stained to reveal *eagle* positive axons (brown). Anterior is to the top. The EW and EG neuron clusters are indicated. In *NetAB* mutants, the EW commissure successfully crosses the midline about 50% of the time (arrowhead). A growth cone can be seen remaining projecting ipsilaterally (arrow). An EW cluster lacking a leading growth cone either due to developmental delay or apoptosis of a neuron can also be seen (asterisk). An overall lack of symmetry of the EW and EG cluster can be seen with the clusters mispositioned and differences in neuron number reflected in differences in staining intensity. The EG commissures are present but not in the plane of focus. (**C**) A *Drosophila NetAB* embryo whose axon guidance defects have been rescued by expression of the *p35* effector caspase inhibitor [25,26]. Growth cones of the EW neuron cluster can be seen crossing the midline (arrowheads), while the growth cones of the contralateral homologues are growing at a slower rate. The growth cones of the most posterior segment in this panel have fasciculated with the contralateral homologue even though the more anterior segments are slightly older in development. In older embryos 90% of the EW axons cross the midline. Images courtesy of G. Newquist.

In flies, the DCC homologue, *frazzled* (*fra*; also called Unc-40 in *C. elegans*), lacks the caspase cleavage site in the cytoplasmic domain, suggesting Fra is not a dependence receptor, though alternative sites could exist [54,56]. Additionally, the **loss** of *fra* activity triggers apoptosis in some tissues, rather than being protective from cell death, as would be expected from the loss of a dependence receptor [57]. The ability to rescue the axon guidance defects of *NetAB* mutants by blocking apoptotic signaling therefore requires an alternative explanation to the dependence receptor hypothesis.

3. How Could Caspase Signaling Operate in Growth Cone Guidance?

Apoptosis requires major changes in the structure of cells via rearrangements of the cytoskeleton, so it is not surprising that caspases cleave a large number of cytoskeletal and structural proteins such as actin, alpha-tubulin, and Spectrin [58–61]. Caspase-3 cleaves Spectrin in growth cones in culture [62]. Changes in levels and localization of cytoskeletal proteins are observed during apoptosis of larval salivary glands, and these changes are prevented by inhibiting caspases [63]. Many proteins that modulate the cytoskeleton are also cleaved during apoptosis such as cofilin, GAP43, and rho kinase (Rock), as well as many cell adhesion molecules [64–66]. These molecules could serve as caspase substrates in the growth cone and their cleavage would likely decrease protrusion (actin driven extension of filopodia and lamellipodia) and inhibit axon outgrowth. Inhibiting caspase activity would protect cytoskeletal components from proteolysis and potentially increase protrusion and forward movement. Migrating border cells in the *Drosophila* ovary provide an apoptosis-independent example for this model. Normal border cell migration requires the activity of the *Diap1* caspase inhibitor [67]. Diap1 forms a complex with the actin cytoskeletal modulators Rac and Profilin, and Diap1 protects these modulators from degradation by Dronc. In ovary border cell migration, Diap1 alters cytoskeletal dynamics independent of apoptosis inhibition, as loss of Diap1 in these cells does not result in cell death and migration phenotypes are not rescued with p35. Similarly, Diap1 promotes F-actin assembly in polarized elongation of sensory organ progenitors by blocking Dronc activation, in a caspase-dependent, apoptosis-independent manner [68,69]. F-actin turnover at the cell margin therefore requires inhibition of Diap1 without any effect on cell survival. An additional example of non-apoptotic functions of caspases comes from the study of so-called "undead" cells in which Hid expression activates apoptosis, but the cells are kept alive by co-expression of p35 [70]. When these cells are created in the anterior-posterior boundary of the wing disc, the undead cells invade and migrate into the posterior domain of the wing disc. This migration is dependent on effector caspases and appears to require intermediate levels of caspase activity [71,72]. Comparable approaches have been taken in vertebrate neurons to separate axon growth from neuronal survival. Blocking cell death through over-expression of Bcl-2 or loss of Bax function was used to demonstrate that axon growth does not appear to be constitutive but likely relies on external signals [73,74].

There are several important implications from these and other studies for non-apoptotic functions of caspase protease activity. The first is that activation of the apoptotic machinery does not necessarily lead to cell death. In vertebrates, molecular mechanisms have been identified that prevent complete activation of the apoptotic cascade [75]. Studies of the synapse suggest that transient and local activation of caspases can remodel the synapse [76], and observations of activated caspase-3 restricted to the growth cone and sites of axon branching are consistent with this model [16,77]. Second, *Drosophila* studies have demonstrated that there is a continuous low level of initiator caspase activation through auto-processing and the role of Diap1 is to counteract this basal caspase activity [43,78–81]. Low levels of caspase activation have been visualized in wing discs using the FRET-based SCAT3 caspase activity probe [42]. A vertebrate Diap1 homologue, X-linked inhibitor of apoptosis (XIAP) plays a role in limiting caspase activation [82,83], and appears central to restricting caspase activation to subcellular compartments of neurons [84].

Returning to the fly CNS and the observation that Netrin mutant phenotypes can be rescued by anti-apoptotic factors, the simplest model to explain the effects of caspase inhibition in *Netrin* mutants is that there is a basal level of caspase activation in growth cones that has to be overcome for maximal forward growth (Figure 2A). This model was first proposed by Gilman and Mattson, after demonstrating that addition of caspase inhibitors to neuronal cultures increases axon outgrowth [76]. The implication is that a basal level of caspase proteolysis keeps normal axon growth below maximum levels, although the molecular targets of caspases in growth cones remain undefined. To support any of the models described, cleavage of specific substrates, such as cytoskeletal components, will have to be demonstrated in the growth cone itself, which may require the study of larger growth cones from other invertebrate or vertebrate species, given the small size of *Drosophila* growth cones.

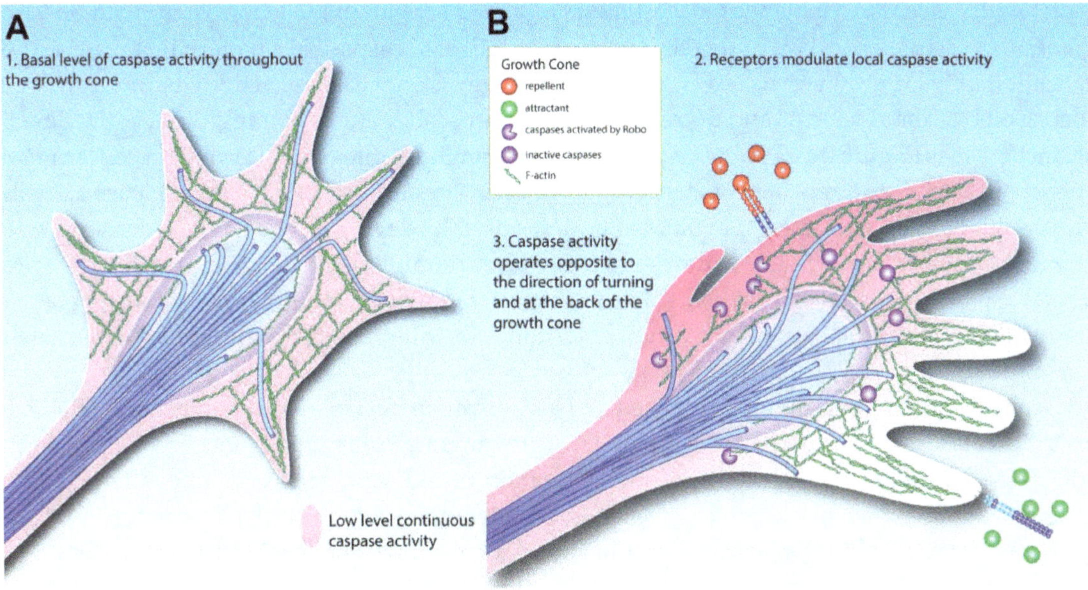

Figure 2. Model for caspase activity in the growth cone. (**A**) Simple low-level activation of caspases throughout the growth cone. (**B**) Attractant and repellent cues modulate caspase activity via cell surface receptors, altering the growth cone trajectory. In this model, repellents increase caspase activity, while attractants decrease caspase activity.

It is worth noting that caspases can also modulate the actin cytoskeleton independently of their protease roles, such as promoting Aip1/cofilin mediated actin polymerization in migrating lymphocytes, though it is not known how widespread this effect is [65].

4. Extracellular Modulation of Caspase Activity in the Growth Cone

An exciting possibility arising from this simple model of caspase-mediated guidance is that caspase activity could be actively modulated in response to external cues (Figure 2B). Mehlen has proposed that axon guidance signaling pathways actively modulate tumor cell survival, explaining why axon guidance molecules are implicated in preventing cancer [85]. Despite a large number of studies implicating Slit/Robo signaling as tumor suppressors in cancer, a direct link of Slit/Robo signals to apoptosis has not yet been shown [86]. In fly screens, the post-translational regulator of Robo, *commissureless* and its paralogue *comm3*, have been shown to suppress apoptotic phenotypes through unidentified mechanisms [87]. DCC has been strongly linked to caspase activity [50], and DCC acts as a tumor suppressor via its dependence receptor activity [52]. In flies, the Netrin receptor *frazzled* appears to act as a tumor suppressor, because while *fra* mutant clones are usually not viable, they can be rescued with p35 expression blocking cell death [57]. Additionally, loss of *fra* activity leads to invasive cell phenotypes reminiscent of metastasis [57,88]. Mutant clones of *fra* have not been made in the embryo, but *fra* homozygotes do not display increased cell death in the developing CNS [15]. This could reflect a tissue-specific difference between the embryo and eye-antennal discs, perhaps due to functional overlap between survival factors in the embryo. Alternatively, the differences between homozygous embryos and mutant clones could be the result of cell competition in which mutant clones are less fit than neighboring cells, a process that interestingly can involve Slit/Robo signaling [89]. In the embryo, NetB likely promotes axon growth by inhibiting caspase activity in the growth cone. Double mutants for the *fra* and *Dscam1* Netrin receptors display an increase in cell death, whereas there is no change in either mutant alone [15], suggesting multiple receptors may mediate this activity.

Emerging evidence reveals that repellent signaling pathways are able to activate caspases, in some cases through direct binding. The Slit/Robo, Eph/Ephrin and Sema/Plexin pathways all recruit and/or activate caspases [17,77,90–94]. Slit/Robo signaling in zebrafish axons has been shown to genetically interact with caspases, in a manner that suggests localized activation [77]. Consistent with Slit/Robo regulation of caspases, we have observed low levels of activated caspase in a pattern that matches Robo localization in the ventral nerve cord (Figure 3). This suggests that caspase activity in the growth cone is increased by axon repellents and decreased by attractants, allowing axon outgrowth in the direction of attractant cues (Figure 2B). *Drosophila* motor neurons similarly integrate information concerning the levels of attractants and repellents emanating from their target muscle [95]. Thus caspase activation and inhibition could sum the input signals to determine lowered caspase activity domains where forward growth occurs, while a widespread basal level of inhibitors like Diap1 restrict the spread of local caspase activation by repellents. In this respect, the model resembles the synapse where the duration and intensity of caspase activation determines the difference between synaptic remodeling and neuronal death [96,97].

Dendritic pruning also requires localized caspase activation but over a much larger subcellular area (dendritic trees can be quite extensive compared to a growth cone), ultimately leading to the degradation of dendrites. Pruning relies heavily on the activation of initiator caspases such as Dronc [47,48]. Expression of p35, which blocks effector caspases, is either ineffective at blocking pruning [47], or only appears to delay pruning [98]. Migration of undead cells also requires initiator but not effector caspase activity, as migration is blocked when Diap1 is expressed (Diap1 inhibits both initiator and effector caspases) [71]. A distinguishing feature of non-apoptotic caspase activity may be a reliance on initiator caspases. Blocking cell death upstream of all caspases in fly embryos has remarkably little effect on the CNS axon scaffold [99], suggesting that caspase activity is dispensable for normal development. However, in sensitized backgrounds such as *NetAB* mutants, blocking effector caspases is sufficient to rescue axon guidance [15]. Overall these results suggest that there may be quantitatively different levels of caspase activation. Growth cones and migrating cells may display low continuous levels of caspase activation. Higher levels may be required for neuronal remodeling events in which structures or parts of the cell such as dendrites are lost, and the highest levels will of course lead to apoptosis. A mechanism for different levels of caspase activation has been proposed from observations of dendritic pruning. The model proposes that there is a threshold for cell death and that, below this threshold, IAPs bound to caspases can be quickly released without activation of the apoptosome [100]. This may represent a general mechanism for non-apoptotic functions of caspases but may need further refinement to account for multiple levels of caspase activation.

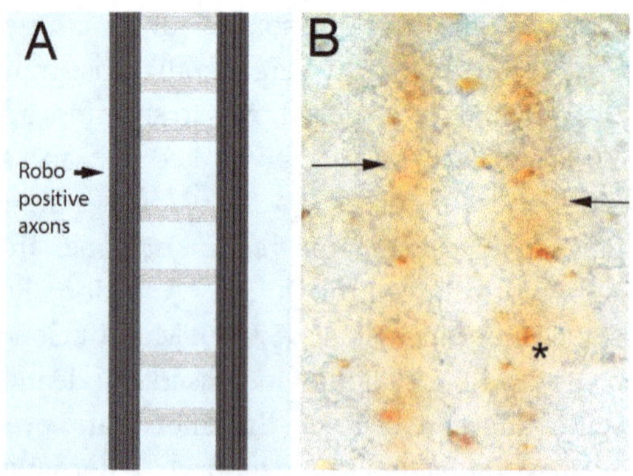

Figure 3. Low-level caspase activity in longitudinal axons. (**A**) Schematic of the CNS axon scaffold in the fly ventral nerve cord. The axons form a ladder-like pattern. The Robo repulsive receptor is only found in the longitudinal portions of CNS axons (arrow). (**B**) *Drosophila* embryonic nerve cord stained with an antibody raised against activated vertebrate Caspase-3 that appears to detect Dronc activation in flies [101]. Dying cells are visible as densely stained regions, usually oval in shape (asterisk). A continuous low level of staining can be seen in the region occupied by the longitudinal axons (arrows). This pattern matches the pattern of Robo localization, with the Robo protein excluded from axons segments crossing the midline while upregulated in axons using active Slit/Robo signaling to avoid the midline [102], suggesting that the Robo and caspase activation pathways may be linked.

If axon repellents activate caspases at low levels, they could mediate contact inhibition or increase sensitivity to survival factors in tumors. Alternatively, it could be that cytoskeletal rearrangements in response to both positive and negative cues require caspase activation, or that caspase activation controls growth cone protein levels [16].

5. Caspase Signaling at the Fly CNS Midline

In fly *NetAB* mutants, most growth cones orient towards the CNS midline but many fail to cross it [103]. The original interpretation of this observation is that Netrins are required for axons to extend across the midline, rather than to attract axons to the midline. The ability of caspase inhibition to rescue axon crossing in *NetAB* mutants suggests that caspase activity inhibits midline crossing or axon outgrowth in general. To explain this observation, we propose that there is a low level of caspase activity in the growth cone that needs to be overcome for forward growth. Alternatively, or in parallel, strict temporal control of caspase activation may allow for the necessary cytoskeletal rearrangements [35]. While many axons do not, a significant number of axons do cross the midline in *NetAB* mutants, revealing that there are undiscovered attractants expressed at the midline. This suggests that caspase inhibition may increase axon outgrowth enough to allow growth cones to use these other cues to locate and grow towards the midline. Interestingly, the fly CNS midline is a source of other neurotrophic factors like the *Drosophila* neurotrophins DNTs [104], acting through neuronal Toll receptors [31]. Classically, the axonal target tissue produces these neurotrophic factors. An increasing number of examples are now known where intermediate targets supply neurotrophic factors—a phenomenon termed *en passant* or pre-target neurotrophic action [54,105–107]. As *NetB* and *DNTs* are expressed by the CNS midline intermediate targets, and required for motor axon targeting [95,108], both gene families may function as *en passant* neurotrophic factors and as guidance cues. Interestingly, midline glia require the axon derived epidermal growth factor Spitz to survive, and when the axons fail to contact the midline glia in *commissureless* mutants, the midline glia migrate to the axons to obtain Spitz [109]. Artificially promoting survival by increasing MAPK signaling, which inhibits Hid, removes the necessity for migration suggesting that Spitz is acting as a caspase-mediated attractive signal. Searching for survival factors could be a general mechanism that influences cell and growth cone migration. MAPK signaling in vertebrate growth cones and commissural axons has been implicated in the response to Netrin supporting this model [16,50]. Neurotrophic factors appear to be required for axon growth independent of their role in cell survival [110], and could do so by inhibiting caspase activation in the growth cone. Together, *Drosophila* axon navigation across the ventral midline suggests that there remains much to be discovered about the functional links between the classical axon guidance problem of midline crossing and caspase signaling.

6. Apoptotic Signaling in Axon Branching

As axons extend, particularly as they enter their target tissues, they also branch, with each branch forming its own growth cone. The process of axon branching is also likely regulated by the apoptotic machinery. One of the most dramatic visualizations of caspase activity in axons is in zebrafish retinal ganglion cells [77]. Caspase activation occurs in a dynamic fashion at branchpoints in developing axonal arbors and genetically interacts with Slit/Robo signaling. Interestingly, Slit is proteolytically cleaved into two fragments, Slit-N and Slit-C. Slit-N stimulates axon branching, whereas full length Slit (Slit-FL) inhibits branching [111–113]. Slit-N is neurotrophic [114], and it will be interesting to see whether Slit-FL can directly activate caspases, perhaps via p38 MAPK signaling as suggesting by zebrafish studies [77]. Additionally, activated caspase activity has been observed in the developing auditory brainstem within several segments of navigating axons as well as their terminal branches within targets where it is proposed to limit unnecessary axonal arborization, because inhibition of caspase activity causes axon branches to spread into inappropriate target tissue [18]. Although much less is known about axon branching within targets, caspase activity appears to be an important regulator of not only primary axons but also their terminal branches. Axon branching may be more important than primary axon growth for regenerative recovery of connections after injury or disease, as branching from spared axons can be major contributors to restoring circuit function [115].

7. Critical Experiments for the Activated Caspase Model

The most important step forward for the proposed model would be to observe asymmetric activation of caspases in growth cones exposed to attractants or repellents. The long axons of zebrafish retinal ganglion cells have been used to successfully demonstrate localized caspase activity using the SCAT3 probe [77]. It seems likely that the same approach could be used in the growth cones of cultured neurons, as asymmetric gradients of phosphorylated Shootin1 (a cytoplasmic axon outgrowth protein) have been observed in growth cones responding to shallow Netrin gradients [116]. Antibodies against activated (cleaved) caspases do not appear to have sufficient sensitivity at present. If Diap1 levels are modulated in response to guidance cues, as seems likely, using split fluorescent protein strategies to tag the endogenous protein and limiting fluorescence to small subsets of neurons may prove invaluable for in vivo analysis [117]. Both cell culture and in vivo approaches would be especially powerful if combined with high-resolution structured illumination microscopy, as has been used for dendritogenesis in *Drosophila* embryos [118]. A second important step for the model would be to define which components of the apoptotic machinery are used in axon guidance and to demonstrate genetic interactions with specific signaling pathways, as has been done for axon degeneration and pruning [119]. A technical problem will be to limit cell death in mutant combinations as this could greatly hinder the interpretation of phenotypes. Dose-sensitive genetic interactions will likely allow sensitization of the cell death machinery without triggering apoptosis. For example, *Diap1* heterozygotes would be predicted to increase caspase activity and enhance *NetAB* axon phenotypes without preventing analysis of axon guidance. Positive results in the proposed experiments would establish the basic model. We also propose that accumulating levels of caspase activation by axon errors will eventually reach the threshold to trigger apoptosis. This model requires an integration system to add up the errors over time, with the simplest mechanism being the accumulation of active caspases. Demonstrating that caspase activation persists and accumulates would require in vivo imaging. The Apoliner caspase reporter in which caspase cleavage results in nuclear translocation of enhanced green fluorescent protein (eGFP) while leaving monomeric red fluorescent protein (mRFP) at the cell membrane could function in this approach [120]. For example, mis-navigating *eagle* commissural axons in *NetAB* mutants would be predicted to have higher levels of eGFP in the nucleus than those that had successfully crossed the midline. A simpler and potentially complementary approach not requiring live imaging would be to use the CD8::Parp::Venus reporter to demonstrate higher levels of caspase activation in the axons of mis-navigating axons [48]. Additional reporters of caspase activity could similarly be employed [121].

8. Conclusions

The neurotrophic hypothesis proposes that competition between neurons for functional connections leads to the correct wiring of the nervous system. Studies of caspases now suggest that the apoptotic machinery plays an active role in forming the connections in the first place. Based on *Drosophila* studies, we propose that there is a continuous low level of basal caspase activation in growth cones that is kept in check by the caspase inhibitor Diap1. Modulation of caspase activity by external signals affects axon growth rates and allows for the integration of multiple, potentially conflicting, inputs to generate a coherent response. These conclusions are summarized:

Highlights of apoptotic signaling in axon guidance

- Localized activation of caspases in the growth cone may modulate axon guidance.
- Axon attractants can promote cell survival, while repellents can promote cell death.
- Neurotrophic factor effects on axon guidance could be through caspase signaling.

- Based on an analogy with systems in which caspase signaling has non-apoptotic roles, we propose that the duration and intensity of caspase activation can modulate growth cone activity, while longer and stronger caspase activity can induce death.
- Crossing the CNS midline is associated with lower caspase activity.
- Correct wiring of the nervous system could result from the elimination of incorrectly navigating neurons due to increased activity of the cell death machinery.

Author Contributions: Writing—original draft preparation: T.K.; Writing—review and editing: L.M.H., R.K., G.S.M., and T.K.

Funding: This work was supported by grants from the National Institutes of Health (P20 GM103554 to T.K.), (EY025205 to G.S.M.) and the National Science Foundation (IOS-1053555 to T.K.).

Acknowledgments: We thank Chaye Propst for preparing the original Adobe Illustrator files used to generate Figure 2 and Gunnar Newquist for the image in Figure 1B. We thank members of the Kidd laboratory for discussing ideas.

Conflicts of Interest: The authors declare no conflict of interest.

References

1. Lowery, L.A.; Van Vactor, D. The trip of the tip: Understanding the growth cone machinery. *Nat. Rev. Mol. Cell Biol.* **2009**, *10*, 332–343. [CrossRef] [PubMed]
2. Kolodkin, A.L.; Tessier-Lavigne, M. Mechanisms and molecules of neuronal wiring: A primer. *Cold Spring Harbor Perspect. Biol.* **2011**, *3*. [CrossRef] [PubMed]
3. Dent, E.W.; Gupton, S.L.; Gertler, F.B. The growth cone cytoskeleton in axon outgrowth and guidance. *Cold Spring Harbor Perspect. Biol.* **2011**, *3*. [CrossRef] [PubMed]
4. Gomez, T.M.; Letourneau, P.C. Actin dynamics in growth cone motility and navigation. *J. Neurochem.* **2014**, *129*, 221–234. [CrossRef] [PubMed]
5. Chacon, M.R.; Fazzari, P. FAK: Dynamic integration of guidance signals at the growth cone. *Cell Adhes. Migr.* **2011**, *5*, 52–55. [CrossRef]
6. Dudanova, I.; Klein, R. Integration of guidance cues: Parallel signaling and crosstalk. *Trends Neurosci.* **2013**, *36*, 295–304. [CrossRef] [PubMed]
7. Varadarajan, S.G.; Kong, J.H.; Phan, K.D.; Kao, T.J.; Panaitof, S.C.; Cardin, J.; Eltzschig, H.; Kania, A.; Novitch, B.G.; Butler, S.J. Netrin1 Produced by Neural Progenitors, Not Floor Plate Cells, Is Required for Axon Guidance in the Spinal Cord. *Neuron* **2017**, *94*, 790–799. [CrossRef] [PubMed]
8. Dominici, C.; Moreno-Bravo, J.A.; Puiggros, S.R.; Rappeneau, Q.; Rama, N.; Vieugue, P.; Bernet, A.; Mehlen, P.; Chedotal, A. Floor-plate-derived netrin-1 is dispensable for commissural axon guidance. *Nature* **2017**, *545*, 350–354. [CrossRef] [PubMed]
9. Tang, X.; Wadsworth, W.G. SAX-3 (Robo) and UNC-40 (DCC) regulate a directional bias for axon guidance in response to multiple extracellular cues. *PLoS ONE* **2014**, *9*, e110031. [CrossRef] [PubMed]
10. Stoeckli, E.T. Understanding axon guidance: Are we nearly there yet? *Development* **2018**, *145*. [CrossRef] [PubMed]
11. Goodhill, G.J. Can Molecular Gradients Wire the Brain? *Trends Neurosci.* **2016**, *39*, 202–211. [CrossRef] [PubMed]
12. Sun, K.L.W.; Correia, J.P.; Kennedy, T.E. Netrins: Versatile extracellular cues with diverse functions. *Development* **2011**, *138*, 2153–2169. [CrossRef] [PubMed]
13. Harris, R.; Sabatelli, L.M.; Seeger, M.A. Guidance cues at the Drosophila CNS midline: Identification and characterization of two Drosophila Netrin/UNC-6 homologs. *Neuron* **1996**, *17*, 217–228. [CrossRef]
14. Mitchell, K.J.; Doyle, J.L.; Serafini, T.; Kennedy, T.E.; Tessier-Lavigne, M.; Goodman, C.S.; Dickson, B.J. Genetic analysis of Netrin genes in Drosophila: Netrins guide CNS commissural axons and peripheral motor axons. *Neuron* **1996**, *17*, 203–215. [CrossRef]
15. Newquist, G.; Drennan, J.M.; Lamanuzzi, M.; Walker, K.; Clemens, J.C.; Kidd, T. Blocking apoptotic signaling rescues axon guidance in Netrin mutants. *Cell Rep.* **2013**, *3*, 595–606. [CrossRef] [PubMed]
16. Campbell, D.S.; Holt, C.E. Apoptotic pathway and MAPKs differentially regulate chemotropic responses of retinal growth cones. *Neuron* **2003**, *37*, 939–952. [CrossRef]

17. Ohsawa, S.; Hamada, S.; Asou, H.; Kuida, K.; Uchiyama, Y.; Yoshida, H.; Miura, M. Caspase-9 activation revealed by semaphorin 7A cleavage is independent of apoptosis in the aged olfactory bulb. *J. Neurosci.* **2009**, *29*, 11385–11392. [CrossRef] [PubMed]
18. Rotschafer, S.E.; Allen-Sharpley, M.R.; Cramer, K.S. Axonal Cleaved Caspase-3 Regulates Axon Targeting and Morphogenesis in the Developing Auditory Brainstem. *Front. Neural Circuits* **2016**, *10*, 84. [CrossRef] [PubMed]
19. Forsthoefel, D.J.; Liebl, E.C.; Kolodziej, P.A.; Seeger, M.A. The Abelson tyrosine kinase, the Trio GEF and Enabled interact with the Netrin receptor Frazzled in Drosophila. *Development* **2005**, *132*, 1983–1994. [CrossRef] [PubMed]
20. O'Donnell, M.P.; Bashaw, G.J. Distinct functional domains of the Abelson tyrosine kinase control axon guidance responses to Netrin and Slit to regulate the assembly of neural circuits. *Development* **2013**, *140*, 2724–2733. [CrossRef] [PubMed]
21. Dorsten, J.N.; Varughese, B.E.; Karmo, S.; Seeger, M.A.; VanBerkum, M.F. In the absence of frazzled over-expression of Abelson tyrosine kinase disrupts commissure formation and causes axons to leave the embryonic CNS. *PLoS ONE* **2010**, *5*, e9822. [CrossRef] [PubMed]
22. Muda, M.; Worby, C.A.; Simonson-Leff, N.; Clemens, J.C.; Dixon, J.E. Use of double-stranded RNA-mediated interference to determine the substrates of protein tyrosine kinases and phosphatases. *Biochem. J.* **2002**, *366*, 73–77. [CrossRef] [PubMed]
23. Schmucker, D.; Clemens, J.C.; Shu, H.; Worby, C.A.; Xiao, J.; Muda, M.; Dixon, J.E.; Zipursky, S.L. Drosophila Dscam is an axon guidance receptor exhibiting extraordinary molecular diversity. *Cell* **2000**, *101*, 671–684. [CrossRef]
24. Sterne, G.R.; Kim, J.H.; Ye, B. Dysregulated Dscam levels act through Abelson tyrosine kinase to enlarge presynaptic arbors. *eLife* **2015**, *4*, e05196. [CrossRef] [PubMed]
25. Worby, C.A.; Simonson-Leff, N.; Clemens, J.C.; Huddler, D., Jr.; Muda, M.; Dixon, J.E. Drosophila Ack targets its substrate, the sorting nexin DSH3PX1, to a protein complex involved in axonal guidance. *J. Biol. Chem.* **2002**, *277*, 9422–9428. [CrossRef] [PubMed]
26. Schoenherr, J.A.; Drennan, J.M.; Martinez, J.S.; Chikka, M.R.; Hall, M.C.; Chang, H.C.; Clemens, J.C. Drosophila activated Cdc42 kinase has an anti-apoptotic function. *PLoS Genet.* **2012**, *8*, e1002725. [CrossRef] [PubMed]
27. Fuchs, Y.; Steller, H. Programmed cell death in animal development and disease. *Cell* **2011**, *147*, 742–758. [CrossRef] [PubMed]
28. Hamburger, V.; Levi-Montalcini, R. Proliferation, differentiation and degeneration in the spinal ganglia of the chick embryo under normal and experimental conditions. *J. Exp. Zool.* **1949**, *111*, 457–501. [CrossRef] [PubMed]
29. Michalak, S.M.; Whitman, M.C.; Park, J.G.; Tischfield, M.A.; Nguyen, E.H.; Engle, E.C. Ocular Motor Nerve Development in the Presence and Absence of Extraocular Muscle. *Investig. Ophthalmol. Vis. Sci.* **2017**, *58*, 2388–2396. [CrossRef]
30. Foldi, I.; Anthoney, N.; Harrison, N.; Gangloff, M.; Verstak, B.; Nallasivan, M.P.; AlAhmed, S.; Zhu, B.; Phizacklea, M.; Losada-Perez, M.; et al. Three-tier regulation of cell number plasticity by neurotrophins and Tolls in Drosophila. *J. Cell Biol.* **2017**, *216*, 1421–1438. [CrossRef] [PubMed]
31. McIlroy, G.; Foldi, I.; Aurikko, J.; Wentzell, J.S.; Lim, M.A.; Fenton, J.C.; Gay, N.J.; Hidalgo, A. Toll-6 and Toll-7 function as neurotrophin receptors in the Drosophila melanogaster CNS. *Nat. Neurosci.* **2013**, *16*, 1248–1256. [CrossRef] [PubMed]
32. Keeler, A.B.; Deppmann, C.D. The evolutionary origins of antagonistic neurotrophin signaling. *J. Cell Biol.* **2017**, *216*, 1223–1225. [CrossRef] [PubMed]
33. Ramirez, M.L.G.; Salvesen, G.S. A primer on caspase mechanisms. *Semin. Cell Dev. Biol.* **2018**. [CrossRef] [PubMed]
34. Clavier, A.; Rincheval-Arnold, A.; Colin, J.; Mignotte, B.; Guenal, I. Apoptosis in Drosophila: Which role for mitochondria? *Apoptosis* **2016**, *21*, 239–251. [CrossRef] [PubMed]
35. Nakajima, Y.I.; Kuranaga, E. Caspase-dependent non-apoptotic processes in development. *Cell Death Differ.* **2017**, *24*, 1422–1430. [CrossRef] [PubMed]
36. Hay, B.A.; Wassarman, D.A.; Rubin, G.M. Drosophila homologs of baculovirus inhibitor of apoptosis proteins function to block cell death. *Cell* **1995**, *83*, 1253–1262. [CrossRef]

37. Hay, B.A.; Wolff, T.; Rubin, G.M. Expression of baculovirus P35 prevents cell death in Drosophila. *Development* **1994**, *120*, 2121–2129. [PubMed]

38. Quinn, L.; Coombe, M.; Mills, K.; Daish, T.; Colussi, P.; Kumar, S.; Richardson, H. Buffy, a Drosophila Bcl-2 protein, has anti-apoptotic and cell cycle inhibitory functions. *EMBO J.* **2003**, *22*, 3568–3579. [CrossRef] [PubMed]

39. Jones, G.; Jones, D.; Zhou, L.; Steller, H.; Chu, Y. Deterin, a new inhibitor of apoptosis from Drosophila melanogaster. *J. Biol. Chem.* **2000**, *275*, 22157–22165. [CrossRef] [PubMed]

40. Vernooy, S.Y.; Chow, V.; Su, J.; Verbrugghe, K.; Yang, J.; Cole, S.; Olson, M.R.; Hay, B.A. Drosophila Bruce can potently suppress Rpr- and Grim-dependent but not Hid-dependent cell death. *Curr. Biol.* **2002**, *12*, 1164–1168. [CrossRef]

41. Bergmann, A. The role of ubiquitylation for the control of cell death in Drosophila. *Cell Death Differ.* **2010**, *17*, 61–67. [CrossRef] [PubMed]

42. Kuranaga, E.; Kanuka, H.; Tonoki, A.; Takemoto, K.; Tomioka, T.; Kobayashi, M.; Hayashi, S.; Miura, M. Drosophila IKK-related kinase regulates nonapoptotic function of caspases via degradation of IAPs. *Cell* **2006**, *126*, 583–596. [CrossRef] [PubMed]

43. Muro, I.; Hay, B.A.; Clem, R.J. The Drosophila DIAP1 protein is required to prevent accumulation of a continuously generated, processed form of the apical caspase DRONC. *J. Biol. Chem.* **2002**, *277*, 49644–49650. [CrossRef] [PubMed]

44. Arama, E.; Agapite, J.; Steller, H. Caspase activity and a specific cytochrome C are required for sperm differentiation in Drosophila. *Dev. Cell* **2003**, *4*, 687–697. [CrossRef]

45. Huh, J.R.; Vernooy, S.Y.; Yu, H.; Yan, N.; Shi, Y.; Guo, M.; Hay, B.A. Multiple apoptotic caspase cascades are required in nonapoptotic roles for Drosophila spermatid individualization. *PLoS Biol.* **2004**, *2*, E15. [CrossRef] [PubMed]

46. Kanuka, H.; Kuranaga, E.; Takemoto, K.; Hiratou, T.; Okano, H.; Miura, M. Drosophila caspase transduces Shaggy/GSK-3beta kinase activity in neural precursor development. *EMBO J.* **2005**, *24*, 3793–3806. [CrossRef] [PubMed]

47. Kuo, C.T.; Zhu, S.; Younger, S.; Jan, L.Y.; Jan, Y.N. Identification of E2/E3 ubiquitinating enzymes and caspase activity regulating Drosophila sensory neuron dendrite pruning. *Neuron* **2006**, *51*, 283–290. [CrossRef] [PubMed]

48. Williams, D.W.; Kondo, S.; Krzyzanowska, A.; Hiromi, Y.; Truman, J.W. Local caspase activity directs engulfment of dendrites during pruning. *Nat. Neurosci.* **2006**, *9*, 1234–1236. [CrossRef] [PubMed]

49. Hong, K.; Hinck, L.; Nishiyama, M.; Poo, M.M.; Tessier-Lavigne, M.; Stein, E. A ligand-gated association between cytoplasmic domains of UNC5 and DCC family receptors converts netrin-induced growth cone attraction to repulsion. *Cell* **1999**, *97*, 927–941. [CrossRef]

50. Forcet, C.; Ye, X.; Granger, L.; Corset, V.; Shin, H.; Bredesen, D.E.; Mehlen, P. The dependence receptor DCC (deleted in colorectal cancer) defines an alternative mechanism for caspase activation. *Proc. Natl. Acad. Sci. USA* **2001**, *98*, 3416–3421. [CrossRef] [PubMed]

51. Mehlen, P.; Rabizadeh, S.; Snipas, S.J.; Assa-Munt, N.; Salvesen, G.S.; Bredesen, D.E. The DCC gene product induces apoptosis by a mechanism requiring receptor proteolysis. *Nature* **1998**, *395*, 801–804. [CrossRef] [PubMed]

52. Castets, M.; Broutier, L.; Molin, Y.; Brevet, M.; Chazot, G.; Gadot, N.; Paquet, A.; Mazelin, L.; Jarrosson-Wuilleme, L.; Scoazec, J.Y.; et al. DCC constrains tumour progression via its dependence receptor activity. *Nature* **2012**, *482*, 534–537. [CrossRef] [PubMed]

53. Williams, M.E.; Lu, X.; McKenna, W.L.; Washington, R.; Boyette, A.; Strickland, P.; Dillon, A.; Kaprielian, Z.; Tessier-Lavigne, M.; Hinck, L. UNC5A promotes neuronal apoptosis during spinal cord development independent of netrin-1. *Nat. Neurosci.* **2006**, *9*, 996–998. [CrossRef] [PubMed]

54. Furne, C.; Rama, N.; Corset, V.; Chedotal, A.; Mehlen, P. Netrin-1 is a survival factor during commissural neuron navigation. *Proc. Natl. Acad. Sci. USA* **2008**, *105*, 14465–14470. [CrossRef] [PubMed]

55. Bin, J.M.; Han, D.; Lai Wing Sun, K.; Croteau, L.P.; Dumontier, E.; Cloutier, J.F.; Kania, A.; Kennedy, T.E. Complete Loss of Netrin-1 Results in Embryonic Lethality and Severe Axon Guidance Defects without Increased Neural Cell Death. *Cell Rep.* **2015**, *12*, 1099–1106. [CrossRef] [PubMed]

56. Neuhaus-Follini, A.; Bashaw, G.J. The Intracellular Domain of the Frazzled/DCC Receptor Is a Transcription Factor Required for Commissural Axon Guidance. *Neuron* **2015**, *87*, 751–763. [CrossRef] [PubMed]

57. VanZomeren-Dohm, A.; Sarro, J.; Flannery, E.; Duman-Scheel, M. The Drosophila Netrin receptor frazzled/DCC functions as an invasive tumor suppressor. *BMC Dev. Biol.* **2011**, *11*, 41. [CrossRef] [PubMed]
58. Igarashi, Y.; Eroshkin, A.; Gramatikova, S.; Gramatikoff, K.; Zhang, Y.; Smith, J.W.; Osterman, A.L.; Godzik, A. CutDB: A proteolytic event database. *Nucleic Acids Res.* **2007**, *35*, D546–D549. [CrossRef] [PubMed]
59. Crawford, E.D.; Seaman, J.E.; Agard, N.; Hsu, G.W.; Julien, O.; Mahrus, S.; Nguyen, H.; Shimbo, K.; Yoshihara, H.A.; Zhuang, M.; et al. The DegraBase: A database of proteolysis in healthy and apoptotic human cells. *Mol. Cell. Proteom.* **2013**, *12*, 813–824. [CrossRef] [PubMed]
60. Luthi, A.U.; Martin, S.J. The CASBAH: A searchable database of caspase substrates. *Cell Death Differ.* **2007**, *14*, 641–650. [CrossRef] [PubMed]
61. Mashima, T.; Naito, M.; Noguchi, K.; Miller, D.K.; Nicholson, D.W.; Tsuruo, T. Actin cleavage by CPP-32/apopain during the development of apoptosis. *Oncogene* **1997**, *14*, 1007–1012. [CrossRef] [PubMed]
62. Westphal, D.; Sytnyk, V.; Schachner, M.; Leshchyns'ka, I. Clustering of the neural cell adhesion molecule (NCAM) at the neuronal cell surface induces caspase-8- and -3-dependent changes of the spectrin meshwork required for NCAM-mediated neurite outgrowth. *J. Biol. Chem.* **2010**, *285*, 42046–42057. [CrossRef] [PubMed]
63. Martin, D.N.; Baehrecke, E.H. Caspases function in autophagic programmed cell death in Drosophila. *Development* **2004**, *131*, 275–284. [CrossRef] [PubMed]
64. Akhter, A.; Caution, K.; Abu Khweek, A.; Tazi, M.; Abdulrahman, B.A.; Abdelaziz, D.H.; Voss, O.H.; Doseff, A.I.; Hassan, H.; Azad, A.K.; et al. Caspase-11 promotes the fusion of phagosomes harboring pathogenic bacteria with lysosomes by modulating actin polymerization. *Immunity* **2012**, *37*, 35–47. [CrossRef] [PubMed]
65. Li, J.; Brieher, W.M.; Scimone, M.L.; Kang, S.J.; Zhu, H.; Yin, H.; von Andrian, U.H.; Mitchison, T.; Yuan, J. Caspase-11 regulates cell migration by promoting Aip1-Cofilin-mediated actin depolymerization. *Nat. Cell Biol.* **2007**, *9*, 276–286. [CrossRef] [PubMed]
66. Han, M.H.; Jiao, S.; Jia, J.M.; Chen, Y.; Chen, C.Y.; Gucek, M.; Markey, S.P.; Li, Z. The novel caspase-3 substrate Gap43 is involved in AMPA receptor endocytosis and long-term depression. *Mol. Cell. Proteom.* **2013**, *12*, 3719–3731. [CrossRef] [PubMed]
67. Geisbrecht, E.R.; Montell, D.J. A role for Drosophila IAP1-mediated caspase inhibition in Rac-dependent cell migration. *Cell* **2004**, *118*, 111–125. [CrossRef] [PubMed]
68. Koto, A.; Kuranaga, E.; Miura, M. Temporal regulation of Drosophila IAP1 determines caspase functions in sensory organ development. *J. Cell Biol.* **2009**, *187*, 219–231. [CrossRef] [PubMed]
69. Oshima, K.; Takeda, M.; Kuranaga, E.; Ueda, R.; Aigaki, T.; Miura, M.; Hayashi, S. IKK epsilon regulates F actin assembly and interacts with Drosophila IAP1 in cellular morphogenesis. *Curr. Biol.* **2006**, *16*, 1531–1537. [CrossRef] [PubMed]
70. Perez-Garijo, A.; Shlevkov, E.; Morata, G. The role of Dpp and Wg in compensatory proliferation and in the formation of hyperplastic overgrowths caused by apoptotic cells in the Drosophila wing disc. *Development* **2009**, *136*, 1169–1177. [CrossRef] [PubMed]
71. Rudrapatna, V.A.; Bangi, E.; Cagan, R.L. Caspase signalling in the absence of apoptosis drives Jnk-dependent invasion. *EMBO Rep.* **2013**, *14*, 172–177. [CrossRef] [PubMed]
72. Portela, M.; Richardson, H.E. Death takes a holiday—Non-apoptotic role for caspases in cell migration and invasion. *EMBO Rep.* **2013**, *14*, 107–108. [CrossRef] [PubMed]
73. Goldberg, J.L.; Espinosa, J.S.; Xu, Y.; Davidson, N.; Kovacs, G.T.; Barres, B.A. Retinal ganglion cells do not extend axons by default: Promotion by neurotrophic signaling and electrical activity. *Neuron* **2002**, *33*, 689–702. [CrossRef]
74. Lentz, S.I.; Knudson, C.M.; Korsmeyer, S.J.; Snider, W.D. Neurotrophins support the development of diverse sensory axon morphologies. *J. Neurosci.* **1999**, *19*, 1038–1048. [CrossRef] [PubMed]
75. Yang, J.Y.; Michod, D.; Walicki, J.; Murphy, B.M.; Kasibhatla, S.; Martin, S.J.; Widmann, C. Partial cleavage of RasGAP by caspases is required for cell survival in mild stress conditions. *Mol. Cell. Biol.* **2004**, *24*, 10425–10436. [CrossRef] [PubMed]
76. Gilman, C.P.; Mattson, M.P. Do apoptotic mechanisms regulate synaptic plasticity and growth-cone motility? *Neuromol. Med.* **2002**, *2*, 197–214. [CrossRef]
77. Campbell, D.S.; Okamoto, H. Local caspase activation interacts with Slit-Robo signaling to restrict axonal arborization. *J. Cell Biol.* **2013**, *203*, 657–672. [CrossRef] [PubMed]
78. Igaki, T.; Yamamoto-Goto, Y.; Tokushige, N.; Kanda, H.; Miura, M. Down-regulation of DIAP1 triggers a novel Drosophila cell death pathway mediated by Dark and DRONC. *J. Biol. Chem.* **2002**, *277*, 23103–23106. [CrossRef] [PubMed]

79. Muro, I.; Means, J.C.; Clem, R.J. Cleavage of the apoptosis inhibitor DIAP1 by the apical caspase DRONC in both normal and apoptotic Drosophila cells. *J. Biol. Chem.* **2005**, *280*, 18683–18688. [CrossRef] [PubMed]

80. Rodriguez, A.; Chen, P.; Oliver, H.; Abrams, J.M. Unrestrained caspase-dependent cell death caused by loss of Diap1 function requires the Drosophila Apaf-1 homolog, Dark. *EMBO J.* **2002**, *21*, 2189–2197. [CrossRef] [PubMed]

81. Zimmermann, K.C.; Ricci, J.E.; Droin, N.M.; Green, D.R. The role of ARK in stress-induced apoptosis in Drosophila cells. *J. Cell Biol.* **2002**, *156*, 1077–1087. [CrossRef] [PubMed]

82. Wright, K.M.; Linhoff, M.W.; Potts, P.R.; Deshmukh, M. Decreased apoptosome activity with neuronal differentiation sets the threshold for strict IAP regulation of apoptosis. *J. Cell Biol.* **2004**, *167*, 303–313. [CrossRef] [PubMed]

83. Erturk, A.; Wang, Y.; Sheng, M. Local pruning of dendrites and spines by caspase-3-dependent and proteasome-limited mechanisms. *J. Neurosci.* **2014**, *34*, 1672–1688. [CrossRef] [PubMed]

84. Hollville, E.; Deshmukh, M. Physiological functions of non-apoptotic caspase activity in the nervous system. *Semin. Cell Dev. Biol.* **2017**. [CrossRef] [PubMed]

85. Mehlen, P.; Delloye-Bourgeois, C.; Chedotal, A. Novel roles for Slits and netrins: Axon guidance cues as anticancer targets? *Nat. Rev. Cancer* **2011**, *11*, 188–197. [CrossRef] [PubMed]

86. Gara, R.K.; Kumari, S.; Ganju, A.; Yallapu, M.M.; Jaggi, M.; Chauhan, S.C. Slit/Robo pathway: A promising therapeutic target for cancer. *Drug Discov. Today* **2015**, *20*, 156–164. [CrossRef] [PubMed]

87. Colin, J.; Garibal, J.; Clavier, A.; Szuplewski, S.; Risler, Y.; Milet, C.; Gaumer, S.; Guenal, I.; Mignotte, B. Screening of suppressors of bax-induced cell death identifies glycerophosphate oxidase-1 as a mediator of debcl-induced apoptosis in Drosophila. *Genes Cancer* **2015**, *6*, 241–253. [CrossRef] [PubMed]

88. Manhire-Heath, R.; Golenkina, S.; Saint, R.; Murray, M.J. Netrin-dependent downregulation of Frazzled/DCC is required for the dissociation of the peripodial epithelium in Drosophila. *Nat. Commun.* **2013**, *4*, 2790. [CrossRef] [PubMed]

89. Vaughen, J.; Igaki, T. Slit-Robo Repulsive Signaling Extrudes Tumorigenic Cells from Epithelia. *Dev. Cell* **2016**, *39*, 683–695. [CrossRef] [PubMed]

90. Ben-Zvi, A.; Manor, O.; Schachner, M.; Yaron, A.; Tessier-Lavigne, M.; Behar, O. The Semaphorin receptor PlexinA3 mediates neuronal apoptosis during dorsal root ganglia development. *J. Neurosci.* **2008**, *28*, 12427–12432. [CrossRef] [PubMed]

91. Wehner, A.B.; Abdesselem, H.; Dickendesher, T.L.; Imai, F.; Yoshida, Y.; Giger, R.J.; Pierchala, B.A. Semaphorin 3A is a retrograde cell death signal in developing sympathetic neurons. *Development* **2016**, *143*, 1560–1570. [CrossRef] [PubMed]

92. Lee, H.; Park, S.; Kang, Y.S.; Park, S. EphA receptors form a complex with caspase-8 to induce apoptotic cell death. *Mol. Cells* **2015**, *38*, 349–355. [CrossRef] [PubMed]

93. Zhang, X.; Li, J.; Liu, J.; Luo, H.; Gou, K.; Cui, S. Prostaglandin F2alpha upregulates Slit/Robo expression in mouse corpus luteum during luteolysis. *J. Endocrinol.* **2013**, *218*, 299–310. [CrossRef] [PubMed]

94. Dickinson, R.E.; Fegan, K.S.; Ren, X.; Hillier, S.G.; Duncan, W.C. Glucocorticoid regulation of SLIT/ROBO tumour suppressor genes in the ovarian surface epithelium and ovarian cancer cells. *PLoS ONE* **2011**, *6*, e27792. [CrossRef] [PubMed]

95. Winberg, M.L.; Mitchell, K.J.; Goodman, C.S. Genetic analysis of the mechanisms controlling target selection: Complementary and combinatorial functions of netrins, semaphorins, and IgCAMs. *Cell* **1998**, *93*, 581–591. [CrossRef]

96. Jiao, S.; Li, Z. Nonapoptotic function of BAD and BAX in long-term depression of synaptic transmission. *Neuron* **2011**, *70*, 758–772. [CrossRef] [PubMed]

97. Gu, Z.; Serradj, N.; Ueno, M.; Liang, M.; Li, J.; Baccei, M.L.; Martin, J.H.; Yoshida, Y. Skilled Movements Require Non-apoptotic Bax/Bak Pathway-Mediated Corticospinal Circuit Reorganization. *Neuron* **2017**, *94*, 626–641. [CrossRef] [PubMed]

98. Williams, D.W.; Truman, J.W. Cellular mechanisms of dendrite pruning in Drosophila: Insights from in vivo time-lapse of remodeling dendritic arborizing sensory neurons. *Development* **2005**, *132*, 3631–3642. [CrossRef] [PubMed]

99. Rogulja-Ortmann, A.; Luer, K.; Seibert, J.; Rickert, C.; Technau, G.M. Programmed cell death in the embryonic central nervous system of Drosophila melanogaster. *Development* **2007**, *134*, 105–116. [CrossRef] [PubMed]

100. Mukherjee, A.; Williams, D.W. More alive than dead: Non-apoptotic roles for caspases in neuronal development, plasticity and disease. *Cell Death Differ.* **2017**, *24*, 1411–1421. [CrossRef] [PubMed]

101. Fan, Y.; Bergmann, A. The cleaved-Caspase-3 antibody is a marker of Caspase-9-like DRONC activity in Drosophila. *Cell Death Differ.* **2010**, *17*, 534–539. [CrossRef] [PubMed]

102. Kidd, T.; Brose, K.; Mitchell, K.J.; Fetter, R.D.; Tessier-Lavigne, M.; Goodman, C.S.; Tear, G. Roundabout controls axon crossing of the CNS midline and defines a novel subfamily of evolutionarily conserved guidance receptors. *Cell* **1998**, *92*, 205–215. [CrossRef]

103. Brankatschk, M.; Dickson, B.J. Netrins guide Drosophila commissural axons at short range. *Nat. Neurosci.* **2006**, *9*, 188–194. [CrossRef] [PubMed]

104. Zhu, B.; Pennack, J.A.; McQuilton, P.; Forero, M.G.; Mizuguchi, K.; Sutcliffe, B.; Gu, C.J.; Fenton, J.C.; Hidalgo, A. Drosophila neurotrophins reveal a common mechanism for nervous system formation. *PLoS Biol.* **2008**, *6*, e284. [CrossRef] [PubMed]

105. Wang, H.; Tessier-Lavigne, M. En passant neurotrophic action of an intermediate axonal target in the developing mammalian CNS. *Nature* **1999**, *401*, 765–769. [CrossRef] [PubMed]

106. Usui, N.; Watanabe, K.; Ono, K.; Tomita, K.; Tamamaki, N.; Ikenaka, K.; Takebayashi, H. Role of motoneuron-derived neurotrophin 3 in survival and axonal projection of sensory neurons during neural circuit formation. *Development* **2012**, *139*, 1125–1132. [CrossRef] [PubMed]

107. Kuruvilla, R.; Zweifel, L.S.; Glebova, N.O.; Lonze, B.E.; Valdez, G.; Ye, H.; Ginty, D.D. A neurotrophin signaling cascade coordinates sympathetic neuron development through differential control of TrkA trafficking and retrograde signaling. *Cell* **2004**, *118*, 243–255. [CrossRef] [PubMed]

108. Sutcliffe, B.; Forero, M.G.; Zhu, B.; Robinson, I.M.; Hidalgo, A. Neuron-type specific functions of DNT1, DNT2 and Spz at the Drosophila neuromuscular junction. *PLoS ONE* **2013**, *8*, e75902. [CrossRef] [PubMed]

109. Bergmann, A.; Tugentman, M.; Shilo, B.Z.; Steller, H. Regulation of cell number by MAPK-dependent control of apoptosis: A mechanism for trophic survival signaling. *Dev. Cell* **2002**, *2*, 159–170. [CrossRef]

110. Goldberg, J.L. How does an axon grow? *Genes Dev.* **2003**, *17*, 941–958. [CrossRef] [PubMed]

111. Wang, K.H.; Brose, K.; Arnott, D.; Kidd, T.; Goodman, C.S.; Henzel, W.; Tessier-Lavigne, M. Biochemical purification of a mammalian slit protein as a positive regulator of sensory axon elongation and branching. *Cell* **1999**, *96*, 771–784. [CrossRef]

112. Nguyen Ba-Charvet, K.T.; Brose, K.; Ma, L.; Wang, K.H.; Marillat, V.; Sotelo, C.; Tessier-Lavigne, M.; Chedotal, A. Diversity and specificity of actions of Slit2 proteolytic fragments in axon guidance. *J. Neurosci.* **2001**, *21*, 4281–4289. [CrossRef] [PubMed]

113. Dascenco, D.; Erfurth, M.-L.; Izadifar, A.; Song, M.; Sachse, S.; Bortnick, R.; Urwyler, O.; Petrovic, M.; Ayaz, D.; He, H.; et al. Slit and Receptor Tyrosine Phosphatase 69D Confer Spatial Specificity to Axon Branching via Dscam1. *Cell* **2015**. [CrossRef] [PubMed]

114. Piper, M.; Nurcombe, V.; Reid, K.; Bartlett, P.; Little, M. N-terminal Slit2 promotes survival and neurite extension in cultured peripheral neurons. *Neuroreport* **2002**, *13*, 2375–2378. [CrossRef] [PubMed]

115. Chen, M.; Zheng, B. Axon plasticity in the mammalian central nervous system after injury. *Trends Neurosci.* **2014**, *37*, 583–593. [CrossRef] [PubMed]

116. Baba, K.; Yoshida, W.; Toriyama, M.; Shimada, T.; Manning, C.F.; Saito, M.; Kohno, K.; Trimmer, J.S.; Watanabe, R.; Inagaki, N. Gradient-reading and mechano-effector machinery for netrin-1-induced axon guidance. *eLife* **2018**, *7*. [CrossRef] [PubMed]

117. Kamiyama, D.; Sekine, S.; Barsi-Rhyne, B.; Hu, J.; Chen, B.; Gilbert, L.A.; Ishikawa, H.; Leonetti, M.D.; Marshall, W.F.; Weissman, J.S.; et al. Versatile protein tagging in cells with split fluorescent protein. *Nat. Commun.* **2016**, *7*, 11046. [CrossRef] [PubMed]

118. Kamiyama, D.; McGorty, R.; Kamiyama, R.; Kim, M.D.; Chiba, A.; Huang, B. Specification of Dendritogenesis Site in Drosophila aCC Motoneuron by Membrane Enrichment of Pak1 through Dscam1. *Dev. Cell* **2015**, *35*, 93–106. [CrossRef] [PubMed]

119. Geden, M.J.; Deshmukh, M. Axon degeneration: Context defines distinct pathways. *Curr. Opin. Neurobiol.* **2016**, *39*, 108–115. [CrossRef] [PubMed]

120. Bardet, P.L.; Kolahgar, G.; Mynett, A.; Miguel-Aliaga, I.; Briscoe, J.; Meier, P.; Vincent, J.P. A fluorescent reporter of caspase activity for live imaging. *Proc. Natl. Acad. Sci. USA* **2008**, *105*, 13901–13905. [CrossRef] [PubMed]
121. Melzer, J.; Broemer, M. Nerve-racking—Apoptotic and non-apoptotic roles of caspases in the nervous system of Drosophila. *Eur. J. Neurosci.* **2016**, *44*, 1683–1690. [CrossRef] [PubMed]

8

Prenatal Neuropathologies in Autism Spectrum Disorder and Intellectual Disability: The Gestation of a Comprehensive Zebrafish Model

Robert A. Kozol

Department of Biology, University of Miami, Coral Gables, FL 33315, USA; r.kozol1@miami.edu;

Abstract: Autism spectrum disorder (ASD) and intellectual disability (ID) are neurodevelopmental disorders with overlapping diagnostic behaviors and risk factors. These include embryonic exposure to teratogens and mutations in genes that have important functions prenatally. Animal models, including rodents and zebrafish, have been essential in delineating mechanisms of neuropathology and identifying developmental critical periods, when those mechanisms are most sensitive to disruption. This review focuses on how the developmentally accessible zebrafish is contributing to our understanding of prenatal pathologies that set the stage for later ASD-ID behavioral deficits. We discuss the known factors that contribute prenatally to ASD-ID and the recent use of zebrafish to model deficits in brain morphogenesis and circuit development. We conclude by suggesting that a future challenge in zebrafish ASD-ID modeling will be to bridge prenatal anatomical and physiological pathologies to behavioral deficits later in life.

Keywords: autism spectrum disorders; ASD; intellectual disability; ID; zebrafish; prenatal; embryonic; fetal; neurodevelopmental disorders; critical periods

---

1. Introduction

Autism spectrum disorder (ASD) and intellectual disability (ID) share diagnostic features, such as deficits in social behavior, and problems with communication and stereotypies [1,2]. As such, individuals with ASD are commonly diagnosed with ID (50–70%) and vis-versa (28–70%) [1,3,4]. These common co-diagnoses are likely a product of environmental and genetic factors that converge on similar molecular pathways [1,5]. Due to the relative diagnostic age (Median ASD diagnosis = 4.5 years; ID = 1–18 years) and enrichment of synaptic gene mutations ASD-ID, animal models initially focused on postnatal time periods [6–9]. However, studies have recently shown that disruptions in fetal development can manifest as childhood behavioral deficits [10,11]. These studies include zebrafish models that are being used to pinpoint critical periods and to provide a more complete picture of the mechanisms that underlie symptoms [12,13].

Zebrafish are increasingly used as a prenatal animal model for neurodevelopmental disorders, due to several anatomical and behavioral characteristics. First, zebrafish are small and transparent during early life stages which facilitates visualization of the nervous system, both anatomically and physiologically [14–17]. Therefore, developmental disruptions that might impact circuits can be investigated morphologically using transgenics to label specific neuronal populations [18,19], or physiologically using genetically encoded calcium indicators to assay neuronal activity [20–22]. Stereotyped swimming behaviors also emerge embryonically and rapidly transition to evasive and predatory behavior [23,24]. Disruptions in critical periods of development can therefore be identified by analyzing these behavioral milestones, such as embryonic tail coiling [25] and larval swimming [26].

Finally, one week old larval zebrafish have a small brain that simplifies studying how circuit-level disruptions affect behaviors at the whole-brain level [22]. Together these attributes produce a cellular-to circuit-level model for establishing functional relationships between prenatal developmental deficits and behavioral phenotypes.

In this paper, we look at how zebrafish are used to study mechanisms by which causal genetic mutations for ASD-ID impact embryonic development. We review recent studies of prenatal factors affecting ASD-ID, that detail critical time periods and the cellular processes responsible for later behavioral deficits. Then, we look at zebrafish proof-of-concept ASD-ID models that investigated embryonic and larval phenotypes related to ASD-ID. Finally, we conclude by discussing how recent advances in methodology are now being employed to investigate how developmental changes in zebrafish ASD-ID models contribute to behavioral phenotypes.

2. What Prenatal Factors Contribute to ASD-ID

The relationship between prenatal development and symptoms of ASD-ID was first established by investigating maternal teratogen exposure and gestationally derived deficits in ASD-ID syndromes. Teratogens are agents, chemical or infectious, that cause malformations to an embryo or fetus [27]. Many teratogenic effects depend on the time of exposure and can be traced back via hospital records or estimated by linking developmental milestones to somatic defects [27,28]. In addition to teratogens, many genes are expressed prenatally and several well-studied ASD-ID syndromes, such as Tuberous Sclerosis, are known to include gestationally derived malformations [10]. Recent research has expanded these prenatal factors to include maternal cytokines, genetic networks and non-coding genetic contributions [29–32]. However many individuals with ASD-ID have a non-etiological diagnosis, with neuropathological roots that may stem from prenatal time periods. In fact, some anatomical features arising from prenatal deficits have been described in adults with ASD-ID, including ectopic superior olivary nucleus neurons and loss of cerebellar Purkinje Cells with intact inferior olivary nuclei [33–36]. These features highlight the need for prenatal animal models to determine whether environmental and genetic risk factors for ASD-ID influence later behavioral deficits.

2.1. *Environmental Risk Factors*

Prenatal critical periods for ASD-ID were initially identified by unintentional population scale exposures to teratogens, that were initially prescribed as medications during pregnancy [37–39]. These studies showed that children exposed to teratogens in utero, such as thalidomide or valproic acid (VPA), developed symptoms of ASD-ID later in life [27]. To narrow down critical periods, researchers matched somatic defects with an accompanying developmental window, such as neural tube defects with neural tube closure [27]. For example, some children of mothers exposed to thalidomide developed autistic symptoms and cranial nerve defects consistent with exposure around 20 to 24 days post conception [27,28]. Despite these temporal signatures exposure can occur throughout the entire pregnancy, making it difficult to ascribe prenatal disruptions to later behavioral deficits. To identify whether these critical periods affect post-natal behavior, animal models have been screened for behavioral deficits, following exposure to teratogens at different time intervals. For example, following exposure to VPA during embryogenesis (E10–E13), rats exhibited learning and social deficits as adults [40,41]. Several zebrafish labs have also modeled the teratogenic effects of thalidomide [42] and VPA [43]. A recent study exposed zebrafish to VPA from gastrulation to neural tube formation, mimicking the window of exposure used in mouse VPA models [44]. VPA exposed larval fish showed decreases in histaminergic, noradrenergic and dopaminergic systems; accompanied by changes to larval sensory behavior and adult social behavior [44]. These models highlight the usefulness of testing teratogenic hypotheses from clinical data, to determine critical embryonic time periods that underlie later ASD-ID symptoms.

Another environmental factor that can increase the risk of ASD-ID are maternal immune responses [32,37,45]. In the 1970's, Chess and colleagues found that mothers exposed to rubella during pregnancy had children with a high incidence of ASD-ID [37]. Recent studies have identified several immune responses, such as higher concentrations of intergerons in ASD-ID versus the general population, or ASD alone [32,45]. Additionally, researchers have shown that inflammatory molecules influence fetal gene expression, including dysregulation of ASD-ID genes associated with mTOR and EIF4E dependent signaling [30]. A recent ASD-ID zebrafish model looked at larval immunological function to determine whether immune responses were affected in a *Methyl-CpG-Binding Protein 2* (*MECP2*) null background [46]. MECP2 mutations cause Rett Syndrome, which includes symptoms associated with inflammation, such as gastrointestinal disturbances and oxidative stress [47]. Mecp2 mutants exhibited decreased expression of *tumor necrosis factor a* (*tnfa)* during early embryogenesis and increased expression of *interleukin-1b* and *interleukin-10* throughout larval stages [46]. Inflammatory molecules from these genes are needed to stimulate the bodies response to a disturbance and promote tissue repair [46]. Therefore, these changes in gene expression may be a response to tissue disturbances or oxidative stress, including a dysregulated inflammatory response of the gastrointestinal tract. These studies provide evidence for an interaction between the environment, inflammation and genetic risk factors in ASD-ID.

2.2. Genetic Risk Factors

Syndromes associated with ASD-ID, including Rett, Fragile X Syndrome (FXS) and Tuberous Sclerosis (TSC) identified the first genetic mutations of large-effect to cause ASD-ID [48]. Causal mutations for all of these conditions occur in genes normally expressed during prenatal periods. The *Fragile X Mental Retardation* (*FMRP*) gene is expressed throughout development and produces many alternatively spliced variants. Because different protein isoforms are expressed at different times and in different cell types [49,50], phenotypes vary depending on the nature and position of *FMRP* mutations. For example, while deletion of *FMRP* in humans and mice is linked to cortical defects, a common *FMRP* repeat mutation in FXS does not cause obvious gross anatomical deficits [50]. None-the-less, pluripotent stem-cells derived from individuals with FXS exhibit developmental delay in neuronal differentiation during embryogenesis [51]. Likewise TSC, caused by mutations in *TSC1/2*, is a ASD-ID syndrome and characterized by large benign cortical tubers and giant cells [1,7]. Individuals with TSC have been shown to have gross anatomical defects during mid-fetal development, including focal lesions [52,53] and giant cells [54]. Finally, recent ASD-ID mutational models have functionally linked deficits in prenatal cellular processes, such as neurogenesis, with postnatal ASD-ID behavioral deficits [11,55,56]. These embryonically expressed single gene mutations point to prenatal time periods as being critical to symptom development in ASD-ID.

In addition to these syndromic causes of ASD-ID, the vast majority of individuals have non-syndromic rare genetic variants, or are diagnosed with idiopathic ASD-ID. Fortunately, more high-confidence ASD-ID genes are being identified, as a result of world-wide consortia, and whole-genome and whole-exome sequencing [57–60]. These studies have shown that ASD-ID gene mutations are heterogenous and rare, implicating hundreds of ASD-ID associated genes. This recent increase in gene discovery, along with transcriptomic and proteomic analysis, has enriched our understanding of molecular networks associated with ASD-ID (recently reviewed in References [9,61,62]. The focus has thus shifted from defining genetic causes of ASD-ID to understanding how biologically relevant networks contribute to ASD-ID symptoms [63–65]. These biologically relevant networks are constructed by analyzing gene expression profiles, including gene ontology terms and protein-protein interactions [31,66,67]. These integrated models continue to identify prenatal ASD-ID clusters with greater temporal and spatial specificity, highlighting commonly impacted structures, like the fetal amygdala and mid-fetal cortical projection neurons [31] and molecular pathways, such as Notch and Wnt signaling [66]. In addition to rare genetic variants, common genetic variation of small-effect size also contribute to the risk and development of ASD-ID [61,68]. Although a single variant confers a small increase in ASD-ID risk, cumulative variation across the genome could account for large contributions (15–50%) towards the genetic etiology of ASD-ID [68]. These common genetic variants may also impact the penetrance of rare genetic variants, influencing the development and severity of ASD-ID phenotypes [69]. The ASD-ID genetic landscape continues to expand, with genetic variants and biological networks explaining a large percentage of the ASD-ID etiology.

Maternal genetics can also contribute to symptoms of ASD-ID via maternal effects. Maternal effects are maternally derived material that can contribute to the development of an offspring [70,71]. Maternal effects include embryonic translation of maternally loaded mRNA transcripts, genetic imprinting and the influence of intrauterine factors [70]. While maternally derived mechanisms including 15q duplications and imprinting in Angelman syndrome are well known, recent studies have discovered novel maternal effects. For example, a recent genome-wide association study linked a parent-of-origin effect with the transfer of a maternal ASD-ID gene *SH3 and multiple ankyrin repeat domain protein 3* (*SHANK3*) variant, suggesting that this parent-of-origin variant increased the likelihood of developing ASD [72]. Finally, maternally loaded genetic variants can also affect developing oocytes. Following long-term storage a Drosophila FXS model exhibited dysregulated neuronal development, due to translational deficiencies of transcripts encoding large proteins [73]. This finding supports previous research that identified deficits in translation of transcripts encoding large proteins as a mechanism plaquing local translation at synapses [73,74]. These studies indicate that mutations in genes linked to ASD-ID can influence neuropathology even before the embryo begins to develop.

3. Comparative Neurodevelopment and Critical Periods: Can Embryonic and Larval Periods in Zebrafish be Compared to Prenatal Development in Humans?

The use of animal models to study human neurodevelopmental disorders requires the ability to translate critical periods across species. However, this can be problematic due to varying rates of development and dramatically different life histories. For instance, the human prenatal period is defined by prolonged in utero gestation and infants are born dependent on parental care [75,76]. In contrast, zebrafish develop externally in a protective chorion and hatch after a few days as fully independent larvae [14]. Moreover, morphogenesis of the zebrafish brain is complete in three days, compared to twenty one days in mice and eight weeks in humans [14,77]. Therefore, internal neurodevelopmental milestones should be used to define critical periods, from the completion of embryogenesis to the maturation of neural circuits essential for complex behaviors.

Currently, a comprehensive guide for comparing zebrafish and mammalian neuronal development is lacking. In mammals a group recently made a resource for comparing critical periods, by using internal and external developmental events. These events included the onset and offset of neurogenesis, cellular processes like myelination and the appearance of behaviors [76,78]. This comparative mammalian study revealed that the order of these events is largely conserved across species, despite differences in gestational time and alternative life histories (e.g., precocial versus altricial development). This order of events may also be conserved in zebrafish; however, a major knowledge gap exists concerning periods of extensive brain growth, between late larval and juvenile stages. Fortunately, research in zebrafish embryos have confirmed that major brain regions, gene expression patterns and cell types remain homologous with mammals (Table 1) [79–81]. This section focuses on comparing zebrafish and mammalian neural development and a strategy for translating prenatal regional or circuit specific critical periods.

Table 1. A timeline of neurogenic events in mammals and zebrafish. Time periods represent approximate dates related to morphological delineation, onset of neurogenesis and onset of behavior. Time periods are approximated as days post-fertilization (dpf).

Structure or Region	Estimated Onset of Development (dpf)		
	Human [75,78]	**Rat [75,78]**	**Zebrafish**
Neural tube	21	10	0.4–0.8 [14]
Telencephalon, Diencephalon, Mesencephalon, Rhombencephalon	28–33	11	0.7–0.75 [14]
Cerebellar primordia	28	12	0.8 [14]
Optic tectum	28	12	1 [82]
Hypothalamus, pineal (epiphysis)	28	12	0.75, 1 [83]
Raphe complex	28 [84]	10.5 [85]	3 [86]
Locus coeruleus	28	11	1 [87,88]
Inferior olive	30	11	3 [89]
Medial longitudinal fasciculus	30	11	1.2 [90]
Thalamus	36	14	2 [91]
Cell type			
Retinal ganglion cells	33	12	1.2 [92]
Mitral cells	33	12	1 [93]
GABAergic neurons	52	15	1.2 [94]
Behavior			
Walking (zebrafish, swimming)	455	48	1.13 [23]
Acoustic startle response	265	29	4 [95]

3.1. Conserved Neuronal Development with Differences in Morphogenesis

Although neuronal development is largely conserved among zebrafish and mammals, several differences exist in morphogenesis and neurogenesis [14,81,96]. Zebrafish and mammals utilize different strategies during the formation of the neural tube: while zebrafish develop a neural thickening or keel that is hollowed by cavitation, mammals create a neural tube by enfolding the neural epithelial sheet [14,97]. Although these mechanisms differ, the underlying genes and signaling pathways responsible are largely conserved [97,98]. Morphogenesis of the telencephalon also differs: the mammalian neural tube evaginates and ventricles develop internally (medial), while the zebrafish telencephalon everts and ventricles develop on the surface of the brain [99,100]. This eversion process rearranges the topographic layout of functionally homologous brain regions, that results in a dorsal-medial amygdala and lateral hippocampus. Furthermore, there is still debate in the fish community over the severity of eversion and the final placement of regions within the forebrain [101,102]. Despite morphogenic differences, regions similar to the amygdala [103,104] and hippocampus [105,106] have been identified by conserved gene expression and function [102]. This provides a rationale for using zebrafish to study neuropsychiatric disorders in relation to emotive, social and memory-based behaviors [107,108].

3.2. Integrated Critical Periods: Gene Expression, Morphogenesis and Function

Currently zebrafish researchers must utilize molecular markers, cellular processes and circuit function to identify translatable critical periods. Some brain regions such as the cerebellum maintain remarkable genetic, cytoarchitectural and functional homology [80,109–111]. As an example of regional developmental staging, milestones of cerebellar development can be used to translate developmentally similar critical periods between zebrafish and mammals (Table 2). First the primordial cerebellum is defined by the expression of *engrailed1/2* in rhombomere 1 of zebrafish (1–2 dpf), mouse (10–12 dpf) and humans (28 dpf) [109,112]. Cerebellar progenitor pools are then specified and can be identified by their expression of the transcription factors *Math1/Atoh1a* and *Ptf1a* [109]. *Math1/Atoh1a* labels the upper rhombic lip that will give rise to excitatory granule cells, while *Ptf1a* labels the ventral cerebellar proliferation zone that gives rise to inhibitory purkinje cells [113,114]. Following neuronal proliferation, the granule, purkinje and deep cerebellar neurons (eurydendroid cells in fish) migrate to assemble a trilaminar structure [113,114]. Critical periods can also be identified as cerebellar regions become compartmentalized for specific functions, similarly to mammals [115,116]. For example, the caudal purkinje cell layer in zebrafish has conserved efferent projections and functions with the dorsal cerebellar vermis in mammals [111,117]. The caudal vermis has indeed been implicated in ASD-ID, including loss of purkinje cells and deficits in saccade adaptation [35,118]. This specific example shows that several criteria can be used to define regional critical periods, including the appearance of a structure, neurogenesis of specific cell types, formation of a functional circuit and emergence of function related to behavior (Table 2).

Table 2. Critical periods in the development of the cerebellum. Time periods represent approximate dates related to onset of neurogenesis, function and formation of structure. Time periods are approximated as days post-fertilization (dpf).

Developmental Event	Estimated Onset of Development (dpf)	
	Rat [112]	**Zebrafish [109,113,119]**
Cerebellar primordia (*En1/2*, *Wnt1*, *Pax2*, *Gbx2*)	12.5	0.8
Progenitor pools specified (*Atoh1a/Ptf1a*)	10.5	1–2
GC/PC migration-differentiation	12.5–15	3
Purkinje cells functionally mature	29.5 [120]	4
Trilaminar structure formed	35–42	5

4. Zebrafish ASD-ID Models: Critical Periods and Developmental Mechanisms

Recently there have been several proof-of-concept studies that have used zebrafish to model inherited neuropsychological disorders [108,121–128]. The discovery of high-confidence ASD-ID genes was initially slow, due to genome-wide studies with small sample sizes and an ever evolving set of diagnostic criteria (Figure 1) [59,129,130]. Many of the original ASD-ID zebrafish models represent genetic variants that, with high-confidence, were deemed causal for ASD-ID (16p11.2 CNV, *SHANK3*, *FMRP*, *MET*, *TSC1*) [127,128,131–134]. Additionally, the majority of ASD-ID zebrafish models were made using morpholino knockdown technology, predating the widespread availability of gene editing techniques to generate site specific knockouts. Interpretations of knockdown phenotypes are limited due to off-targeting concerns and the transient nature of knockdown technologies, restricting analyses to early stages [135]. Comparisons between knockdowns and knockouts of the same gene are complicated by off-target effects induced by morpholinos and genetic compensation seen in mutants ([136] discussed in Section 5.2). Given the recent rate of discovery of causal genetic mutations [9,64], future studies should produce a rich understanding of critical developmental time periods and mechanisms. None-the-less, past ASD-ID zebrafish models illustrate the importance of neuronal critical periods and developmental check points; when initiation of gene expression, activity dependent growth and synaptic refinement are required to build behavioral circuits [137].

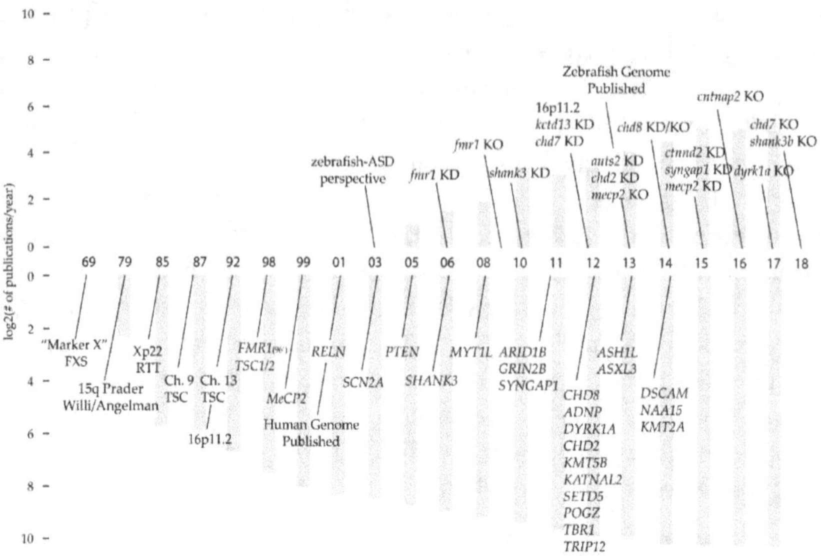

Figure 1. A timeline of ASD-ID genetic discoveries and ASD-ID modeling in zebrafish. Names represent the characterization of a gene associated with ASD-ID or the first report of a genetic variant associated with ASD-ID (bottom). ASD-ID genes where chosen from the Simons Autism Research Initiative (tier 1 genes) and common ASD-ID syndromes. Zebrafish genetic models denote the first reported embryonic or larval model and the method of genetic manipulation, KD = morpholino knock-down, KO = mutant allele. Histograms describe publications per year for PubMed keywords ASD, ID, Genes, Genetics, Loci, Chromosome (top zebrafish, bottom human). Non-linear intervals of years were averaged. Human timeline references: "Marker X" [138]; 15q Prader Wili/Angelman [139]; Xp22 RTT [140]; TSC Ch. 9 [141], Ch. 16 [142]; *FMR1* [143]; *TSC1/2* [144,145]; *MeCP2* [47]; *RELN* [146]; *SCN2A* [147]; PTEN [148]; *SHANK3* [149]; *MYT1L* [150]; *ARID1B* [151]; *GRIN2B* [152]; *CHD8, ADNP, DYRK1A, TBR1* [57]; *KMT5B, KATNAL2, SETD5, POGZ* [153,154]; *TRIP12* [155]; *ASH1L* [31]; *ASXL3* [156]; *DSCAM, NAA15, KMT2A* [157]. Zebrafish timeline references: zebrafish-ASD perspective [121]; *fmr1* KD [132], *shank3* KD [133], 16p11.2 KD [127], *kctd13* KD [134], *chd7* KD, *auts2* KD [158], *chd2* KD [159], *mecp2* KO, *chd8* KD/KO [160], *ctnnd2* KD [161], *syngap1b* KD [126], *mecp2* KD, *cntnap2ab* KO [125], *dyrk1a* KO, *shank3b* KO [162], *chd7* KO.

4.1. Disruptions in Morphogenesis and Circuit Development of the Embryonic Brain

Early in development the embryo must undergo major morphological changes to specify neural ectoderm, establish the divisions of the nervous system and grow a brain anteriorly. During gastrulation cells of the blastodisc migrate to engulf the yolk, while cells at the leading edge involute to establish the epiblast (ectoderm) and the hypoblast (endo-mesoderm) [163–165]. As gastrulation and neurulation progress, molecules such as *noggin* and *chordin* are expressed to establish anterior brain regions [166]. Due to embryonic accessibility, seminal work in zebrafish helped to further elucidate morphogenic mechanisms that lead to the specification and formation of the forebrain, midbrain and hindbrain [167–169]. These studies illustrate how transcription factors and signaling molecules are expressed to pattern the neural tube, including feedback loops to restrict brain regions and produce a myriad of neuronal subtypes necessary to build diverse neural circuits [79,81]. For instance, the mid-hindbrain boundary (MHB) is demarcated by midbrain *Orthodenticle Homeobox 2* (*Otx2*) and hindbrain *Gastrulation Brain Homeobox 2* (*Gbx2*) expression. This MHB is further maintained by the mid-hindbrain program consisting of *Fibroblast growth factor 8* (*Fgf8*), *Wingless* (*Wnt1*), *Engrailed 1/2* (*En1/2*) and *Paired Box* (*Pax*) genes [81]. These genes form a complex network of regional signaling that differentially restrict gene expression and promote the generation of the midbrain optic tectum and the hindbrain cerebellum [81]. Many early studies focused on genetic mutants that exhibited the loss of a brain region, such as *no isthmus (noi)* and *acerebellar (ace)*, that lacked the mid-hindbrain boundary and cerebellum, respectively [170]. These studies exemplify the importance of tightly controlled cellular signaling to establish boundaries and provide lineage restricted differential growth of brain regions.

4.1.1. The Dysregulation of Morphogenic Genes and Conserved Signaling Pathways Influences Brain Size

Some ASD-ID genes are expressed during embryogenesis, when loss-of-function has the potential to affect the spatial and temporal growth of neuronal tissue [12,72,73]. These changes in growth can be caused by dysregulated expression of genes involved in conserved signaling pathways, many of which are responsible for delineating the boundaries of tissues and brain regions. The morphogenic gene *chordin* encodes a BMP antagonist necessary for neurulation, that is instrumental in determining the size of anterior brain regions [171]. Zebrafish ASD-ID gene models have shown disrupted *chordin* expression during gastrulation that greatly influences the size of brain regions. For instance, knocking-down or knocking-out the ASD gene *chromodomain helicase DNA binding protein 8 (chd8)* in zebrafish causes an expanded expression of *chordin* and the forebrain marker *otx2* [160]. These changes in expression increase the tissue that will give rise to the forebrain, that the authors suggest may underlie macrocephaly and increased brain size seen in a subset of individuals carrying *CHD8* mutations [160]. In contrast to *chordin* expansion seen in *chd8* models, knocking-down D*elta Catenin-2* (*CTNND2*), another gene linked to ASD-ID, causes a shortening and widening of *chordin* expression during epiboly and a reduction in presumptive forebrain tissue [172]. Direct interactions between *CHD8* and *CTNND2* with Wnt signaling components may therefore explain changes in expression of *chordin* and other brain region specific markers [173]. For instance, *CHD8* indirectly influences embryonic cortical progenitors through chromatin remodeling, affecting the expression of several Wnt signaling genes [55]. Additionally, inactivation of beta catenin is known to cause *dlx2* upregulation and *dlx2* expression was increased in the prethalamus of *chd8* morphants [160,174]. Furthermore, upregulated *dlx2* expression coincides with an enlarged midbrain that, in *chd8* morphants, is likely due to an earlier increase in neuronal proliferation [175]. Finally, Ctnnd1 also interacts directly with beta-catenin and *ctnnd1* morphants have reduced *axin* expression that may underlie the changes seen in *chordin* expression [57]. These studies demonstrate how gene mutations can affect gastrulation and neurulation through convergent developmental mechanisms, producing diametrically opposite phenotypes.

Zebrafish knockdown and knockout models of *autism susceptibility candidate 2 (auts2)* and *shank3* also suggest ASD-ID genes play essential roles in brain morphogenesis. *Auts2* morphants are microcephalic, have increased apoptosis and a decreased number of neurons in the forebrain [158,176]. Morphant and mutant models of *shank3* are also microcephalic and appear developmentally delayed [126,162]. *Shank3* microcephaly is likely due to increased cell death or decreased neuronal proliferation. Mid-hindbrain apoptosis was increased in *shank3a* morphants, while *shank3b* homozygous mutants exhibited brain-wide decreases in *HuC:RFP* fluorescence [126,162]. Furthermore, olfactory placodal neurons derived from humans with *SHANK3* microdeletions exhibited developmental phenotypes [177]. These young post-mitotic neurons had smaller cell bodies, more extensively branched neurites and motility deficits, that likely impact early morphogenesis [177]. While these studies provide evidence that ASD-ID genes play important roles during brain morphogenesis, more research is needed to identify the molecular mechanisms underlying these early developmental phenotypes.

Zebrafish knockdown and knockout strategies have also been used to model the ASD associated 16p11.2 deletion syndrome. These 16p11.2 models have identified several genes that may contribute to changes in brain size. With 29 core genes affected, a 16p11.2 knockdown screen identified multiple genes that may contribute to brain growth [127]. A follow up study used genetic knockouts to further investigate developmental phenotypes and found that homozygote *fam57ba* mutations contribute specifically to macrocephaly [178]. In contrast, two 16p11.2 models have focused on *potassium channel tetramerization domain containing 13 (kctd13)*, with morphant and mutant models producing different results [134,179]. A knockdown and overexpression study found that *Kctd13* caused brain morphogenesis phenotypes in a dosage-dependent manner [134]. *Kctd13* morphants exhibited increased proliferation and macrocephaly, while overexpression of *kctd13* mRNA caused increased apoptosis, decreased proliferation and microcephaly [134]. In contrast, a more recent *kctd13* gene deletion model found no phenotypes associated with head size or neuronal proliferation [179]. These researchers suggest that the discordance in knockdown and knockout phenotypes could either be due to off-target effects that enhance phenotypes in morphants or genetic compensation that masks phenotypes in *kctd13* mutants [179]. These studies highlight the challenges in comparing mutant and morphant phenotypes (see Section 5.2).

4.1.2. Deficits in Neurogenesis Negatively Impact Axonogenesis and Circuit Connectivity

During neurogenesis primary neurons extend axons throughout the nervous system that act as a foundation for future circuits. Primary neurons are born within the first 24 h after fertilization, repeat serially and have large cell bodies with long axons [180,181]. Because later-born neurons navigate to their targets along the primary neuron axons, changes in primary neuron axonogenesis can have a significant impact on the function of mature circuits. As these axons grow, synchronized developmental steps establish a delicate balance of circuit components, eventually supporting a flexible repertoire of behaviors. These steps include the differential expression of signaling pathway genes that guide them to correct targets, activity dependent neurite growth and synaptogenesis, and both neuronal and synaptic refinement [137,182,183]. These events are especially important for maintaining balance of excitatory and inhibitory (E/I) inputs, whose homeostasis is necessary to coordinate complex behaviors [184]. Although circuits are resilient, disruptions to E/I balance are common in ASD-ID, underlying comorbid symptoms such as epilepsy and hypotonia [184,185].

ASD-ID mutations have been shown to impact axonogenesis, including several zebrafish models investigating embryonic axonogenesis. The X-linked intellectual disability gene *FGF13* was found to play a role in stabilizing microtubules to control proper extension of growth cones [186]. When *FGF13* is mutated, growth cones overextend past their targets, which likely explains defects in *FGF13* null cortical layering. This axonogenic deficit may also underlie learning and memory deficits seen in mice carrying *Fgf13* LOF mutations [186]. Several zebrafish ASD-ID gene models show similar deficits, for example, the *Disrupted In Schizophrenia 1* (*DISC1*) gene is a negative regulator of Wnt signaling that influences cellular migration, axonal transport and axonogenesis [187]. Both mutant and morphant zebrafish *disc1* models exhibit axonogenesis phenotypes, including the absence of supraoptic tracts and deficits in rhombomeric axon development [122,187]. Importantly, these *zdisc1* morphant phenotypes could be rescued by full-length human *DISC1* mRNA but not by mRNA encoding Bipolar/Schizophrenia causing *DISC1* variants [122]. Axon tract phenotypes have also phenocopied anatomical deficits seen in ASD-ID patients. The RAC1 pathway has been implicated as a common molecular pathway affecting individuals with ASD-ID and *RAC1* gene mutations have been implicated in several developmental disorders [188]. Cerebellar hypoplasia is common in individuals with *RAC1* mutations, while over-expression of patient *RAC1* mutations in zebrafish caused disruptions in cerebellar axonogenesis [188]. As mentioned previously, a recent study knocking-down 16p11.2 genes found that 80% of genes tested showed deficits in axonogenesis, including reduced fasciculation, disorganized morphology and supernumerary axon tracts that could be rescued by the corresponding human mRNA [127]. These results suggest that several 16p11.2 CNV genes are necessary for early brain development and contribute to neurological phenotypes. Furthermore, because axon tracts are shaped by environmental cues, disruptions to morphogenesis and axonogenesis are likely mechanistically linked.

Developing circuits can also be impacted by disruptions in the balance of excitatory and inhibitory (E/I) inputs [184]. Evidence for increased or decreased excitatory (glutamatergic) and inhibitory (GABAergic) neurons have been reported in different zebrafish ASD-ID gene models. Recent zebrafish ASD-ID studies have used markers that specify E/I neuronal populations to assess changes in the number and distribution of E/I components. The ASD-ID genes *SHANK3* and *SYNGAP1* encode post-synaptic density proteins, that are essential for maintaining synaptic strength of excitatory glutamatergic neurons via dendritic spine growth and AMPA receptor expression [189,190]. Morphant models of *shank3a* and *syngap1b* exhibited severe behavioral deficits, including seizure-like bouts, unproductive body undulations and decreased sensitivity to touch [126]. *Shank3a* and *syngap1b* morphants showed broad decreases in inhibitory GABAergic neurons, suggesting a shift towards an increased ratio of excitatory neurons [126]. A zebrafish mutant loss-of-function *Contactin Associated Protein 2 (CNTNAP2)* ASD-ID model also quantified E/I balance [125]. In contrast to *SHANK3* and *SYNGAP1*, CNTNAP2 is expressed in inhibitory GABAergic neurons, localizes voltage-gated potassium channels and plays a role in migration of GABAergic neurons [125]. *Cntnap2ab* mutant zebrafish exhibited a decrease in forebrain GABAergic neurons, night-time hyperactivity and increased seizure susceptibility [125]. To determine the possible neurological basis for these behavioral deficits, a drug screen was performed confirming that *cntnap2ab* mutants were sensitive to GABAergic agonists. Surprisingly, NMDA receptor antagonists in wild-type larvae strongly correlated with mutant phenotypes suggesting that *cntnap2ab* mutations also affect behaviorally relevant excitatory components. This combination of systems level neuronal deficits, behavioral phenotypes and drug screening provide a compelling model for future ASD-ID studies.

Future studies investing changes in E/I balance will require the identification of ASD-ID relevant neuronal subtypes in zebrafish. Many ASD-ID genes are expressed in glutamatergic excitatory or inhibitory GABAergic neurons, including *Neurexin* (*NRX*) and SHANK3, or *Sodium Voltage-Gated Channels Alpha Subunit 1* (*SCN1A*) and *CNTNAP2*, respectively [185,191]. Several of these genes have been selectively knocked-out in specific neuronal subtypes to demonstrate that genetic mutations in those neuronal subtypes are sufficient to cause ASD-ID behavioral deficits [185,192]. For example, knocking-out *Scn1a* in forebrain GABAergic neurons recapitulated several Dravet syndrome phenotypes, including seizure susceptibility [192]. These phenotypes are likely due to *Scn1a* loss-of-function in fast-spiking Parvalbumin positive cell types [192]. Zebrafish *scn1lLab* mutants were recently used to discover novel anti-convulsant molecules. This study provided evidence for a GABAergic component, however cellular level and subtype specific neuropathology was not investigated [193]. Identifying similar Parvalbumin positive GABAergic neurons in zebrafish forebrain could provide a better understanding of whether *scn1Lab* mutations affect the same cell types, while determining how those neurons relate to seizure susceptibility and deficits in behavior. Studies looking at peptidergic cell types in the hypothalamus [194,195] and dopaminergic neurons [196,197] have been well characterized by specific molecular markers. Further studies are needed to define other neuronal subtypes, including a thorough molecular census of glutamatergic and GABAergic sub-populations, to further validate zebrafish ASD-ID models.

5. Current Limitations for Using Zebrafish to Study Human Disease and Disorder Genes

Although zebrafish provide many advantages for modeling human genetic disorders, the retention of many duplicate genes after a whole genome duplication and the differences in morphant and mutant phenotypes provide challenges for researchers. For example, duplicate genes in zebrafish require researchers to study two genes in parallel, due to the likelihood that subfunctionalization has split ancestral functions between duplicate pairs [198]. In addition due to off targeting and toxicity with morpholinos and genetic compensation in mutants, knockdown and knockout models of the same gene can present different phenotypes [135,199]. Fortunately these issues are not barriers to using zebrafish for disease models and can be tackled via careful consideration while planning projects.

5.1. The Teleost Whole Genome Duplication

A whole genome duplication occurred during the evolution of teleosts (zebrafish infraclass) and resulted in the retention of many disorder related genes in zebrafish [200]. When a duplicate is retained, the duplicate pair either share ancestral functions (functional redundancy), split functions (subfunctionalization), or develop new functions (neofunctionalization) [201]. In an effort to resolve the ancestral state of zebrafish genes, recent studies have compared genes in the spotted gar (diverged before the Teleostei duplication) with zebrafish, to determine which gene within a pair is most similar to the ancestral orthologue [198,202]. This method can also be combined with gene expression data (temporal and spatial patterns) and genetic manipulation, to determine whether a gene duplicate is relevant to a human disorder [202]. A recent study showed that a majority of duplicate pairs experience subfunctionalization, requiring double knockouts for disorder modeling [203]. Although only a minority of zebrafish duplicates have been retained (~13%), [203] developmental disorder duplicates have been retained at a much higher rate (~60%) [12]. Therefore, the majority of ASD-ID gene orthologues are likely duplicated and will require studying two genes. Although studying two genes becomes a daunting task, clear advantageous can be garnered from studying a duplicate pair. For example, subfunctionalization could provide a means for teasing out the relationship between phenotypes. One gene may retain a role in ASD-ID social behavior while the duplicate retains a role important for a comorbid phenotype, such as intestinal distress. While subfunctionalization may be advantageous for studying certain genotype-phenotype relationship, gene duplicates remain challenging to study, and researchers need to utilize improved methodology for evaluating ancestral functions when starting a project.

5.2. Gene Knockdowns Versus Knockouts

Genetic knockdown and knockout technologies have strengths and weaknesses when modeling human disorders [199]. Gene knockdown techniques rely on anti-sense morpholinos or short-hairpin RNA molecules, to either inhibit mRNA translation or disrupt pre-mRNA splicing [135]. Gene knockdown technologies have been used extensively to functionally validate human disease genes [204–208]. However, in some cases, morpholinos cause severe phenotypes due to off-target effects, activation of p53 cell death pathways and toxicity [135,209,210]. Alternatively, some morpholinos faithfully phenocopy stable mutant phenotypes that can be rescued by injecting mRNA encoding the targeted gene [170,211,212]. In these cases morpholinos can complement mutant studies, allowing loss-of-function phenotypes to be studied rapidly in multiple transgenic zebrafish lines. Because gene knockdowns are transient, phenotypes must be studied in the embryo [135,213]. Therefore, gene knockdowns could be useful for future embryonic models of ASD-ID and researchers should follow the guidelines for using morpholino knockdown technology. These guidelines detail the proper control experiments needed for morphant studies and suggest using morphants as a complement to genetic mutants [213].

More recently, CRISPR-Cas9, TALENS and zinc finger endonucleases have provided a powerful tool, not only to produce gene knockouts but to induce site-specific mutations and knock-ins [214,215]. In some cases, phenotypes are subtler than expected due to genetic compensation, though the pervasiveness of such compensation has yet to be determined [136,216]. None-the-less, targeted knockouts and tailored mutants provide a better representation of the genetic state in human disorders, in comparison to morpholinos [209]. Additionally, when mutagenesis efficiencies are high, phenotypes can be assessed in embryos injected with endonucleases [217,218]. This provides the same expediency and versatility as morpholinos, but without the drawbacks of toxicity and transient phenotypes [217,218]. Therefore when phenotypes are present, knockout models should be used as the primary model for studying genetic causes of human disorders.

6. The Integrated Model: Utilizing Emerging Technology to Integrate Anatomy, Physiology and Behavior

These initial ASD-ID developmental models provide windows into the anatomical changes, mechanisms and critical periods that relate to prenatal neuropathology. However, zebrafish neuropsychiatric models can reach their full potential by using an integrative methodological approach. Future studies can bridge the gap between anatomy, physiology and behavior by combining systems level developmental changes (e.g., decreased proliferation), with specificity to neuronal subclasses (e.g., decreased dopaminergic neurons) and behavior (e.g., decreased dopaminergic drive affects motivated behavior). Ideally the next generation of models will utilize innovative methods including, brain atlases providing molecular marker libraries to interrogate anatomical and physiological causes of behavioral phenotypes, and small molecule screens for drug discovery.

6.1. Characteristics of Behaviors during Zebrafish Embryonic and Larval Development

Zebrafish are a powerful prenatal model because precocious embryonic and larval behavior provide an early indicator for disruptions of neural circuits. Prenatal disruptions of neural circuits may go undetected in humans, because ASD-ID diagnostic behaviors emerge slowly during post-natal development. Therefore, a delay between circuit disruption and the emergence of diagnostic behavior might result in prenatal etiological knowledge gaps. These gaps can easily be addressed by studying emerging behavior in zebrafish gene models of ASD-ID. During embryogenesis zebrafish tails begin to bend spontaneously (18 hpf), due to large calcium depolarizations flowing from head to tail [23,219]. Early synchronous behavior then transitions into fully functional swimming by 28 hpf, including touch evoked evasive movements [23]. Importantly, these behaviors utilize conserved midbrain, hindbrain and spinal cord circuits that drive ambulatory movement in humans [220,221]. Zebrafish begin to wire and integrate sensory-motor systems during larval development needed to avoid predators and capture prey [24]. These sensory provoked behaviors have extensive background literature, including functional studies detailing the relationship of sensory input to motor output [22,111,222,223]. This attribute could provide the most compelling translational research from zebrafish ASD-ID models, as sensory based deficits are common in ASD and were recently added to the diagnostic criteria [224,225]. Therefore, we argue that emergent behavioral phenotypes can be used in zebrafish ASD-ID models as a means to assay human prenatal developmental milestones.

Zebrafish also develop social and memory-based behaviors during larval development [24,226]. Social behaviors are important for modeling ASD-ID and several groups have established larval assays, that include shoaling and kin recognition [107,227,228]. However, these behaviors emerge at later stages (>9 DPF) and, although likely conserved with mammals, the neural circuitry remains largely unknown [229]. Studies in adult zebrafish have established assays and provided insight into oxytocinergic and dopaminergic circuits involved in social interactions [230]. Additionally, a novel social interaction paradigm was recently developed that identified forebrain cholinergic neurons essential for social orientating [231]. These emerging behaviors also evoke similar comparative questions discussed in Section 3, such as, what are the neurodevelopmental events and circuits that correspond to these complex behaviors? The zebrafish community will need to address these knowledge gaps in order to interpret circuit-level changes in larval social behavior.

6.2. Brain Atlases and High Resolution Morphometry

Research focusing on structural and functional mapping of the brain has recently made impressive progress. Subtle anatomical changes are common in ASD-ID and will require analyzing microstructure of brain circuits in zebrafish ASD-ID gene models [33,232,233]. Zebrafish may be well suited for studying these subtle anatomical changes due to their smaller brains. Several zebrafish groups have established community brain atlases and libraries using brain registration methods [19,234–237]. Confocal stacks of larval brains are orientated to a reference brain in a specified x, y, z axis using non-linear registration [238]. These registration methods provide sub-cellular accuracy and ensure that brain regions and neuronal populations overlap properly. Reference brains are then used to map libraries of molecular markers, to delineate neuronal subpopulations and circuits within the brain. Therefore, users can register confocal stacks of neuronal activity (e.g., genetically encoded calcium indicator, GECI) or transgenic backgrounds (e.g., vesicular glutamate transporter 2, *vGluT2a*:DsRed transgenic line [239]), to determine discrete sources of physiological and anatomical phenotypes. Furthermore, some of these analytical methods provide fine structural detail that could be missed analyzing gross anatomical regions by hand. For example, many studies looking at macro- and microcephaly measure changes to the entire brain or conspicuous regions (e.g., between the optic tecta [160]). However, innovations in mapping allow users to analyze specific somas or neuropil in fine detail. A new registration and brain segmentation method was recently used to provide a neuroanatomical map based on transgene expression profiles [235]. This method was able to determine significant differences in microstructure of larval brain regions exposed to VPA [235]. VPA exposed larvae displayed several anatomical deficits, including a decrease in excitatory components of the statoacoustic ganglion. Consistent with this structural alteration, VPA-exposed larvae showed decreased responsiveness following an acoustic startle stimulus. These resources have overcome challenges in analyzing circuits, allowing the community to link anatomical and physiological changes to behavioral phenotypes.

6.3. Whole-Brain Visual Outputs of Neuronal Activity

Larval zebrafish have a small brain (~100,000 neuron at 6 dpf) in comparison to adult humans (80,000,000,000), allowing for brain-wide neuronal activity monitoring using a neuronal activity indicator [22]. This includes immediate early genes (IEGs) and genetically encoded calcium indicators (GECIs). Although both are readily available for any lab to use, the former (IEGs) are more accessible to be captured on a typical confocal microscope and analyzed using a community brain atlas.

IEGs are genes and proteins whose expression increases exponentially following an action potential [240]. IEGs, such as cFOS, are commonly used as a proxy for neuronal firing. However, cFOS expression peaks around an hour following an action potential, making it impossible to use for transient stimulus provoked behavior [241]. In contrast, ERK protein is phosphorylated (pERK) within seconds of neuronal firing and phosphorylation peaks within five minutes [242]. pERK expression can provide information on activity changes following a temporally short stimulus, such as a light flash or vibration. Recently, pERK expression was coupled with a brain mapping technique to provide regional discrimination of changes in neuronal activity [234]. This MAP-Mapping technique takes normalized pERK expression in two groups and computes the difference in expression between comparable anatomical regions [234]. For example, the GABA potentiator pentylenetetrazol (PTZ) causes uninhibited excitatory activity throughout the brain that can be visualized in MAP-Map showing an increase in pERK throughout the brain in comparison to pERK expression with no PTZ. This method provides an unbiased way to survey activity changes in the entire brain during a disorder relevant behavior.

Calcium indicators become more fluorescent upon binding calcium and are used to study nervous system activity. Calcium indicator dyes have been utilized by zebrafish researchers for decades, because dyes can be injected into blastomeres and backfilled in the spinal cord to label neurons [243–245]. This system has advantages over Genetically Encoded Calcium Indicators (GECI), because mutant backgrounds do not have to be outcrossed to GECIs. However, GECI methods have been refined and are now sensitive enough to record whole-brain activity in awake behaving fish [22,246]. Labs have provided a wealth of information on zebrafish brain activity underlying specific behaviors during early larval behavior (6–7 dpf), including response to visual and auditory stimuli [17,20,22,25]. Typically, zebrafish need to be paralyzed or restrained in low melt agarose for recording, however two labs have recently built robotic light-sheet microscopes that can follow freely swimming zebrafish larvae in a dish, while recording GCaMP whole-brain imaging [247,248]. Future studies will be able to follow zebrafish ASD-ID models during complex sensory or social behavior, allowing researchers to define, in real-time, the neural circuit deficits underlying ASD-ID behavioral phenotypes. These methods will be revolutionary in determining whether conserved brain regions are utilized during larval social behavior and memory to model core ASD-ID deficits.

Finally, a new approach called Multi-MAP has combined fixed and live imaging techniques to assess modulations of brain activity by arousal pathways [249]. This approach uses GCaMP6s transgenic fish to record neuronal activity in live fish, followed by fixation and immunostaining to identify changes in different neuromodulatory neuronal sub-classes. A looming stimulus was used to measure reaction time and determine alertness [249]. Fast and slow alertness was found to be correlated with the firing of specific subtypes of neuromodulatory neurons in the locus coeruleus, tegmentum and hypothalamus [249]. Importantly, alertness correlated brain regions showed similar activity patterns in mice, suggesting that the relationship between alertness circuits and behavior are conserved [249]. Furthermore, these conserved circuits and behaviors are important for ASD-ID research because arousal pathways have been implicated in ASD-ID [191,250,251]. This method could be strengthened by functionally testing circuits using lines from a Gal4 or Cre-Lox library [19,252]. The multi-MAP approach leverages the power of live whole-brain activity monitoring and molecularly defined circuits to discover cellular elements underlying behavior.

6.4. *Small Molecule Screening and Drug Discovery*

Zebrafish larvae are small and take up small molecules through the water column, macking zebrafish attractive for medium-throughput drug screening. Small molecules can be dissolved into water or embryo media, and automated behavioral rigs can test 96 fish simultaneously [253]. Additionally, outbred zebrafish lines provide a heterogeneous animal model that reflects the genetic diversity seen in humans [254]. A potential drug discovery should therefore be attractive for preclinical trials when a phenotype-drug relationship is clearly reproducible. Seminal work in small molecule screens have discovered novel compounds that affect sleep [125,255], the photomotor response [256], seizures [193] and behavioral effects of neuropsychological drugs [257]. A drug screen recently correlated changes in behavior between known neuropsychiatric drugs and novel molecules [257]. By using the drug haloperidol, this method validated the therapeutic efficacy of structurally similar compounds [257]. Identifying alternative molecules to treat neuropsychiatric disorders is important, because many neuropsychiatric drugs target multiple neuromodulatory systems that results in unwanted side-effects. Using this method, zebrafish ASD-ID gene models could be screened for small molecules that recapitulate the therapeutic effect of a marketed drug, while lessoning the effect of unwanted side effects. This strategy could be important for currently prescribed ASD medications, such as risperidone and lithium, that cause severe side effects [258,259]. Finally, whole-brain activity methods could be combined with drug screens to determine relationships between drug influenced circuits and improvements in behavioral deficits.

7. Conclusions

Past proof-of-concept studies have established zebrafish as a powerful model for studying prenatal neuropathologies underlying ASD-ID. Prenatal risk factors for developing ASD-ID have been well-documented for over a half century, yet there is still little known about the earliest developmental origins of these disorders. In addition, geneticists are beginning to identify prenatally expressed genes that constitute risk factors for ASD-ID. The current body of zebrafish ASD-ID literature has provided a groundwork for studying broad mechanisms underlying ASD-ID neuropathology, such as brain morphogenesis and circuit development. However, the majority of these studies involved gene knockdowns and future genetic mutants will be needed to determine the link between ASD-ID gene mutations and developmental phenotypes. These mutant studies will also need to integrate the current technological leaps, linking anatomical phenotypes to physiological mechanisms underlying ASD-ID relevant behavior.

Funding: University of Miami's Department of Biology Teaching Assistantship.

Acknowledgments: I would like to thank Julia Dallman for feedback and guidance during the writing and editing processes.

Conflicts of Interest: The author declares no conflict of interest.

References

1. Srivastava, A.K.; Schwartz, C.E. Intellectual disability and autism spectrum disorders: Causal genes and molecular mechanisms. *Neurosci. Biobehav. Rev.* **2014**, *46*, 161–174. [CrossRef] [PubMed]
2. Association, A.P. Intellectual Disability. In *Diagnostic and Statistical Manual of Mental Disorders*, 5th ed.; DSM: Heerlen, The Netherlands, 2013.
3. Fombonne, E. Epidemiological surveys of autism and other pervasive developmental disorders: An update. *J. Autism. Dev. Disord.* **2003**, *33*, 365–382. [CrossRef] [PubMed]
4. Matson, J.L.; Shoemaker, M. Intellectual disability and its relationship to autism spectrum disorders. *Res. Dev. Disabil.* **2009**, *30*, 1107–1114. [CrossRef] [PubMed]
5. Kou, Y.; Betancur, C.; Xu, H.; Buxbaum, J.D.; Ma'ayan, A. Network- and attribute-based classifiers can prioritize genes and pathways for autism spectrum disorders and intellectual disability. *Am. J. Med. Genet. C Semin. Med. Genet.* **2012**, *160C*, 130–142. [CrossRef] [PubMed]
6. Baio, J.; Wiggins, L.; Christensen, D.L.; Maenner, M.J.; Daniels, J.; Warren, Z.; Kurzius-Spencer, M.; Zahorodny, W.; Rosenberg, C.R.; White, T.; et al. Prevalence of Autism Spectrum Disorder Among Children Aged 8 Years—Autism and Developmental Disabilities Monitoring Network, 11 Sites, United States, 2014. *MMWR Surveillance Summer* **2018**, *67*, 1–23. [CrossRef] [PubMed]
7. Berg, J.M.; Geschwind, D.H. Autism genetics: Searching for specificity and convergence. *Genome Biol.* **2012**, *13*, 247. [CrossRef] [PubMed]
8. Peca, J.; Feng, G. Cellular and synaptic network defects in autism. *Curr. Opin. Neurobiol.* **2012**, *22*, 866–872. [CrossRef] [PubMed]
9. Chen, J.A.; Penagarikano, O.; Belgard, T.G.; Swarup, V.; Geschwind, D.H. The emerging picture of autism spectrum disorder: Genetics and pathology. *Annu. Rev. Pathol.* **2015**, *10*, 111–144. [CrossRef] [PubMed]
10. Ehninger, D.; Sano, Y.; de Vries, P.J.; Dies, K.; Franz, D.; Geschwind, D.H.; Kaur, M.; Lee, Y.S.; Li, W.; Lowe, J.K.; et al. Gestational immune activation and Tsc2 haploinsufficiency cooperate to disrupt fetal survival and may perturb social behavior in adult mice. *Mol. Psychiatry* **2012**, *17*, 62–70. [CrossRef] [PubMed]
11. Belinson, H.; Nakatani, J.; Babineau, B.A.; Birnbaum, R.Y.; Ellegood, J.; Bershteyn, M.; McEvilly, R.J.; Long, J.M.; Willert, K.; Klein, O.D.; et al. Prenatal beta-catenin/Brn2/Tbr2 transcriptional cascade regulates adult social and stereotypic behaviors. *Mol. Psychiatry* **2016**, *21*, 1417–1433. [CrossRef] [PubMed]
12. Kozol, R.A.; Abrams, A.J.; James, D.M.; Buglo, E.; Yan, Q.; Dallman, J.E. Function Over Form: Modeling Groups of Inherited Neurological Conditions in Zebrafish. *Front. Mol. Neurosci.* **2016**, *9*, 55. [CrossRef] [PubMed]
13. Sakai, C.; Ijaz, S.; Hoffman, E.J. Zebrafish Models of Neurodevelopmental Disorders: Past, Present, and Future. *Front. Mol. Neurosci.* **2018**, *11*, 294. [CrossRef] [PubMed]
14. Kimmel, C.B.; Ballard, W.W.; Kimmel, S.R.; Ullmann, B.; Schilling, T.F. Stages of embryonic development of the zebrafish. *Dev. Dyn.* **1995**, *203*, 253–310. [CrossRef] [PubMed]
15. Kimmel, C.B.; Hatta, K.; Metcalfe, W.K. Early axonal contacts during development of an identified dendrite in the brain of the zebrafish. *Neuron* **1990**, *4*, 535–545. [CrossRef]
16. Higashijima, S.; Masino, M.A.; Mandel, G.; Fetcho, J.R. Imaging neuronal activity during zebrafish behavior with a genetically encoded calcium indicator. *J. Neurophysiol.* **2003**, *90*, 3986–3997. [CrossRef] [PubMed]
17. Ahrens, M.B.; Orger, M.B.; Robson, D.N.; Li, J.M.; Keller, P.J. Whole-brain functional imaging at cellular resolution using light-sheet microscopy. *Nat. Methods* **2013**, *10*, 413–420. [CrossRef] [PubMed]
18. Satou, C.; Kimura, Y.; Hirata, H.; Suster, M.L.; Kawakami, K.; Higashijima, S. Transgenic tools to characterize neuronal properties of discrete populations of zebrafish neurons. *Development* **2013**, *140*, 3927–3931. [CrossRef] [PubMed]
19. Marquart, G.D.; Tabor, K.M.; Brown, M.; Strykowski, J.L.; Varshney, G.K.; LaFave, M.C.; Mueller, T.; Burgess, S.M.; Higashijima, S.; Burgess, H.A. A 3D Searchable Database of Transgenic Zebrafish Gal4 and Cre Lines for Functional Neuroanatomy Studies. *Front. Neural Circuits* **2015**, *9*, 78. [CrossRef] [PubMed]
20. Akerboom, J.; Chen, T.W.; Wardill, T.J.; Tian, L.; Marvin, J.S.; Mutlu, S.; Calderon, N.C.; Esposti, F.; Borghuis, B.G.; Sun, X.R.; et al. Optimization of a GCaMP calcium indicator for neural activity imaging. *J. Neurosci.* **2012**, *32*, 13819–13840. [CrossRef] [PubMed]
21. Fosque, B.F.; Sun, Y.; Dana, H.; Yang, C.T.; Ohyama, T.; Tadross, M.R.; Patel, R.; Zlatic, M.; Kim, D.S.;

Ahrens, M.B.; et al. Neural circuits. Labeling of active neural circuits in vivo with designed calcium integrators. *Science* **2015**, *347*, 755–760. [CrossRef] [PubMed]

22. Ahrens, M.B.; Li, J.M.; Orger, M.B.; Robson, D.N.; Schier, A.F.; Engert, F.; Portugues, R. Brain-wide neuronal dynamics during motor adaptation in zebrafish. *Nature* **2012**, *485*, 471–477. [CrossRef] [PubMed]
23. Brustein, E.; Saint-Amant, L.; Buss, R.R.; Chong, M.; McDearmid, J.R.; Drapeau, P. Steps during the development of the zebrafish locomotor network. *J. Physiol. Paris* **2003**, *97*, 77–86. [CrossRef] [PubMed]
24. Fero, K.Y.T.; Burgess, H.A. The Behavioral Repertoire of Larval Zebrafish. In *Zebrafish Models in Neurobehavioral Research*; Humana Press: New York, NY, USA, 2011; pp. 249–291.
25. Warp, E.; Agarwal, G.; Wyart, C.; Friedmann, D.; Oldfield, C.S.; Conner, A.; Del Bene, F.; Arrenberg, A.B.; Baier, H.; Isacoff, E.Y. Emergence of patterned activity in the developing zebrafish spinal cord. *Curr. Biol.* **2012**, *22*, 93–102. [CrossRef] [PubMed]
26. Orger, M.B.; de Polavieja, G.G. Zebrafish Behavior: Opportunities and Challenges. *Annu. Rev. Neurosci.* **2017**, *40*, 125–147. [CrossRef] [PubMed]
27. Arndt, T.L.; Stodgell, C.J.; Rodier, P.M. The teratology of autism. *Int. J. Dev. Neurosci.* **2005**, *23*, 189–199. [CrossRef] [PubMed]
28. Miller, M.T.; Stromland, K.; Ventura, L.; Johansson, M.; Bandim, J.M.; Gillberg, C. Autism associated with conditions characterized by developmental errors in early embryogenesis: A mini review. *Int. J. Dev. Neurosci.* **2005**, *23*, 201–219. [CrossRef] [PubMed]
29. Birnbaum, R.; Jaffe, A.E.; Hyde, T.M.; Kleinman, J.E.; Weinberger, D.R. Prenatal expression patterns of genes associated with neuropsychiatric disorders. *Am. J. Psychiatry* **2014**, *171*, 758–767. [CrossRef] [PubMed]
30. Lombardo, M.V.; Moon, H.M.; Su, J.; Palmer, T.D.; Courchesne, E.; Pramparo, T. Maternal immune activation dysregulation of the fetal brain transcriptome and relevance to the pathophysiology of autism spectrum disorder. *Mol. Psychiatry* **2018**, *23*, 1001–1013. [CrossRef] [PubMed]
31. Willsey, A.J.; Sanders, S.J.; Li, M.; Dong, S.; Tebbenkamp, A.T.; Muhle, R.A.; Reilly, S.K.; Lin, L.; Fertuzinhos, S.; Miller, J.A.; et al. Coexpression networks implicate human midfetal deep cortical projection neurons in the pathogenesis of autism. *Cell* **2013**, *155*, 997–1007. [CrossRef] [PubMed]
32. Jones, K.L.; Croen, L.A.; Yoshida, C.K.; Heuer, L.; Hansen, R.; Zerbo, O.; DeLorenze, G.N.; Kharrazi, M.; Yolken, R.; Ashwood, P.; et al. Autism with intellectual disability is associated with increased levels of maternal cytokines and chemokines during gestation. *Mol. Psychiatry* **2017**, *22*, 273–279. [CrossRef] [PubMed]
33. Bauman, M.L. Microscopic neuroanatomic abnormalities in autism. *Pediatrics* **1991**, *87*, 791–796. [PubMed]
34. Bauman, M.L.; Kemper, T.L. Neuroanatomic observations of the brain in autism: A review and future directions. *Int. J. Dev. Neurosci.* **2005**, *23*, 183–187. [CrossRef] [PubMed]
35. Fatemi, S.H.; Aldinger, K.A.; Ashwood, P.; Bauman, M.L.; Blaha, C.D.; Blatt, G.J.; Chauhan, A.; Chauhan, V.; Dager, S.R.; Dickson, P.E.; et al. Consensus paper: Pathological role of the cerebellum in autism. *Cerebellum* **2012**, *11*, 777–807. [CrossRef] [PubMed]
36. Wegiel, J.; Kuchna, I.; Nowicki, K.; Imaki, H.; Wegiel, J.; Marchi, E.; Ma, S.Y.; Chauhan, A.; Chauhan, V.; Bobrowicz, T.W.; et al. The neuropathology of autism: Defects of neurogenesis and neuronal migration, and dysplastic changes. *Acta Neuropathol.* **2010**, *119*, 755–770. [CrossRef] [PubMed]
37. Chess, S.; Fernandez, P. Neurologic damage and behavior disorder in rubella children. *Am. Ann. Deaf* **1980**, *125*, 998–1001. [CrossRef] [PubMed]
38. Moore, S.J.; Turnpenny, P.; Quinn, A.; Glover, S.; Lloyd, D.J.; Montgomery, T.; Dean, J.C. A clinical study of 57 children with fetal anticonvulsant syndromes. *J. Med. Genet.* **2000**, *37*, 489–497. [CrossRef] [PubMed]
39. Stromland, K.; Nordin, V.; Miller, M.; Akerstrom, B.; Gillberg, C. Autism in thalidomide embryopathy: A population study. *Dev. Med. Child. Neurol.* **1994**, *36*, 351–356. [CrossRef] [PubMed]
40. Nicolini, C.; Fahnestock, M. The valproic acid-induced rodent model of autism. *Exp. Neurol.* **2018**, *299*, 217–227. [CrossRef] [PubMed]
41. Kim, J.W.; Seung, H.; Kwon, K.J.; Ko, M.J.; Lee, E.J.; Oh, H.A.; Choi, C.S.; Kim, K.C.; Gonzales, E.L.; You, J.S.; et al. Subchronic treatment of donepezil rescues impaired social, hyperactive, and stereotypic behavior in valproic acid-induced animal model of autism. *PLoS ONE* **2014**, *9*, e104927. [CrossRef] [PubMed]
42. Ito, T.; Ando, H.; Suzuki, T.; Ogura, T.; Hotta, K.; Imamura, Y.; Yamaguchi, Y.; Handa, H. Identification of a primary target of thalidomide teratogenicity. *Science* **2010**, *327*, 1345–1350. [CrossRef] [PubMed]
43. Chen, J.; Lei, L.; Tian, L.; Hou, F.; Roper, C.; Ge, X.; Zhao, Y.; Chen, Y.; Dong, Q.; Tanguay, R.L.; et al.

Developmental and behavioral alterations in zebrafish embryonically exposed to valproic acid (VPA): An aquatic model for autism. *Neurotoxicol. Teratol.* **2018**, *66*, 8–16. [CrossRef] [PubMed]

44. Baronio, D.; Puttonen, H.A.J.; Sundvik, M.; Semenova, S.; Lehtonen, E.; Panula, P. Embryonic exposure to valproic acid affects the histaminergic system and the social behaviour of adult zebrafish (*Danio rerio*). *Br. J. Pharmacol.* **2018**, *175*, 797–809. [CrossRef] [PubMed]
45. Krakowiak, P.; Goines, P.E.; Tancredi, D.J.; Ashwood, P.; Hansen, R.L.; Hertz-Picciotto, I.; Van de Water, J. Neonatal Cytokine Profiles Associated With Autism Spectrum Disorder. *Biol. Psychiatry* **2017**, *81*, 442–451. [CrossRef] [PubMed]
46. van der Vaart, M.; Svoboda, O.; Weijts, B.G.; Espin-Palazon, R.; Sapp, V.; Pietri, T.; Bagnat, M.; Muotri, A.R.; Traver, D. Mecp2 regulates tnfa during zebrafish embryonic development and acute inflammation. *Dis. Model Mech.* **2017**, *10*, 1439–1451. [CrossRef] [PubMed]
47. Amir, R.E.; Van den Veyver, I.B.; Wan, M.; Tran, C.Q.; Francke, U.; Zoghbi, H.Y. Rett syndrome is caused by mutations in X-linked MECP2, encoding methyl-CpG-binding protein 2. *Nat. Genet.* **1999**, *23*, 185–188. [CrossRef] [PubMed]
48. Folstein, S.E.; Piven, J. Etiology of autism: Genetic influences. *Pediatrics* **1991**, *87*, 767–773. [PubMed]
49. Tassone, F.; De Rubeis, S.; Carosi, C.; La Fata, G.; Serpa, G.; Raske, C.; Willemsen, R.; Hagerman, P.J.; Bagni, C. Differential usage of transcriptional start sites and polyadenylation sites in FMR1 premutation alleles. *Nucleic Acids Res.* **2011**, *39*, 6172–6185. [CrossRef] [PubMed]
50. Banerjee, P.; Schoenfeld, B.P.; Bell, A.J.; Choi, C.H.; Bradley, M.P.; Hinchey, P.; Kollaros, M.; Park, J.H.; McBride, S.M.; Dockendorff, T.C. Short- and long-term memory are modulated by multiple isoforms of the fragile X mental retardation protein. *J. Neurosci.* **2010**, *30*, 6782–6792. [CrossRef] [PubMed]
51. Boland, M.J.; Nazor, K.L.; Tran, H.T.; Szucs, A.; Lynch, C.L.; Paredes, R.; Tassone, F.; Sanna, P.P.; Hagerman, R.J.; Loring, J.F. Molecular analyses of neurogenic defects in a human pluripotent stem cell model of fragile X syndrome. *Brain* **2017**, *140*, 582–598. [CrossRef] [PubMed]
52. Park, S.-H.; Pepkowitz, S.H.; Kerfoot, C.; de Rosa, M.J.; Poukens, V.; Wienecke, R.; de Clue, J.E.; Vinters, H.V. Tuberous sclerosis in a 20-week gestation fetus: Immunohistochemical study. *Acta Neuropathol.* **1997**, *94*, 180–186. [CrossRef] [PubMed]
53. Wei, J.; Li, P.; Chiriboga, L.; Mizuguchi, M.; Yee, H.; Miller, D.C.; Greco, M.A. Tuberous sclerosis in a 19-week fetus: Immunohistochemical and molecular study of hamartin and tuberin. *Pediatr. Dev. Pathol.* **2002**, *5*, 448–464. [CrossRef] [PubMed]
54. Bordarier, C.; Lellouch-Tubiana, A.; Robain, O. Cardiac rhabdomyoma and tuberous sclerosis in three fetuses: A neuropathological study. *Brain Dev.* **1994**, *16*, 467–471. [CrossRef]
55. Durak, O.; Gao, F.; Kaeser-Woo, Y.J.; Rueda, R.; Martorell, A.J.; Nott, A.; Liu, C.Y.; Watson, L.A.; Tsai, L.-H. Chd8 mediates cortical neurogenesis via transcriptional regulation of cell cycle and Wnt signaling. *Nat. Neurosci.* **2016**, *19*, 1477–1488. [CrossRef] [PubMed]
56. Durak, O.; de Anda, F.C.; Singh, K.K.; Leussis, M.P.; Petryshen, T.L.; Sklar, P.; Tsai, L.-H. Ankyrin-G regulates neurogenesis and Wnt signaling by altering the subcellular localization of beta-catenin. *Mol. Psychiatry* **2015**, *20*, 388–397. [CrossRef] [PubMed]
57. O'Roak, B.J.; Vives, L.; Fu, W.; Egertson, J.D.; Stanaway, I.B.; Phelps, I.G.; Carvill, G.; Kumar, A.; Lee, C.; Ankenman, K.; et al. Multiplex targeted sequencing identifies recurrently mutated genes in autism spectrum disorders. *Science* **2012**, *338*, 1619–1622. [CrossRef] [PubMed]
58. O'Roak, B.J.; Vives, L.; Girirajan, S.; Karakoc, E.; Krumm, N.; Coe, B.P.; Levy, R.; Ko, A.; Lee, C.; Smith, J.D.; et al. Sporadic autism exomes reveal a highly interconnected protein network of de novo mutations. *Nature* **2012**, *485*, 246–250. [CrossRef] [PubMed]
59. Sanders, S.J.; Murtha, M.T.; Gupta, A.R.; Murdoch, J.D.; Raubeson, M.J.; Willsey, A.J.; Ercan-Sencicek, A.G.; DiLullo, N.M.; Parikshak, N.N.; Stein, J.L.; et al. De novo mutations revealed by whole-exome sequencing are strongly associated with autism. *Nature* **2012**, *485*, 237–241. [CrossRef] [PubMed]
60. Werling, D.M.; Brand, H.; An, J.Y.; Stone, M.R.; Zhu, L.; Glessner, J.T.; Collins, R.L.; Dong, S.; Layer, R.M.; Markenscoff-Papadimitriou, E.; et al. An analytical framework for whole-genome sequence association studies and its implications for autism spectrum disorder. *Nat. Genet.* **2018**, *50*, 727–736. [CrossRef] [PubMed]
61. Geschwind, D.H.; State, M.W. Gene hunting in autism spectrum disorder: On the path to precision medicine. *Lancet Neurol.* **2015**, *14*, 1109–1120. [CrossRef]

62. De Rubeis, S.; Buxbaum, J.D. Genetics and genomics of autism spectrum disorder: Embracing complexity. *Hum. Mol. Genet.* **2015**, *24*, R24–R31. [CrossRef] [PubMed]
63. Ionita-Laza, I.; Capanu, M.; De Rubeis, S.; McCallum, K.; Buxbaum, J.D. Identification of rare causal variants in sequence-based studies: Methods and applications to VPS13B, a gene involved in Cohen syndrome and autism. *PLoS Genet.* **2014**, *10*, e1004729. [CrossRef] [PubMed]
64. Parikshak, N.N.; Luo, R.; Zhang, A.; Won, H.; Lowe, J.K.; Chandran, V.; Horvath, S.; Geschwind, D.H. Integrative functional genomic analyses implicate specific molecular pathways and circuits in autism. *Cell* **2013**, *155*, 1008–1021. [CrossRef] [PubMed]
65. Cristino, A.S.; Williams, S.M.; Hawi, Z.; An, J.Y.; Bellgrove, M.A.; Schwartz, C.E.; Costa Lda, F.; Claudianos, C. Neurodevelopmental and neuropsychiatric disorders represent an interconnected molecular system. *Mol. Psychiatry* **2014**, *19*, 294–301. [CrossRef] [PubMed]
66. Hormozdiari, F.; Penn, O.; Borenstein, E.; Eichler, E.E. The discovery of integrated gene networks for autism and related disorders. *Genome Res.* **2015**, *25*, 142–154. [CrossRef] [PubMed]
67. Voineagu, I.; Wang, X.; Johnston, P.; Lowe, J.K.; Tian, Y.; Horvath, S.; Mill, J.; Cantor, R.M.; Blencowe, B.J.; Geschwind, D.H. Transcriptomic analysis of autistic brain reveals convergent molecular pathology. *Nature* **2011**, *474*, 380–384. [CrossRef] [PubMed]
68. Vorstman, J.A.S.; Parr, J.R.; Moreno-De-Luca, D.; Anney, R.J.L.; Nurnberger, J.I., Jr.; Hallmayer, J.F. Autism genetics: Opportunities and challenges for clinical translation. *Nat. Rev. Genet.* **2017**, *18*, 362–376. [CrossRef] [PubMed]
69. Niemi, M.E.K.; Martin, H.C.; Rice, D.L.; Gallone, G.; Gordon, S.; Kelemen, M.; McAloney, K.; McRae, J.; Radford, E.J.; Yu, S.; et al. Common genetic variants contribute to risk of rare severe neurodevelopmental disorders. *Nature* **2018**, *562*, 268–271. [CrossRef] [PubMed]
70. Wolf, J.B.; Wade, M.J. What are maternal effects (and what are they not)? *Philos. Trans. R. Soc. Lond. B Biol. Sci.* **2009**, *364*, 1107–1115. [CrossRef] [PubMed]
71. Guilmatre, A.; Sharp, A.J. Parent of origin effects. *Clin. Genet.* **2012**, *81*, 201–209. [CrossRef] [PubMed]
72. Connolly, S.; Anney, R.; Gallagher, L.; Heron, E.A. A genome-wide investigation into parent-of-origin effects in autism spectrum disorder identifies previously associated genes including SHANK3. *Eur. J. Hum. Genet.* **2017**, *25*, 234–239. [CrossRef] [PubMed]
73. Greenblatt, E.J.; Spradling, A.C. Fragile X mental retardation 1 gene enhances the translation of large autism-related proteins. *Science* **2018**, *361*, 709–712. [CrossRef] [PubMed]
74. Kindler, S.; Kreienkamp, H.J. Dendritic mRNA targeting and translation. *Adv. Exp. Med. Biol.* **2012**, *970*, 285–305. [PubMed]
75. Rice, D.; Barone, S., Jr. Critical periods of vulnerability for the developing nervous system: Evidence from humans and animal models. *Environ. Health Perspect.* **2000**, *108* (Suppl. 3), 511–533. [PubMed]
76. Clancy, B.; Darlington, R.B.; Finlay, B.L. Translating developmental time across mammalian species. *Neuroscience* **2001**, *105*, 7–17. [CrossRef]
77. Vaillancourt, C.; Lafond, J. Human embryogenesis: Overview. *Methods Mol. Biol.* **2009**, *550*, 3–7. [PubMed]
78. Workman, A.D.; Charvet, C.J.; Clancy, B.; Darlington, R.B.; Finlay, B.L. Modeling transformations of neurodevelopmental sequences across mammalian species. *J. Neurosci.* **2013**, *33*, 7368–7383. [CrossRef] [PubMed]
79. Moens, C.B.; Prince, V.E. Constructing the hindbrain: Insights from the zebrafish. *Dev. Dyn.* **2002**, *224*, 1–17. [CrossRef] [PubMed]
80. Hashimoto, M.; Hibi, M. Development and evolution of cerebellar neural circuits. *Dev. Growth Differ.* **2012**, *54*, 373–389. [CrossRef] [PubMed]
81. Wurst, W.; Bally-Cuif, L. Neural plate patterning: Upstream and downstream of the isthmic organizer. *Nat. Rev. Neurosci.* **2001**, *2*, 99–108. [CrossRef] [PubMed]
82. Wilson, S.W.; Ross, L.S.; Parrett, T.; Easter, S.S., Jr. The development of a simple scaffold of axon tracts in the brain of the embryonic zebrafish, *Brachydanio rerio*. *Development* **1990**, *108*, 121–145. [PubMed]
83. Ross, L.S.; Parrett, T.; Easter, S.S., Jr. Axonogenesis and morphogenesis in the embryonic zebrafish brain. *J. Neurosci.* **1992**, *12*, 467–482. [CrossRef] [PubMed]
84. Golden, G.S. A review of the neuroembryology of monoamine systems. *Brain Res. Bull.* **1982**, *9*, 553–558. [CrossRef]

85. Bonnin, A.; Levitt, P. Fetal, maternal, and placental sources of serotonin and new implications for developmental programming of the brain. *Neuroscience* **2011**, *197*, 1–7. [CrossRef] [PubMed]
86. McLean, D.L.; Fetcho, J.R. Relationship of tyrosine hydroxylase and serotonin immunoreactivity to sensorimotor circuitry in larval zebrafish. *J. Comp. Neurol.* **2004**, *480*, 57–71. [CrossRef] [PubMed]
87. Guo, S.; Wilson, S.W.; Cooke, S.; Chitnis, A.B.; Driever, W.; Rosenthal, A. Mutations in the zebrafish unmask shared regulatory pathways controlling the development of catecholaminergic neurons. *Dev. Biol.* **1999**, *208*, 473–487. [CrossRef] [PubMed]
88. Holzschuh, J.; Ryu, S.; Aberger, F.; Driever, W. Dopamine transporter expression distinguishes dopaminergic neurons from other catecholaminergic neurons in the developing zebrafish embryo. *Mech. Dev.* **2001**, *101*, 237–243. [CrossRef]
89. Bae, Y.K.; Kani, S.; Shimizu, T.; Tanabe, K.; Nojima, H.; Kimura, Y.; Higashijima, S.; Hibi, M. Anatomy of zebrafish cerebellum and screen for mutations affecting its development. *Dev. Biol.* **2009**, *330*, 406–426. [CrossRef] [PubMed]
90. Chitnis, A.B.; Kuwada, J.Y. Axonogenesis in the brain of zebrafish embryos. *J. Neurosci.* **1990**, *10*, 1892–1905. [CrossRef] [PubMed]
91. Mueller, T. What is the Thalamus in Zebrafish? *Front. Neurosci.* **2012**, *6*, 64. [CrossRef] [PubMed]
92. Laessing, U.; Stuermer, C.A. Spatiotemporal pattern of retinal ganglion cell differentiation revealed by the expression of neurolin in embryonic zebrafish. *J. Neurobiol.* **1996**, *29*, 65–74. [CrossRef]
93. Sakata, Y.; Olson, J.K.; Michel, W.C. Assessment of neuronal maturation and acquisition of functional competence in the developing zebrafish olfactory system. *Methods Cell Sci.* **2003**, *25*, 39–48. [CrossRef] [PubMed]
94. Mueller, T.; Vernier, P.; Wullimann, M.F. A phylotypic stage in vertebrate brain development: GABA cell patterns in zebrafish compared with mouse. *J. Comp. Neurol.* **2006**, *494*, 620–634. [CrossRef] [PubMed]
95. Kimmel, C.B.; Patterson, J.; Kimmel, R.O. The development and behavioral characteristics of the startle response in the zebra fish. *Dev. Psychobiol.* **1974**, *7*, 47–60. [CrossRef] [PubMed]
96. Lumsden, A.; Krumlauf, R. Patterning the vertebrate neuraxis. *Science* **1996**, *274*, 1109–1115. [CrossRef] [PubMed]
97. Schmidt, R.; Strahle, U.; Scholpp, S. Neurogenesis in zebrafish—From embryo to adult. *Neural Dev.* **2013**, *8*, 3. [CrossRef] [PubMed]
98. Strahle, U.; Blader, P. Early neurogenesis in the zebrafish embryo. *FASEB J.* **1994**, *8*, 692–698. [CrossRef] [PubMed]
99. Folgueira, M.; Bayley, P.; Navratilova, P.; Becker, T.S.; Wilson, S.W.; Clarke, J.D. Morphogenesis underlying the development of the everted teleost telencephalon. *Neural Dev.* **2012**, *7*, 32. [CrossRef] [PubMed]
100. Mueller, T.; Wullimann, M.F. An evolutionary interpretation of teleostean forebrain anatomy. *Brain Behav. Evol.* **2009**, *74*, 30–42. [CrossRef] [PubMed]
101. Nieuwenhuys, R. The forebrain of actinopterygians revisited. *Brain Behav. Evol.* **2009**, *73*, 229–252. [CrossRef] [PubMed]
102. Ganz, J.; Kroehne, V.; Freudenreich, D.; Machate, A.; Geffarth, M.; Braasch, I.; Kaslin, J.; Brand, M. Subdivisions of the adult zebrafish pallium based on molecular marker analysis. *F1000Res* **2014**, *3*, 308. [CrossRef] [PubMed]
103. Portavella, M.; Torres, B.; Salas, C.; Papini, M.R. Lesions of the medial pallium, but not of the lateral pallium, disrupt spaced-trial avoidance learning in goldfish (*Carassius auratus*). *Neurosci. Lett.* **2004**, *362*, 75–78. [CrossRef] [PubMed]
104. Lal, P.; Tanabe, H.; Suster, M.L.; Ailani, D.; Kotani, Y.; Muto, A.; Itoh, M.; Iwasaki, M.; Wada, H.; Yaksi, E.; et al. Identification of a neuronal population in the telencephalon essential for fear conditioning in zebrafish. *BMC Biol.* **2018**, *16*, 45. [CrossRef] [PubMed]
105. Broglio, C.; Gomez, A.; Duran, E.; Ocana, F.M.; Jimenez-Moya, F.; Rodriguez, F.; Salas, C. Hallmarks of a common forebrain vertebrate plan: Specialized pallial areas for spatial, temporal and emotional memory in actinopterygian fish. *Brain Res. Bull.* **2005**, *66*, 277–281. [CrossRef] [PubMed]
106. Broglio, C.; Rodriguez, F.; Gomez, A.; Arias, J.L.; Salas, C. Selective involvement of the goldfish lateral pallium in spatial memory. *Behav. Brain Res.* **2010**, *210*, 191–201. [CrossRef] [PubMed]

107. Stewart, A.M.; Ullmann, J.F.; Norton, W.H.; Parker, M.O.; Brennan, C.H.; Gerlai, R.; Kalueff, A.V. Molecular psychiatry of zebrafish. *Mol. Psychiatry* **2015**, *20*, 2–17. [CrossRef] [PubMed]

108. Norton, W.H. Toward developmental models of psychiatric disorders in zebrafish. *Front. Neural Circuits* **2013**, *7*, 79. [CrossRef] [PubMed]

109. Wullimann, M.F.; Mueller, T.; Distel, M.; Babaryka, A.; Grothe, B.; Koster, R.W. The long adventurous journey of rhombic lip cells in jawed vertebrates: A comparative developmental analysis. *Front. Neuroanat* **2011**, *5*, 27. [CrossRef] [PubMed]

110. Takeuchi, M.; Yamaguchi, S.; Sakakibara, Y.; Hayashi, T.; Matsuda, K.; Hara, Y.; Tanegashima, C.; Shimizu, T.; Kuraku, S.; Hibi, M. Gene expression profiling of granule cells and Purkinje cells in the zebrafish cerebellum. *J. Comp. Neurol.* **2017**, *525*, 1558–1585. [CrossRef] [PubMed]

111. Matsui, H.; Namikawa, K.; Babaryka, A.; Koster, R.W. Functional regionalization of the teleost cerebellum analyzed in vivo. *Proc. Natl. Acad. Sci. USA* **2014**, *111*, 11846–11851. [CrossRef] [PubMed]

112. Sillitoe, R.V.; Joyner, A.L. Morphology, molecular codes, and circuitry produce the three-dimensional complexity of the cerebellum. *Annu. Rev. Cell Dev. Biol.* **2007**, *23*, 549–577. [CrossRef] [PubMed]

113. Kani, S.; Bae, Y.K.; Shimizu, T.; Tanabe, K.; Satou, C.; Parsons, M.J.; Scott, E.; Higashijima, S.; Hibi, M. Proneural gene-linked neurogenesis in zebrafish cerebellum. *Dev. Biol.* **2010**, *343*, 1–17. [CrossRef] [PubMed]

114. Hibi, M.; Matsuda, K.; Takeuchi, M.; Shimizu, T.; Murakami, Y. Evolutionary mechanisms that generate morphology and neural-circuit diversity of the cerebellum. *Dev. Growth Differ.* **2017**, *59*, 228–243. [CrossRef] [PubMed]

115. Sylvester, S.J.G.; Lee, M.M.; Ramirez, A.D.; Lim, S.; Goldman, M.S.; Aksay, E.R.F. Population-scale organization of cerebellar granule neuron signaling during a visuomotor behavior. *Sci. Rep.* **2017**, *7*, 16240. [CrossRef] [PubMed]

116. Severi, K.E.; Bohm, U.L.; Wyart, C. Investigation of hindbrain activity during active locomotion reveals inhibitory neurons involved in sensorimotor processing. *Sci. Rep.* **2018**, *8*, 13615. [CrossRef] [PubMed]

117. Kheradmand, A.; Zee, D.S. Cerebellum and ocular motor control. *Front. Neurol.* **2011**, *2*, 53. [CrossRef] [PubMed]

118. Schmitt, L.M.; Cook, E.H.; Sweeney, J.A.; Mosconi, M.W. Saccadic eye movement abnormalities in autism spectrum disorder indicate dysfunctions in cerebellum and brainstem. *Mol. Autism* **2014**, *5*, 47. [CrossRef] [PubMed]

119. Hsieh, J.Y.; Ulrich, B.; Issa, F.A.; Wan, J.; Papazian, D.M. Rapid development of Purkinje cell excitability, functional cerebellar circuit, and afferent sensory input to cerebellum in zebrafish. *Front. Neural Circuits* **2014**, *8*, 147. [CrossRef] [PubMed]

120. McKay, B.E.; Turner, R.W. Physiological and morphological development of the rat cerebellar Purkinje cell. *J. Physiol.* **2005**, *567*, 829–850. [CrossRef] [PubMed]

121. Tropepe, V.; Sive, H.L. Can zebrafish be used as a model to study the neurodevelopmental causes of autism? *Genes Brain Behav.* **2003**, *2*, 268–281. [CrossRef] [PubMed]

122. De Rienzo, G.; Bishop, J.A.; Mao, Y.; Pan, L.; Ma, T.P.; Moens, C.B.; Tsai, L.H.; Sive, H. Disc1 regulates both beta-catenin-mediated and noncanonical Wnt signaling during vertebrate embryogenesis. *FASEB J.* **2011**, *25*, 4184–4197. [CrossRef] [PubMed]

123. Kalueff, A.V.; Stewart, A.M.; Gerlai, R. Zebrafish as an emerging model for studying complex brain disorders. *Trends Pharmacol. Sci.* **2014**, *35*, 63–75. [CrossRef] [PubMed]

124. Ijaz, S.; Hoffman, E.J. Zebrafish: A Translational Model System for Studying Neuropsychiatric Disorders. *J. Am. Acad. Child. Adolesc. Psychiatry* **2016**, *55*, 746–748. [CrossRef] [PubMed]

125. Hoffman, E.J.; Turner, K.J.; Fernandez, J.M.; Cifuentes, D.; Ghosh, M.; Ijaz, S.; Jain, R.A.; Kubo, F.; Bill, B.R.; Baier, H.; et al. Estrogens Suppress a Behavioral Phenotype in Zebrafish Mutants of the Autism Risk Gene, CNTNAP2. *Neuron* **2016**, *89*, 725–733. [CrossRef] [PubMed]

126. Kozol, R.A.; Cukier, H.N.; Zou, B.; Mayo, V.; De Rubeis, S.; Cai, G.; Griswold, A.J.; Whitehead, P.L.; Haines, J.L.; Gilbert, J.R.; et al. Two knockdown models of the autism genes SYNGAP1 and SHANK3 in zebrafish produce similar behavioral phenotypes associated with embryonic disruptions of brain morphogenesis. *Hum. Mol. Genet.* **2015**, *24*, 4006–4023. [CrossRef] [PubMed]

127. Blaker-Lee, A.; Gupta, S.; McCammon, J.M.; De Rienzo, G.; Sive, H. Zebrafish homologs of genes within

16p11.2, a genomic region associated with brain disorders, are active during brain development, and include two deletion dosage sensor genes. *Dis. Model Mech.* **2012**, *5*, 834–851. [CrossRef] [PubMed]
128. Elsen, G.E.; Choi, L.Y.; Prince, V.E.; Ho, R.K. The autism susceptibility gene met regulates zebrafish cerebellar development and facial motor neuron migration. *Dev. Biol.* **2009**, *335*, 78–92. [CrossRef] [PubMed]
129. Psychiatric, G.C.C.C.; Cichon, S.; Craddock, N.; Daly, M.; Faraone, S.V.; Gejman, P.V.; Kelsoe, J.; Lehner, T.; Levinson, D.F.; Moran, A.; et al. Genomewide association studies: History, rationale, and prospects for psychiatric disorders. *Am. J. Psychiatry* **2009**, *166*, 540–556.
130. Volkmar, F.R.; McPartland, J.C. From Kanner to DSM-5: Autism as an evolving diagnostic concept. *Annu. Rev. Clin. Psychol.* **2014**, *10*, 193–212. [CrossRef] [PubMed]
131. DiBella, L.M.; Park, A.; Sun, Z. Zebrafish Tsc1 reveals functional interactions between the cilium and the TOR pathway. *Hum. Mol. Genet.* **2009**, *18*, 595–606. [CrossRef] [PubMed]
132. Tucker, B.; Richards, R.I.; Lardelli, M. Contribution of mGluR and Fmr1 functional pathways to neurite morphogenesis, craniofacial development and fragile X syndrome. *Hum. Mol. Genet.* **2006**, *15*, 3446–3458. [CrossRef] [PubMed]
133. Gauthier, J.; Champagne, N.; Lafreniere, R.G.; Xiong, L.; Spiegelman, D.; Brustein, E.; Lapointe, M.; Peng, H.; Cote, M.; Noreau, A.; et al. De novo mutations in the gene encoding the synaptic scaffolding protein SHANK3 in patients ascertained for schizophrenia. *Proc. Natl. Acad. Sci. USA* **2010**, *107*, 7863–7868. [CrossRef] [PubMed]
134. Golzio, C.; Willer, J.; Talkowski, M.E.; Oh, E.C.; Taniguchi, Y.; Jacquemont, S.; Reymond, A.; Sun, M.; Sawa, A.; Gusella, J.F.; et al. KCTD13 is a major driver of mirrored neuroanatomical phenotypes of the 16p11.2 copy number variant. *Nature* **2012**, *485*, 363–367. [CrossRef] [PubMed]
135. Eisen, J.S.; Smith, J.C. Controlling morpholino experiments: Don't stop making antisense. *Development* **2008**, *135*, 1735–1743. [CrossRef] [PubMed]
136. Rossi, A.; Kontarakis, Z.; Gerri, C.; Nolte, H.; Holper, S.; Kruger, M.; Stainier, D.Y. Genetic compensation induced by deleterious mutations but not gene knockdowns. *Nature* **2015**, *524*, 230–233. [CrossRef] [PubMed]
137. Ben-Ari, Y.; Spitzer, N.C. Phenotypic checkpoints regulate neuronal development. *Trends Neurosci.* **2010**, *33*, 485–492. [CrossRef] [PubMed]
138. Lubs, H.A. A marker X chromosome. *Am. J. Hum. Genet.* **1969**, *21*, 231–244. [PubMed]
139. Lejeune, J.; Maunoury, C.; Prieur, M.; Van den Akker, J. A jumping translocation (5p;15q), (8q;15q), and (12q;15q) (author's transl). *Ann. Genet.* **1979**, *22*, 210–213. [PubMed]
140. Archidiacono, N.; Rett, A.; Rocchi, M.; Rolando, S.; Lugaresi, E.; Romeo, G. Rett syndrome and fragile site in Xp22. *Lancet* **1985**, *2*, 1242–1243. [CrossRef]
141. Fryer, A.E.; Chalmers, A.; Connor, J.M.; Fraser, I.; Povey, S.; Yates, A.D.; Yates, J.R.; Osborne, J.P. Evidence that the gene for tuberous sclerosis is on chromosome 9. *Lancet* **1987**, *1*, 659–661. [CrossRef]
142. Kandt, R.S.; Haines, J.L.; Smith, M.; Northrup, H.; Gardner, R.J.; Short, M.P.; Dumars, K.; Roach, E.S.; Steingold, S.; Wall, S.; et al. Linkage of an important gene locus for tuberous sclerosis to a chromosome 16 marker for polycystic kidney disease. *Nat. Genet.* **1992**, *2*, 37–41. [CrossRef] [PubMed]
143. Vincent, J.B.; Konecki, D.S.; Munstermann, E.; Bolton, P.; Poustka, A.; Poustka, F.; Gurling, H.M. Point mutation analysis of the FMR-1 gene in autism. *Mol. Psychiatry* **1996**, *1*, 227–231. [PubMed]
144. van Slegtenhorst, M.; de Hoogt, R.; Hermans, C.; Nellist, M.; Janssen, B.; Verhoef, S.; Lindhout, D.; van den Ouweland, A.; Halley, D.; Young, J.; et al. Identification of the tuberous sclerosis gene TSC1 on chromosome 9q34. *Science* **1997**, *277*, 805–808. [CrossRef] [PubMed]
145. van Slegtenhorst, M.; Nellist, M.; Nagelkerken, B.; Cheadle, J.; Snell, R.; van den Ouweland, A.; Reuser, A.; Sampson, J.; Halley, D.; van der Sluijs, P. Interaction between hamartin and tuberin, the TSC1 and TSC2 gene products. *Hum. Mol. Genet.* **1998**, *7*, 1053–1057. [CrossRef] [PubMed]
146. Persico, A.M.; D'Agruma, L.; Maiorano, N.; Totaro, A.; Militerni, R.; Bravaccio, C.; Wassink, T.H.; Schneider, C.; Melmed, R.; Trillo, S.; et al. Reelin gene alleles and haplotypes as a factor predisposing to autistic disorder. *Mol. Psychiatry* **2001**, *6*, 150–159. [CrossRef] [PubMed]
147. Weiss, L.A.; Escayg, A.; Kearney, J.A.; Trudeau, M.; MacDonald, B.T.; Mori, M.; Reichert, J.; Buxbaum, J.D.; Meisler, M.H. Sodium channels SCN1A, SCN2A and SCN3A in familial autism. *Mol. Psychiatry* **2003**, *8*, 186–194. [CrossRef] [PubMed]
148. Butler, M.G.; Dasouki, M.J.; Zhou, X.P.; Talebizadeh, Z.; Brown, M.; Takahashi, T.N.; Miles, J.H.; Wang, C.H.;

Stratton, R.; Pilarski, R.; et al. Subset of individuals with autism spectrum disorders and extreme macrocephaly associated with germline PTEN tumour suppressor gene mutations. *J. Med. Genet.* **2005**, *42*, 318–321. [CrossRef] [PubMed]

149. Bonaglia, M.C.; Giorda, R.; Mani, E.; Aceti, G.; Anderlid, B.M.; Baroncini, A.; Pramparo, T.; Zuffardi, O. Identification of a recurrent breakpoint within the SHANK3 gene in the 22q13.3 deletion syndrome. *J. Med. Genet.* **2006**, *43*, 822–828. [CrossRef] [PubMed]
150. Vrijenhoek, T.; Buizer-Voskamp, J.E.; van der Stelt, I.; Strengman, E.; Genetic, R.; Outcome in Psychosis, C.; Sabatti, C.; Geurts van Kessel, A.; Brunner, H.G.; Ophoff, R.A.; et al. Recurrent CNVs disrupt three candidate genes in schizophrenia patients. *Am. J. Hum. Genet.* **2008**, *83*, 504–510. [CrossRef] [PubMed]
151. Nord, A.S.; Roeb, W.; Dickel, D.E.; Walsh, T.; Kusenda, M.; O'Connor, K.L.; Malhotra, D.; McCarthy, S.E.; Stray, S.M.; Taylor, S.M.; et al. Reduced transcript expression of genes affected by inherited and de novo CNVs in autism. *Eur. J. Hum. Genet.* **2011**, *19*, 727–731. [CrossRef] [PubMed]
152. O'Roak, B.J.; Deriziotis, P.; Lee, C.; Vives, L.; Schwartz, J.J.; Girirajan, S.; Karakoc, E.; Mackenzie, A.P.; Ng, S.B.; Baker, C.; et al. Exome sequencing in sporadic autism spectrum disorders identifies severe de novo mutations. *Nat. Genet.* **2011**, *43*, 585–589. [CrossRef] [PubMed]
153. Iossifov, I.; Ronemus, M.; Levy, D.; Wang, Z.; Hakker, I.; Rosenbaum, J.; Yamrom, B.; Lee, Y.H.; Narzisi, G.; Leotta, A.; et al. De novo gene disruptions in children on the autistic spectrum. *Neuron* **2012**, *74*, 285–299. [CrossRef] [PubMed]
154. Neale, B.M.; Kou, Y.; Liu, L.; Ma'ayan, A.; Samocha, K.E.; Sabo, A.; Lin, C.F.; Stevens, C.; Wang, L.S.; Makarov, V.; et al. Patterns and rates of exonic de novo mutations in autism spectrum disorders. *Nature* **2012**, *485*, 242–245. [CrossRef] [PubMed]
155. Chahrour, M.H.; Yu, T.W.; Lim, E.T.; Ataman, B.; Coulter, M.E.; Hill, R.S.; Stevens, C.R.; Schubert, C.R.; Collaboration, A.A.S.; Greenberg, M.E.; et al. Whole-exome sequencing and homozygosity analysis implicate depolarization-regulated neuronal genes in autism. *PLoS Genet.* **2012**, *8*, e1002635. [CrossRef] [PubMed]
156. Dinwiddie, D.L.; Soden, S.E.; Saunders, C.J.; Miller, N.A.; Farrow, E.G.; Smith, L.D.; Kingsmore, S.F. De novo frameshift mutation in ASXL3 in a patient with global developmental delay, microcephaly, and craniofacial anomalies. *BMC Med. Genom.* **2013**, *6*, 32. [CrossRef] [PubMed]
157. Iossifov, I.; O'Roak, B.J.; Sanders, S.J.; Ronemus, M.; Krumm, N.; Levy, D.; Stessman, H.A.; Witherspoon, K.T.; Vives, L.; Patterson, K.E.; et al. The contribution of de novo coding mutations to autism spectrum disorder. *Nature* **2014**, *515*, 216–221. [CrossRef] [PubMed]
158. Beunders, G.; Voorhoeve, E.; Golzio, C.; Pardo, L.M.; Rosenfeld, J.A.; Talkowski, M.E.; Simonic, I.; Lionel, A.C.; Vergult, S.; Pyatt, R.E.; et al. Exonic deletions in AUTS2 cause a syndromic form of intellectual disability and suggest a critical role for the C terminus. *Am. J. Hum. Genet.* **2013**, *92*, 210–220. [CrossRef] [PubMed]
159. Suls, A.; Jaehn, J.A.; Kecskes, A.; Weber, Y.; Weckhuysen, S.; Craiu, D.C.; Siekierska, A.; Djemie, T.; Afrikanova, T.; Gormley, P.; et al. De novo loss-of-function mutations in CHD2 cause a fever-sensitive myoclonic epileptic encephalopathy sharing features with Dravet syndrome. *Am. J. Hum. Genet.* **2013**, *93*, 967–975. [CrossRef] [PubMed]
160. Bernier, R.; Golzio, C.; Xiong, B.; Stessman, H.A.; Coe, B.P.; Penn, O.; Witherspoon, K.; Gerdts, J.; Baker, C.; Vulto-van Silfhout, A.T.; et al. Disruptive CHD8 mutations define a subtype of autism early in development. *Cell* **2014**, *158*, 263–276. [CrossRef] [PubMed]
161. Hofmeister, W.; Nilsson, D.; Topa, A.; Anderlid, B.M.; Darki, F.; Matsson, H.; Tapia Paez, I.; Klingberg, T.; Samuelsson, L.; Wirta, V.; et al. CTNND2-a candidate gene for reading problems and mild intellectual disability. *J. Med. Genet.* **2015**, *52*, 111–122. [CrossRef] [PubMed]
162. Liu, C.X.; Li, C.Y.; Hu, C.C.; Wang, Y.; Lin, J.; Jiang, Y.H.; Li, Q.; Xu, X. CRISPR/Cas9-induced shank3b mutant zebrafish display autism-like behaviors. *Mol. Autism* **2018**, *9*, 23. [CrossRef] [PubMed]
163. Warga, R.M.; Kimmel, C.B. Cell movements during epiboly and gastrulation in zebrafish. *Development* **1990**, *108*, 569–580. [PubMed]
164. Ho, R.K.; Kimmel, C.B. Commitment of cell fate in the early zebrafish embryo. *Science* **1993**, *261*, 109–111. [CrossRef] [PubMed]
165. Feldman, B.; Gates, M.A.; Egan, E.S.; Dougan, S.T.; Rennebeck, G.; Sirotkin, H.I.; Schier, A.F.; Talbot, W.S. Zebrafish organizer development and germ-layer formation require nodal-related signals. *Nature* **1998**, *395*, 181–185. [CrossRef] [PubMed]

166. Bachiller, D.; Klingensmith, J.; Kemp, C.; Belo, J.A.; Anderson, R.M.; May, S.R.; McMahon, J.A.; McMahon, A.P.; Harland, R.M.; Rossant, J.; et al. The organizer factors Chordin and Noggin are required for mouse forebrain development. *Nature* **2000**, *403*, 658–661. [CrossRef] [PubMed]

167. Houart, C.; Westerfield, M.; Wilson, S.W. A small population of anterior cells patterns the forebrain during zebrafish gastrulation. *Nature* **1998**, *391*, 788–792. [CrossRef] [PubMed]

168. Lun, K.; Brand, M. A series of no isthmus (noi) alleles of the zebrafish pax2.1 gene reveals multiple signaling events in development of the midbrain-hindbrain boundary. *Development* **1998**, *125*, 3049–3062. [PubMed]

169. Grinblat, Y.; Gamse, J.; Patel, M.; Sive, H. Determination of the zebrafish forebrain: Induction and patterning. *Development* **1998**, *125*, 4403–4416. [PubMed]

170. Brand, M.; Heisenberg, C.P.; Jiang, Y.J.; Beuchle, D.; Lun, K.; Furutani-Seiki, M.; Granato, M.; Haffter, P.; Hammerschmidt, M.; Kane, D.A.; et al. Mutations in zebrafish genes affecting the formation of the boundary between midbrain and hindbrain. *Development* **1996**, *123*, 179–190. [PubMed]

171. Piccolo, S.; Sasai, Y.; Lu, B.; De Robertis, E.M. Dorsoventral patterning in Xenopus: Inhibition of ventral signals by direct binding of chordin to BMP-4. *Cell* **1996**, *86*, 589–598. [CrossRef]

172. Turner, T.N.; Sharma, K.; Oh, E.C.; Liu, Y.P.; Collins, R.L.; Sosa, M.X.; Auer, D.R.; Brand, H.; Sanders, S.J.; Moreno-De-Luca, D.; et al. Loss of delta-catenin function in severe autism. *Nature* **2015**, *520*, 51–56. [CrossRef] [PubMed]

173. Itoh, K.; Sokol, S.Y. Graded amounts of Xenopus dishevelled specify discrete anteroposterior cell fates in prospective ectoderm. *Mech. Dev.* **1997**, *61*, 113–125. [CrossRef]

174. Backman, M.; Machon, O.; Mygland, L.; van den Bout, C.J.; Zhong, W.; Taketo, M.M.; Krauss, S. Effects of canonical Wnt signaling on dorso-ventral specification of the mouse telencephalon. *Dev. Biol.* **2005**, *279*, 155–168. [CrossRef] [PubMed]

175. Sugathan, A.; Biagioli, M.; Golzio, C.; Erdin, S.; Blumenthal, I.; Manavalan, P.; Ragavendran, A.; Brand, H.; Lucente, D.; Miles, J.; et al. CHD8 regulates neurodevelopmental pathways associated with autism spectrum disorder in neural progenitors. *Proc. Natl. Acad. Sci. USA* **2014**, *111*, E4468–4477. [CrossRef] [PubMed]

176. Oksenberg, N.; Stevison, L.; Wall, J.D.; Ahituv, N. Function and regulation of AUTS2, a gene implicated in autism and human evolution. *PLoS Genet.* **2013**, *9*, e1003221. [CrossRef] [PubMed]

177. Kathuria, A.; Nowosiad, P.; Jagasia, R.; Aigner, S.; Taylor, R.D.; Andreae, L.C.; Gatford, N.J.F.; Lucchesi, W.; Srivastava, D.P.; Price, J. Stem cell-derived neurons from autistic individuals with SHANK3 mutation show morphogenetic abnormalities during early development. *Mol. Psychiatry* **2018**, *23*, 735–746. [CrossRef] [PubMed]

178. McCammon, J.M.; Blaker-Lee, A.; Chen, X.; Sive, H. The 16p11.2 homologs fam57ba and doc2a generate certain brain and body phenotypes. *Hum. Mol. Genet.* **2017**, *26*, 3699–3712. [CrossRef] [PubMed]

179. Escamilla, C.O.; Filonova, I.; Walker, A.K.; Xuan, Z.X.; Holehonnur, R.; Espinosa, F.; Liu, S.; Thyme, S.B.; Lopez-Garcia, I.A.; Mendoza, D.B.; et al. Kctd13 deletion reduces synaptic transmission via increased RhoA. *Nature* **2017**, *551*, 227–231. [CrossRef] [PubMed]

180. Kuwada, J.Y.; Bernhardt, R.R. Axonal outgrowth by identified neurons in the spinal cord of zebrafish embryos. *Exp. Neurol.* **1990**, *109*, 29–34. [CrossRef]

181. Metcalfe, W.K.; Myers, P.Z.; Trevarrow, B.; Bass, M.B.; Kimmel, C.B. Primary neurons that express the L2/HNK-1 carbohydrate during early development in the zebrafish. *Development* **1990**, *110*, 491–504. [PubMed]

182. Borodinsky, L.N.; Root, C.M.; Cronin, J.A.; Sann, S.B.; Gu, H.L.; Spitzer, N.C. Activity-dependent homeostatic specification of transmitter expression in embryonic neurons. *Nature* **2004**, *429*, 523–530. [CrossRef] [PubMed]

183. Spitzer, N.C. Electrical activity in early neuronal development. *Nature* **2006**, *444*, 707–712. [CrossRef] [PubMed]

184. Gatto, C.L.; Broadie, K. Genetic controls balancing excitatory and inhibitory synaptogenesis in neurodevelopmental disorder models. *Front. Synaptic Neurosci.* **2010**, *2*, 4. [CrossRef] [PubMed]

185. Nelson, S.B.; Valakh, V. Excitatory/Inhibitory Balance and Circuit Homeostasis in Autism Spectrum Disorders. *Neuron* **2015**, *87*, 684–698. [CrossRef] [PubMed]

186. Wu, Q.F.; Yang, L.; Li, S.; Wang, Q.; Yuan, X.B.; Gao, X.; Bao, L.; Zhang, X. Fibroblast growth factor 13 is a microtubule-stabilizing protein regulating neuronal polarization and migration. *Cell* **2012**, *149*, 1549–1564. [CrossRef] [PubMed]

187. Singh, K.K.; De Rienzo, G.; Drane, L.; Mao, Y.; Flood, Z.; Madison, J.; Ferreira, M.; Bergen, S.; King, C.; Sklar, P.; et al. Common DISC1 polymorphisms disrupt Wnt/GSK3beta signaling and brain development. *Neuron* **2011**, *72*, 545–558. [CrossRef] [PubMed]

188. Reijnders, M.R.F.; Ansor, N.M.; Kousi, M.; Yue, W.W.; Tan, P.L.; Clarkson, K.; Clayton-Smith, J.; Corning, K.; Jones, J.R.; Lam, W.W.K.; et al. RAC1 Missense Mutations in Developmental Disorders with Diverse Phenotypes. *Am. J. Hum. Genet.* **2017**, *101*, 466–477. [CrossRef] [PubMed]

189. Peca, J.; Feliciano, C.; Ting, J.T.; Wang, W.; Wells, M.F.; Venkatraman, T.N.; Lascola, C.D.; Fu, Z.; Feng, G. Shank3 mutant mice display autistic-like behaviours and striatal dysfunction. *Nature* **2011**, *472*, 437–442. [CrossRef] [PubMed]

190. Clement, J.P.; Aceti, M.; Creson, T.K.; Ozkan, E.D.; Shi, Y.; Reish, N.J.; Almonte, A.G.; Miller, B.H.; Wiltgen, B.J.; Miller, C.A.; et al. Pathogenic SYNGAP1 mutations impair cognitive development by disrupting maturation of dendritic spine synapses. *Cell* **2012**, *151*, 709–723. [CrossRef] [PubMed]

191. Golden, C.E.; Buxbaum, J.D.; De Rubeis, S. Disrupted circuits in mouse models of autism spectrum disorder and intellectual disability. *Curr. Opin. Neurobiol.* **2018**, *48*, 106–112. [CrossRef] [PubMed]

192. Dutton, S.B.; Makinson, C.D.; Papale, L.A.; Shankar, A.; Balakrishnan, B.; Nakazawa, K.; Escayg, A. Preferential inactivation of Scn1a in parvalbumin interneurons increases seizure susceptibility. *Neurobiol. Dis.* **2013**, *49*, 211–220. [CrossRef] [PubMed]

193. Baraban, S.C.; Dinday, M.T.; Hortopan, G.A. Drug screening in Scn1a zebrafish mutant identifies clemizole as a potential Dravet syndrome treatment. *Nat. Commun.* **2013**, *4*, 2410. [CrossRef] [PubMed]

194. Herget, U.; Gutierrez-Triana, J.A.; Salazar Thula, O.; Knerr, B.; Ryu, S. Single-Cell Reconstruction of Oxytocinergic Neurons Reveals Separate Hypophysiotropic and Encephalotropic Subtypes in Larval Zebrafish. *eNeuro* **2017**, *4*. [CrossRef] [PubMed]

195. Herget, U.; Wolf, A.; Wullimann, M.F.; Ryu, S. Molecular neuroanatomy and chemoarchitecture of the neurosecretory preoptic-hypothalamic area in zebrafish larvae. *J. Comp. Neurol.* **2014**, *522*, 1542–1564. [CrossRef] [PubMed]

196. Filippi, A.; Mueller, T.; Driever, W. vglut2 and gad expression reveal distinct patterns of dual GABAergic versus glutamatergic cotransmitter phenotypes of dopaminergic and noradrenergic neurons in the zebrafish brain. *J. Comp. Neurol.* **2014**, *522*, 2019–2037. [CrossRef] [PubMed]

197. Haehnel-Taguchi, M.; Fernandes, A.M.; Bohler, M.; Schmitt, I.; Tittel, L.; Driever, W. Projections of the Diencephalospinal Dopaminergic System to Peripheral Sense Organs in Larval Zebrafish (*Danio rerio*). *Front. Neuroanat.* **2018**, *12*, 20. [CrossRef] [PubMed]

198. Braasch, I.; Gehrke, A.R.; Smith, J.J.; Kawasaki, K.; Manousaki, T.; Pasquier, J.; Amores, A.; Desvignes, T.; Batzel, P.; Catchen, J.; et al. The spotted gar genome illuminates vertebrate evolution and facilitates human-teleost comparisons. *Nat. Genet.* **2016**, *48*, 427–437. [CrossRef] [PubMed]

199. Stainier, D.Y.; Kontarakis, Z.; Rossi, A. Making sense of anti-sense data. *Dev. Cell* **2015**, *32*, 7–8. [CrossRef] [PubMed]

200. Howe, K.; Clark, M.D.; Torroja, C.F.; Torrance, J.; Berthelot, C.; Muffato, M.; Collins, J.E.; Humphray, S.; McLaren, K.; Matthews, L.; et al. The zebrafish reference genome sequence and its relationship to the human genome. *Nature* **2013**, *496*, 498–503. [CrossRef] [PubMed]

201. Postlethwait, J.; Amores, A.; Cresko, W.; Singer, A.; Yan, Y.L. Subfunction partitioning, the teleost radiation and the annotation of the human genome. *Trends Genet.* **2004**, *20*, 481–490. [CrossRef] [PubMed]

202. Catchen, J.M.; Braasch, I.; Postlethwait, J.H. Conserved synteny and the zebrafish genome. *Methods Cell. Biol.* **2011**, *104*, 259–285. [PubMed]

203. Pasquier, J.; Braasch, I.; Batzel, P.; Cabau, C.; Montfort, J.; Nguyen, T.; Jouanno, E.; Berthelot, C.; Klopp, C.; Journot, L.; et al. Evolution of gene expression after whole-genome duplication: New insights from the spotted gar genome. *J. Exp. Zool. B Mol. Dev. Evol.* **2017**, *328*, 709–721. [CrossRef] [PubMed]

204. McWhorter, M.L.; Monani, U.R.; Burghes, A.H.; Beattie, C.E. Knockdown of the survival motor neuron (Smn) protein in zebrafish causes defects in motor axon outgrowth and pathfinding. *J. Cell Biol.* **2003**, *162*, 919–931. [CrossRef] [PubMed]

205. Shepard, J.L.; Amatruda, J.F.; Stern, H.M.; Subramanian, A.; Finkelstein, D.; Ziai, J.; Finley, K.R.; Pfaff, K.L.; Hersey, C.; Zhou, Y.; et al. A zebrafish bmyb mutation causes genome instability and increased cancer susceptibility. *Proc. Natl. Acad. Sci. USA* **2005**, *102*, 13194–13199. [CrossRef] [PubMed]

206. Jeong, J.Y.; Einhorn, Z.; Mercurio, S.; Lee, S.; Lau, B.; Mione, M.; Wilson, S.W.; Guo, S. Neurogenin1 is a determinant of zebrafish basal forebrain dopaminergic neurons and is regulated by the conserved zinc finger protein Tof/Fezl. *Proc. Natl. Acad. Sci. USA* **2006**, *103*, 5143–5148. [CrossRef] [PubMed]
207. Omran, H.; Kobayashi, D.; Olbrich, H.; Tsukahara, T.; Loges, N.T.; Hagiwara, H.; Zhang, Q.; Leblond, G.; O'Toole, E.; Hara, C.; et al. Ktu/PF13 is required for cytoplasmic pre-assembly of axonemal dyneins. *Nature* **2008**, *456*, 611–616. [CrossRef] [PubMed]
208. Simons, M.; Gloy, J.; Ganner, A.; Bullerkotte, A.; Bashkurov, M.; Kronig, C.; Schermer, B.; Benzing, T.; Cabello, O.A.; Jenny, A.; et al. Inversin, the gene product mutated in nephronophthisis type II, functions as a molecular switch between Wnt signaling pathways. *Nat. Genet.* **2005**, *37*, 537–543. [CrossRef] [PubMed]
209. Kok, F.O.; Shin, M.; Ni, C.W.; Gupta, A.; Grosse, A.S.; van Impel, A.; Kirchmaier, B.C.; Peterson-Maduro, J.; Kourkoulis, G.; Male, I.; et al. Reverse genetic screening reveals poor correlation between morpholino-induced and mutant phenotypes in zebrafish. *Dev. Cell* **2015**, *32*, 97–108. [CrossRef] [PubMed]
210. Robu, M.E.; Larson, J.D.; Nasevicius, A.; Beiraghi, S.; Brenner, C.; Farber, S.A.; Ekker, S.C. p53 activation by knockdown technologies. *PLoS Genet.* **2007**, *3*, e78. [CrossRef] [PubMed]
211. Nasevicius, A.; Ekker, S.C. Effective targeted gene 'knockdown' in zebrafish. *Nat. Genet.* **2000**, *26*, 216–220. [CrossRef] [PubMed]
212. Karlen, S.; Rebagliati, M. A morpholino phenocopy of the cyclops mutation. *Genesis* **2001**, *30*, 126–128. [CrossRef] [PubMed]
213. Stainier, D.Y.R.; Raz, E.; Lawson, N.D.; Ekker, S.C.; Burdine, R.D.; Eisen, J.S.; Ingham, P.W.; Schulte-Merker, S.; Yelon, D.; Weinstein, B.M.; et al. Guidelines for morpholino use in zebrafish. *PLoS Genet.* **2017**, *13*, e1007000. [CrossRef] [PubMed]
214. Li, M.; Zhao, L.; Page-McCaw, P.S.; Chen, W. Zebrafish Genome Engineering Using the CRISPR-Cas9 System. *Trends Genet.* **2016**, *32*, 815–827. [CrossRef] [PubMed]
215. Simone, B.W.; Martinez-Galvez, G.; WareJoncas, Z.; Ekker, S.C. Fishing for understanding: Unlocking the zebrafish gene editor's toolbox. *Methods* **2018**, *150*, 3–10. [CrossRef] [PubMed]
216. El-Brolosy, M.A.; Stainier, D.Y.R. Genetic compensation: A phenomenon in search of mechanisms. *PLoS Genet.* **2017**, *13*, e1006780. [CrossRef] [PubMed]
217. Shankaran, S.S.; Dahlem, T.J.; Bisgrove, B.W.; Yost, H.J.; Tristani-Firouzi, M. CRISPR/Cas9-Directed Gene Editing for the Generation of Loss-of-Function Mutants in High-Throughput Zebrafish F0 Screens. *Curr. Protoc. Mol. Biol.* **2017**, *119*, 31–39. [PubMed]
218. Shah, A.N.; Moens, C.B.; Miller, A.C. Targeted candidate gene screens using CRISPR/Cas9 technology. *Methods Cell Biol.* **2016**, *135*, 89–106. [PubMed]
219. Saint-Amant, L.; Drapeau, P. Time course of the development of motor behaviors in the zebrafish embryo. *J. Neurobiol.* **1998**, *37*, 622–632. [CrossRef]
220. Grillner, S. The motor infrastructure: From ion channels to neuronal networks. *Nat. Rev. Neurosci.* **2003**, *4*, 573–586. [CrossRef] [PubMed]
221. Goulding, M.D. Circuits controlling vertebrate locomotion: Moving in a new direction. *Nat. Rev. Neurosci.* **2009**, *10*, 507–518. [CrossRef] [PubMed]
222. Burgess, H.A.; Granato, M. Sensorimotor gating in larval zebrafish. *J. Neurosci.* **2007**, *27*, 4984–4994. [CrossRef] [PubMed]
223. Portugues, R.; Feierstein, C.E.; Engert, F.; Orger, M.B. Whole-brain activity maps reveal stereotyped, distributed networks for visuomotor behavior. *Neuron* **2014**, *81*, 1328–1343. [CrossRef] [PubMed]
224. Tavassoli, T.; Bellesheim, K.; Siper, P.M.; Wang, A.T.; Halpern, D.; Gorenstein, M.; Grodberg, D.; Kolevzon, A.; Buxbaum, J.D. Measuring Sensory Reactivity in Autism Spectrum Disorder: Application and Simplification of a Clinician-Administered Sensory Observation Scale. *J. Autism. Dev. Disord.* **2016**, *46*, 287–293. [CrossRef] [PubMed]
225. American Psychiatric Association; American Psychiatric Association. DSM-5 Task Force. *Diagnostic and Statistical Manual of Mental Disorders: DSM-5*, 5th ed. American Psychiatric Association: Washington, DC, USA, 2013.
226. Engeszer, R.E.; Barbiano, L.A.; Ryan, M.J.; Parichy, D.M. Timing and plasticity of shoaling behaviour in the zebrafish, *Danio rerio· Anim. Behav.* **2007,** *74*, 1269–1275. [CrossRef] [PubMed]

227. Dreosti, E.; Lopes, G.; Kampff, A.R.; Wilson, S.W. Development of social behavior in young zebrafish. *Front. Neural Circuits* **2015**, *9*, 39. [CrossRef] [PubMed]
228. Gerlach, G.; Hodgins-Davis, A.; Avolio, C.; Schunter, C. Kin recognition in zebrafish: A 24-hour window for olfactory imprinting. *Proc. Biol. Sci.* **2008**, *275*, 2165–2170. [CrossRef] [PubMed]
229. Wircer, E.; Blechman, J.; Borodovsky, N.; Tsoory, M.; Nunes, A.R.; Oliveira, R.F.; Levkowitz, G. Homeodomain protein Otp affects developmental neuropeptide switching in oxytocin neurons associated with a long-term effect on social behavior. *Elife* **2017**, *6*, e22170. [CrossRef] [PubMed]
230. Engeszer, R.E.; Ryan, M.J.; Parichy, D.M. Learned social preference in zebrafish. *Curr. Biol.* **2004**, *14*, 881–884. [CrossRef] [PubMed]
231. Stednitz, S.J.; McDermott, E.M.; Ncube, D.; Tallafuss, A.; Eisen, J.S.; Washbourne, P. Forebrain Control of Behaviorally Driven Social Orienting in Zebrafish. *Curr. Biol.* **2018**, *28*, 2445–2451. [CrossRef] [PubMed]
232. Myers, L.; Anderlid, B.M.; Nordgren, A.; Willfors, C.; Kuja-Halkola, R.; Tammimies, K.; Bolte, S. Minor physical anomalies in neurodevelopmental disorders: A twin study. *Child. Adolesc. Psychiatry Ment. Health* **2017**, *11*, 57. [CrossRef] [PubMed]
233. Nicolson, R.; DeVito, T.J.; Vidal, C.N.; Sui, Y.; Hayashi, K.M.; Drost, D.J.; Williamson, P.C.; Rajakumar, N.; Toga, A.W.; Thompson, P.M. Detection and mapping of hippocampal abnormalities in autism. *Psychiatry Res.* **2006**, *148*, 11–21. [CrossRef] [PubMed]
234. Randlett, O.; Wee, C.L.; Naumann, E.A.; Nnaemeka, O.; Schoppik, D.; Fitzgerald, J.E.; Portugues, R.; Lacoste, A.M.; Riegler, C.; Engert, F.; et al. Whole-brain activity mapping onto a zebrafish brain atlas. *Nat. Methods* **2015**, *12*, 1039–1046. [CrossRef] [PubMed]
235. Gupta, T.; Marquart, G.D.; Horstick, E.J.; Tabor, K.M.; Pajevic, S.; Burgess, H.A. Morphometric analysis and neuroanatomical mapping of the zebrafish brain. *Methods* **2018**. [CrossRef] [PubMed]
236. Marquart, G.D.; Tabor, K.M.; Horstick, E.J.; Brown, M.; Geoca, A.K.; Polys, N.F.; Nogare, D.D.; Burgess, H.A. High-precision registration between zebrafish brain atlases using symmetric diffeomorphic normalization. *Gigascience* **2017**, *6*, 1–15. [CrossRef] [PubMed]
237. Forster, D.; Arnold-Ammer, I.; Laurell, E.; Barker, A.J.; Fernandes, A.M.; Finger-Baier, K.; Filosa, A.; Helmbrecht, T.O.; Kolsch, Y.; Kuhn, E.; et al. Genetic targeting and anatomical registration of neuronal populations in the zebrafish brain with a new set of BAC transgenic tools. *Sci. Rep.* **2017**, *7*, 5230. [CrossRef] [PubMed]
238. Jefferis, G.S.; Potter, C.J.; Chan, A.M.; Marin, E.C.; Rohlfing, T.; Maurer, C.R., Jr.; Luo, L. Comprehensive maps of Drosophila higher olfactory centers: Spatially segregated fruit and pheromone representation. *Cell* **2007**, *128*, 1187–1203. [CrossRef] [PubMed]
239. Satou, C.; Kimura, Y.; Higashijima, S. Generation of multiple classes of V0 neurons in zebrafish spinal cord: Progenitor heterogeneity and temporal control of neuronal diversity. *J. Neurosci.* **2012**, *32*, 1771–1783. [CrossRef] [PubMed]
240. Guzowski, J.F.; Timlin, J.A.; Roysam, B.; McNaughton, B.L.; Worley, P.F.; Barnes, C.A. Mapping behaviorally relevant neural circuits with immediate-early gene expression. *Curr. Opin. Neurobiol.* **2005**, *15*, 599–606. [CrossRef] [PubMed]
241. Barros, V.N.; Mundim, M.; Galindo, L.T.; Bittencourt, S.; Porcionatto, M.; Mello, L.E. The pattern of c-Fos expression and its refractory period in the brain of rats and monkeys. *Front. Cell Neurosci.* **2015**, *9*, 72. [CrossRef] [PubMed]
242. Xia, Z.; Dudek, H.; Miranti, C.K.; Greenberg, M.E. Calcium influx via the NMDA receptor induces immediate early gene transcription by a MAP kinase/ERK-dependent mechanism. *J. Neurosci.* **1996**, *16*, 5425–5436. [CrossRef] [PubMed]
243. Cox, K.J.; Fetcho, J.R. Labeling blastomeres with a calcium indicator: A non-invasive method of visualizing neuronal activity in zebrafish. *J. Neurosci. Methods* **1996**, *68*, 185–191. [CrossRef]
244. Takahashi, M.; Narushima, M.; Oda, Y. In vivo imaging of functional inhibitory networks on the mauthner cell of larval zebrafish. *J. Neurosci.* **2002**, *22*, 3929–3938. [CrossRef] [PubMed]
245. Sumbre, G.; Muto, A.; Baier, H.; Poo, M.M. Entrained rhythmic activities of neuronal ensembles as perceptual memory of time interval. *Nature* **2008**, *456*, 102–106. [CrossRef] [PubMed]
246. Miyawaki, A.; Llopis, J.; Heim, R.; McCaffery, J.M.; Adams, J.A.; Ikura, M.; Tsien, R.Y. Fluorescent indicators for Ca^{2+} based on green fluorescent proteins and calmodulin. *Nature* **1997**, *388*, 882–887. [CrossRef] [PubMed]

247. Kim, D.H.; Kim, J.; Marques, J.C.; Grama, A.; Hildebrand, D.G.C.; Gu, W.; Li, J.M.; Robson, D.N. Pan-neuronal calcium imaging with cellular resolution in freely swimming zebrafish. *Nat. Methods* **2017**, *14*, 1107–1114. [CrossRef] [PubMed]

248. Cong, L.; Wang, Z.; Chai, Y.; Hang, W.; Shang, C.; Yang, W.; Bai, L.; Du, J.; Wang, K.; Wen, Q. Rapid whole brain imaging of neural activity in freely behaving larval zebrafish (*Danio rerio*). *Elife* **2017**, *6*, e28158. [CrossRef] [PubMed]

249. Lovett-Barron, M.; Andalman, A.S.; Allen, W.E.; Vesuna, S.; Kauvar, I.; Burns, V.M.; Deisseroth, K. Ancestral Circuits for the Coordinated Modulation of Brain State. *Cell* **2017**, *171*, 1411–1423. [CrossRef] [PubMed]

250. Jawinski, P.; Kirsten, H.; Sander, C.; Spada, J.; Ulke, C.; Huang, J.; Burkhardt, R.; Scholz, M.; Hensch, T.; Hegerl, U. Human brain arousal in the resting state: A genome-wide association study. *Mol. Psychiatry* **2018**. [CrossRef] [PubMed]

251. Fernandez, M.; Mollinedo-Gajate, I.; Penagarikano, O. Neural Circuits for Social Cognition: Implications for Autism. *Neuroscience* **2018**, *370*, 148–162. [CrossRef] [PubMed]

252. Kawakami, K.; Asakawa, K.; Hibi, M.; Itoh, M.; Muto, A.; Wada, H. Gal4 Driver Transgenic Zebrafish: Powerful Tools to Study Developmental Biology, Organogenesis, and Neuroscience. *Adv. Genet.* **2016**, *95*, 65–87. [PubMed]

253. Wiley, D.S.; Redfield, S.E.; Zon, L.I. Chemical screening in zebrafish for novel biological and therapeutic discovery. *Methods Cell Biol.* **2017**, *138*, 651–679. [PubMed]

254. Svenson, K.L.; Gatti, D.M.; Valdar, W.; Welsh, C.E.; Cheng, R.; Chesler, E.J.; Palmer, A.A.; McMillan, L.; Churchill, G.A. High-resolution genetic mapping using the Mouse Diversity outbred population. *Genetics* **2012**, *190*, 437–447. [CrossRef] [PubMed]

255. Rihel, J.; Prober, D.A.; Arvanites, A.; Lam, K.; Zimmerman, S.; Jang, S.; Haggarty, S.J.; Kokel, D.; Rubin, L.L.; Peterson, R.T.; et al. Zebrafish behavioral profiling links drugs to biological targets and rest/wake regulation. *Science* **2010**, *327*, 348–351. [CrossRef] [PubMed]

256. Kokel, D.; Peterson, R.T. Using the zebrafish photomotor response for psychotropic drug screening. *Methods Cell. Biol.* **2011**, *105*, 517–524. [PubMed]

257. Bruni, G.; Rennekamp, A.J.; Velenich, A.; McCarroll, M.; Gendelev, L.; Fertsch, E.; Taylor, J.; Lakhani, P.; Lensen, D.; Evron, T.; et al. Zebrafish behavioral profiling identifies multitarget antipsychotic-like compounds. *Nat. Chem. Biol.* **2016**, *12*, 559–566. [CrossRef] [PubMed]

258. Siegel, M.; Beresford, C.A.; Bunker, M.; Verdi, M.; Vishnevetsky, D.; Karlsson, C.; Teer, O.; Stedman, A.; Smith, K.A. Preliminary investigation of lithium for mood disorder symptoms in children and adolescents with autism spectrum disorder. *J. Child Adolesc. Psychopharmacol.* **2014**, *24*, 399–402. [CrossRef] [PubMed]

259. Lemmon, M.E.; Gregas, M.; Jeste, S.S. Risperidone use in autism spectrum disorders: A retrospective review of a clinic-referred patient population. *J. Child Neurol.* **2011**, *26*, 428–432. [CrossRef] [PubMed]

9

Yeast Gup1(2) Proteins Are Homologues of the Hedgehog Morphogens Acyltransferases HHAT(L): Facts and Implications

Cândida Lucas [1,*], Célia Ferreira [1,2], Giulia Cazzanelli [1], Ricardo Franco-Duarte [1] and Joana Tulha [1]

1 CBMA—Centre of Molecular and Environmental Biology, University of Minho, Campus de Gualtar, 4710-054 Braga, Portugal; giulia.cazzanelli88@gmail.com (G.C.); ricardofrancoduarte@gmail.com (R.F.-D.); tulha.joana@gmail.com (J.T.)
2 School of Biology, Faculty of Biological Sciences, University of Leeds, Leeds LS2 9JT, UK; c.m.ferreira@leeds.ac.uk
* Correspondence: clucas@bio.uminho.pt

Academic Editors: Henk Roelink and Simon J. Conway

Abstract: In multiple tissues, the Hedgehog secreted morphogen activates in the receiving cells a pathway involved in cell fate, proliferation and differentiation in the receiving cells. This pathway is particularly important during embryogenesis. The protein HHAT (Hedgehog *O*-acyltransferase) modifies Hh morphogens prior to their secretion, while HHATL (Hh *O*-acyltransferase-like) negatively regulates the pathway. HHAT and HHATL are homologous to *Saccharomyces cerevisiae* Gup2 and Gup1, respectively. In yeast, Gup1 is associated with a high number and diversity of biological functions, namely polarity establishment, secretory/endocytic pathway functionality, vacuole morphology and wall and membrane composition, structure and maintenance. Phenotypes underlying death, morphogenesis and differentiation are also included. Paracrine signalling, like the one promoted by the Hh pathway, has not been shown to occur in microbial communities, despite the fact that large aggregates of cells like biofilms or colonies behave as proto-tissues. Instead, these have been suggested to sense the population density through the secretion of quorum-sensing chemicals. This review focuses on Gup1/HHATL and Gup2/HHAT proteins. We review the functions and physiology associated with these proteins in yeasts and higher eukaryotes. We suggest standardisation of the presently chaotic Gup-related nomenclature, which includes KIAA117, c3orf3, RASP, Skinny, Sightless and Central Missing, in order to avoid the disclosure of otherwise unnoticed information.

Keywords: yeast; GUP; HHAT; morphogenesis; Hedgehog

1. Introduction

The yeast *Saccharomyces cerevisiae* is a recognised model to enlighten higher eukaryotic molecular processes, including the ones underlying human pathologies, which are very far from microbial life [1]. Intuitively, we associate yeasts, as microbes, to a planktonic form of life. However, microbes in the wild live mostly as large communities of cells forming biofilms or colonies, the biology of which remains largely unknown [2]. A colony or a biofilm displays a *proto-tissue* complex behaviour [3]. In these communities, cells organise spatially, morphologically and functionally to ensure the survival of the group. This implies the outlying orchestrated differentiation and death of cells to complete an efficient colonisation of the substrate, be it the pulp of a fruit or the surface of medical devices.

Within the large multicellular communities of yeast, cells behave similarly to their higher eukaryotic counterparts. They are born, grow larger and age while replicating until they senesce and die. Alternatively, they may die young, following an apoptotic cell death programme [4,5], allowing the supply of nutrients to the inner layers of the group [4], located further from the nutrient richer environment. Additionally, yeast cells may differentiate, shifting from yeast into true or pseudo-hyphae. These differ morphologically and physiologically [6–8]. The differentiation shift promotes an invasive behaviour generally associated with strain virulence [9]. Therefore, in contrast to planktonic growth, the survival strategy is collective [2] and resembles tissues from higher eukaryotes. Accordingly, large populations of yeast cells are imbedded in an extracellular matrix (ECM) composed of polysaccharides [10] and a large proteome [11], including many representatives of higher eukaryote ECM key proteins [11].

In higher eukaryotes, long distance communication between cells is achieved by secreted signalling proteins, namely from the Hedgehog (Hh) or Wingless (Wnt) pathways. These ligands interact with specific membrane-resident receptors that trigger a downstream signalling pathway in the receiving cell. No yeast cell-to-cell molecular communication vehicle has been recognised so far, although ammonia gradients have been suggested to perform a role in the collective orchestration of cell behaviour [12]. Moreover, the secretion of quorum-sensing small chemicals has been described in yeasts and suggested to perform a role in the control of population density [13,14]. Concomitantly, neither Hh- or Wnt-like pathways have been described in yeasts.

The maturation of the Hedgehog secreted morphogens involves intein self-splicing and C- and N-terminal lipidation (reviewed by [15]). In *S. cerevisiae*, there is only one recognised self-splicing protein that has a Hedgehog-like Hint domain, the endonuclease VDE (PI-SceI), which is required for gene conversion during meiosis [16]. VDE derives from the splicing of Vma1, the alpha subunit of the V-ATPase [17], whose human homologue has a role in human breast cancer cell invasiveness [18]. More importantly, the highly homologous yeast proteins Gup1 and Gup2, discovered by our group in 2000 [19], have two homologues in higher eukaryotes that are responsible for the Hh-secreted morphogen N-terminal palmitoylation (HHAT/Gup2) [20,21] and the negative regulation of the pathway (HHATL/Gup1) [22]. Gup1 and Gup2 are members of the MBOAT superfamily of membrane bound *O*-acyltransferases [23,24] and are present in all eukaryotic genomes sequenced to date. Their designation as GUP (Glycerol Uptake Protein) comes from their influence on the performance of glycerol active transport [19,25], which is actually accomplished by Stl1 [26], meaning that Gup proteins were often mistaken for the actual permease (e.g., [27]). In animal cells, the nomenclature is not consistent. In mouse, human or fly cells, both Gup1 and Gup2 are known by numerous aliases (Table 1). Such ambiguity generated the dispersion of experimental data on these proteins, as well as the under-interpretation of their roles. In yeasts, the Gup proteins have been implicated in a vast number of phenotypes from very diverse but fundamental biological/molecular processes, which will be revised in this work. In mammals and fly, the experimental data are scarcer and more recent. Here, we review the information available from *S. cerevisiae* and from another model yeast, the human commensal/opportunistic pathogen *Candida albicans*, and complement this with the role of Gup homologues in the regulation of Hedgehog ligand secretion in higher eukaryotes. Moreover, data on amino acid sequence comparison throughout model organisms are presented and a harmonising nomenclature is suggested.

Table 1. Gup1 and Gup2 aliases in the literature and databases.

Aliases		Organism	Key References
Gup1	Gup2	*Saccharomyces cerevisiae*	[19,28]
Gup1	-	*Candida albicans*	[29]
-	RASP	*Drosophila melanogaster*	[30]
	Skinny Hedgehog		[20]
	Sightless		[31]
	Central Missing		[32]
HHATL	HHAT	*Mus musculus*	[22]
HHATL	HHAT	*Homo sapiens*	[33]
c3orf3			[34]
KIAA117			[35,36]
MBOAT3 *			[33]
MSTP002 #			[33]
OACT3 *			[33]

* Nomenclature shared with *Xenopus laevis* and *X. tropicalis*; # Nomenclature shared with *X. tropicalis*.

2. Yeast Gup1 and Gup2 Proteins

2.1. *Gup1 and Gup2 Are Members of the MBOAT Superfamily*

S. cerevisiae GUP1 was first associated with the recovery of glycerol from the medium for osmoregulation purposes [19]. In accordance, glycerol active transport is deficient in *GUP1*-deleted mutants [19,25]. However, this is an indirect effect, resulting from the involvement of Gup1 in lipid raft formation [37]. Raft disassembly causes Pma1 H^+-ATPase plasma membrane misdistribution [27], decreasing the appropriate proton motive force for active transporters like Stl1 to work properly [37]. Gup1 has a close homologue Gup2. The two proteins share a high degree of similarity (77%) and identity (57%) [19]. Gup1 and Gup2 are members of the MBOAT superfamily of multispanning membrane-bound *O*-acyltransferases, a superfamily that was first suggested based exclusively on sequence similarity between a small group of proteins [23]. Besides Gup proteins from *S. cerevisiae*, the MBOAT superfamily includes proteins as diverse as mammalian acyl-CoA:cholesterol acyltransferases (ACAT), fly palmitoyltransferases (Porcn and Nessy), *Arabidopsis thaliana* diglyceride acyltransferases (DGAT), bacteria *Bacillus subtilis*, *Staphylococcus aureus* and *Pseudomonas aeruginosa* D-alanyltransferases (DltB) and alginate *O*-acetyltransferases (AlgI) [23]. All of these proteins share a highly conserved histidine residue localised in a hydrophobic domain (His^{447} in yeast Gup1), as well as another residue of histidine, asparagine or aspartic acid, localised 30–50 amino acids upstream, within a hydrophilic region (His^{411} in the yeast Gup1). These conserved positions specify the active centre of these enzymes. The MBOAT superfamily is presently subdivided into three functionally different subgroups. One group includes enzymes involved in neutral lipid biosynthesis (ACATs and DGATs), another group includes proteins involved in phospholipid remodelling (lysophosphatidate acyltransferases—LPATs) and a third group includes the enzymes implicated in protein acylation (Porcupine (from the Wnt pathway), HHAT(L) (from the Hh pathway) and Ghrelin *O*-acyltransferase—GOAT (from the insulin regulatory pathway)) [38]. Gup1, although included in the third group, has also been included in the LPATs group due to the multiple biological and molecular roles assigned to it. In the mammalian HHAT(L), the highly conserved MBOAT signature amino acid residues are different. The yeast Gup1 His^{411}/His^{447} and Gup2 His^{461}/His^{498} positions are occupied by aspartic acid and histidine in HHAT and by aspartic acid and leucine in HHATL, respectively. Interestingly, the substitution of His^{447} by a leucine in *S. cerevisiae* Gup1 caused loss of phenotype [39], raising the question whether mammalian HHATL actually functions as an acyltransferase [22]. The roles attributed to Gup1 in all cellular models are for now exclusively biological/phenomenological, not biochemical. In contrast, the HHAT from high eukaryotes has a fully recognised enzymatic function in the N-palmitoylation of the Hh-secreted signals that can be reproduced in vivo and in vitro [21,40]. The yeast Gup2 shares with Gup1 the two MBOAT conserved histidines in the amino acid sequence positions 460 and 496.

According to the most common algorithms used to calculate the hydrophobicity of amino acid sequences, the Gup1 and Gup2 proteins both have 10 well-defined transmembrane domains (TMD). This number of TMD implicates that Gup1 terminals are both located on the same side of the membrane. However, the Gup1 C- and N-terminal domains are localised, respectively, in the periplasmic and cytosolic sides of the membrane [41], which is a rather uncommon topology. HHAT from mammals has a re-entrant loop stabilised by a lipid attachment, a palmitoylation of the MBOAT aspartic acid signature (Asp^{339}) [42]. This re-entrant loop actually comprises both MBOAT signature amino acids (*M. musculus* Asp^{339} and His^{374}; *H. sapiens* Asp^{340} and His^{377}). *S. cerevisiae* Gup1 and Gup2 and *C. albicans* Gup1 present a partially hydrophobic region, located between TMD 7 and 8 (using the TMD prediction engine available at [43]). The correspondent sequence aligns perfectly with the sequence from the *M. musculus* HHAT re-entrant loop, showing several conserved several amino acid residues. In silico palmitoylation prediction of Gup amino acid sequence (using the prediction engine available at [44]) indicates a fully conserved cysteine residue in that region: *S. cerevisiae* Gup1 Cys^{396} and Gup2 Cys^{445}; *C. albicans* Gup1 Cys^{417}; *M. musculus* HHAT Cys^{324} and HHATL Cys^{326}; *H. sapiens* HHAT Cys^{325} and HHATL Cys^{326}; and *D. melanogaster* RASP Cys^{326}. This degree of conservation, in addition to the possibility that a palmitoylation is involved, suggests that Gup proteins might require the MBOAT signature residues to be embedded in or at the surface of the membrane, whether this is obtained by Cys or Asp palmitoylation as in mammalian HHAT [42]. This possibility will have to be studied in the future.

2.2. Gup1 and Gup2 Subcellular Localisation

In *S. cerevisiae*, Gup1 protein co-localises with markers of the plasma membrane, the endoplasmic reticulum (ER) and the mitochondria [19]. Gup2 shares the first two locations [45]. The more prominent localisation of Gup1 appears to be the plasma membrane, as revealed by GFP tagging at either N- or -C-terminal [41]. The plasma membrane localisation of Gup1 is dependent on the secretory pathway proteins Sec6-4 [41] and End-3-mediated excision [26,41]. In opposition, in mammalian cells (mouse myocytes cell line), the Gup1 orthologue localises predominantly in the cytosol and only residually in the ER closer to the nucleus [34].

2.3. *GUP1* and *GUP2* Expression in Yeast

The yeast *GUP2* promoter displays consensus sequences for 65 transcription factors (TF), and *GUP1* for 19 TF (Table 2; [46]). These include major players in yeast transcription control, like the general stress regulators Msn2 and Msn4, the glucose-repression controller Mig1, Ste12 from pheromone response and mating and Gcn4 from nitrogen-associated regulation. Three TFs are predicted to regulate both promoters identically (Tec1, Mig3 and Rap1), and one—Ash1—is predicted to activate *GUP1* and inhibits *GUP2* transcription (Table 2).

The presence of numerous TF consensus sequences in yeast *GUP1/2* promoters could reflect a complex transcription regulation, yet *GUP1* and *GUP2* are almost invariantly expressed in several different conditions. This is the case for cells actively growing on glucose (repression conditions—fermentation) and on glycerol or ethanol (de-repression conditions—respiration) [47], as well as in cells adapted to osmotic stress conditions created by high amounts of salt [47,48], or subjected to osmotic shock by high amounts of sorbitol [49]. These results suggest that the expression of these genes might be constitutive [47]. Nevertheless, this cannot be the case since *GUP1* transcription duplicates upon NaCl shock [50]. Additionally, at the level of the protein, Gup1 is stably internalised in an End3-dependent manner upon sudden shift from glycerol to glucose [26,41]. Glycerol has multiple and complex roles in yeast cells. It is a respired poor carbon source whose metabolism requires induction and relief from glucose-repression. It functions as an osmolyte (reviewed by [51]), a lipid synthesis precursor [52], a *pendulum* of the NAD(H) redox balancing cytosol/mitochondrial shuttle [53], and is necessary for maintaining the signalling competent state of the cells [54]. The fact that Gup1 is retrieved from the plasma membrane upon glycerol-to-glucose transition, and overexpressed upon salt shock when glycerol accumulation is required for cell survival [51], establishes a connection between Gup1 and glycerol-associated regulation, which nature keeps unknown.

Table 2. Documented transcription factors that function as regulators of Gup1 and Gup2 [46]. Activ.: activator; Inhib.: inhibitor.

Predicted Function in *GUP1* and *GUP2* Transcription Regulation											
GUP1			*GUP2*				Identical Regulation			Opposite Regulation	
Activ.	Inhib.	Both	Activ.		Inhib.	Both	Activ.	Inhib.	Both	Activ. *GUP1* + Inhib. *GUP2*	Inhib. *GUP1* + Activ. *GUP2*
Ash1	Abf1	Rap1	Ecm22	Sfp1	Ace2	Aft1	Tec1	Mig3	Rap1	Ash1	none
Met4	Cac2	Spt23	Gat4	Snf2	Ash1	Cbf1					
Tec1	Cup9	Yrm1	Gcr2	Snf6	Bas1	Cin5					
	Gal4		Gis1	Sok2	Fhl1	Plm2					
	Hir3		Gln3	Spt10	Gcn4	Pho4					
	Mig3		Gzf3	Spt2	Gcr2	Rap1					
	Sfp1		Hap5	Ste12	Mig3	Rfx1					
	Sum1		Hda1	Sut1	Msn4	Sko1					
	Uls1		Hms1	Swi3	Rif2	Xbp1					
			Isw2	Taf14	Set2	Yap6					
			Mbp1	Tec1	Swi5	Msn2					
			Mcm1	Tup1	Tbf1	Swi4					
			Mga1	Uls1							
			Mig1	Uc2							
			Mot3	Urc2							
			Nrg1	Xbp1							
			Rgm1	Yhp1							
			Rox1	Yox1							
			YR015C								

Observations: Ash1: TF inhibitor of HO transcription; Mig3: TR with a major role in catabolite repression and ethanol response; inactivated by phosphorylation by Snf1 or Mec1 pathway; Rap1: Essential TR for many loci: transcription activation and repression, chromatin silencing, and telomere length maintenance; Tec1: TF targeting filamentation genes and Ty1 expression; activity dependent on association with Ste12; Ste12 activation of most filamentation gene promoters depends on Tec1.

GUP2 is generally less expressed than *GUP1* [47], and *gup2*Δ null mutants do not present any marked phenotype in response to carbon sources (glucose/glycerol) or stress (salts, ethanol, weak acids and high temperature). In contrast, *gup1*Δ mutants are moderately sensitive to all stresses [28] with the exception of high temperature, to which *gup1*Δ is highly sensitive [28,55]. The expression of *GUP2* (not *GUP1*) has been shown to increase considerably upon ultraviolet irradiation [56], leading the authors to suggest Gup2 as part of the Early Genotoxic Response, i.e., the response to metal poisoning.

Up-regulation of Gup1 induces the up-regulation of several genes resident in the ER and Golgi, from the secretory pathway and lipid synthesis [39]. Kar2, a key protein from the Unfolded Protein Response (UPR) signalling pathway, stands out for its 24-fold increased expression [38]. Up-regulation of Gup1 further induces proliferation of membranous cisternal structures, with 1–7 per cell, physically separated from each other. This phenomenon is dependent on Ire1, another protein from the UPR pathway. ER, Golgi and itinerant proteins are present in these structures. Gup1 was also found, appearing to diffuse in and out of these structures [39]. Remarkably, the production of these membrane structures is dependent on the H^{447} MBOAT signature amino acid.

3. Yeast Phenotypes Associated with the Deletion of *GUP1*

3.1. Phenotypes Emerging from Genome-Wide Yeast Screenings

The extensive list of phenotypes associated with the deletion of *GUP1* comprises many that were obtained in genome-wide screening. These include cytoskeleton polarisation and bud site selection patterns [57,58], secretory and endocytic pathways [59], vacuole morphology [59] and anaerobic growth [60]. Other high throughput studies cannot be presently rationalised, namely Askree et al. [61], which mentions Gup1 as an important protein in the maintenance of proper telomere length. Importantly, many of these studies show the resistance of the *gup1Δ* mutant to pharmaceutical drugs such as the anti-tumoral imatinib [62], the cytostatic cisplatin [63], the anti-inflammatory ibuprofen [64], the immunosuppressant FK506 [65] and the antimicrobial thymol [66]. Further resistance associated with *GUP1* deletion includes lithium [67] and arsenic [68]. This large list, by itself, suggests an equally large complexity of Gup1 functions. In contrast, *GUP2* has been referred to in only two HTP screenings [69,70]. The few evidences show that *GUP2* expression is increased in response to ethanol stress in wine fermentative conditions/strains.

3.2. Cell Wall Integrity and Biogenesis

As mentioned above, *gup1Δ* is very sensitive to high temperature [28,55]. Generally, the sensitivity of yeast to high temperature is correlated with major defects in cell wall. Consistently, the deletion of *GUP1* in *S. cerevisiae* causes an increase in the cell wall content of β-1,3-glucans (+25%) [28] and chitin (+90%) [28,71], and a decrease in the amount of mannoproteins (−70%) [28]. These large differences are responsible for a deficient wall morphology and sensitivity to wall-perturbing agents [28]. In spite of these defects, the Cell Wall Integrity (CWI)/PKC pathway is signalling normally in the *gup1Δ* mutant, as seen by MAPK (Slt2) (Figure 1A) dually phosphorylated state [28,72] after induction of the pathway by hypo-osmotic shock [28] or high temperature, or by NaCl or ethanol stress [73].

The CWI/PKC pathway controls not only the biogenesis of cell wall constituents and their remodelling, but also the associated organisation of the actin cytoskeleton and secretory pathway (reviewed by [74]). Unlike mammalian cells, *S. cerevisiae* only has one protein kinase C (Pkc1) that responds to the G-protein Rho1, the yeast homologue of RhoA. Rho1 activates Pkc1 besides other effectors. This allows the coordinated control of the actin cytoskeleton, exocytosis, membrane fluidity and the synthesis of wall glucans through transcription regulation in response to nutrients and stress [75]. Rho1 in turn is activated by Rom2/1 GDP/GTP exchange factors. These respond directly at least to a couple of wall stress sensors, Wsc1 and Mid2, and are further activated by the ligation of PI(4,5)P2 (phosphoinositol biphosphate) and/or of Slm1/2 proteins. The latter are targets for TORC2 (Target of Rapamycin 2 (TOR2) Complex) phosphorylation [76]. TORC2 also directly activates Pkc1, inducing the PKC pathway response. Importantly, Rho1 and consequently the PKC pathway are also activated by the α-factor plasma membrane receptor Ste2, a physical partner of Gup2 (Saccharomyces Genome Database). Rho1 is a signalling node, regulating many proteins and being under the effect of many others (including Kog1 from TORC1 (Target of Rapamycin 1 (TOR1) Complex) [77]), and it acts on actin cytoskeleton, thus causing cellular shape/polarity and filamentation [78].

TORC2 is redundant with TORC1 in the control of major cellular functions, but the cell wall composition and integrity and the polarisation of the actin cytoskeleton associated with cell growth, division and morphogenesis is only dependent on TORC2 [79,80]. Even though TOR proteins are very similar phosphatidylinositol kinase (PIK)-related kinases, only TOR1 is sensitive to inhibition by caffeine or the immunosuppressant rapamycin (reviewed by [80]). The *gup1Δ* mutant, both alone and in combination with *gup2Δ*, is resistant to rapamycin [73]. On the other hand, *gup1Δ* is equally sensitive to caffeine as the wild type strain but, unlike this last, its sensitivity is remediable with sorbitol [28].

The mechanisms of action of caffeine and rapamycin are poorly understood. Both induce Rho1 activation and depend on this induction to inactivate the TORC1 pathway [77]. Caffeine induces the activation of Slt2 in a Tor1/Rom2 dependent manner [81] and elicits a transient decrease in the intracellular levels of cAMP, concomitantly inhibiting the Ras/cAMP pathway [81]. Moreover, caffeine supposedly activates the Pkc1 cascade through Tor1-mediated signalling [81]. Rapamycin, on the other hand, represses rRNA transcription and induces arrest of translation and cell cycle at G1. Furthermore, it causes glycogen accumulation, sporulation and autophagy [82]. These effects demand the complexation of rapamycin with Fpr1 (yeast equivalent of the immunophilin FKBP12) and subsequently with TOR1 [83]. Resistance to rapamycin in *S. cerevisiae*, can be provided by Fap1, which competes with Fpr1 [83,84]. However, the way in which Gup1 might be involved remains obscure.

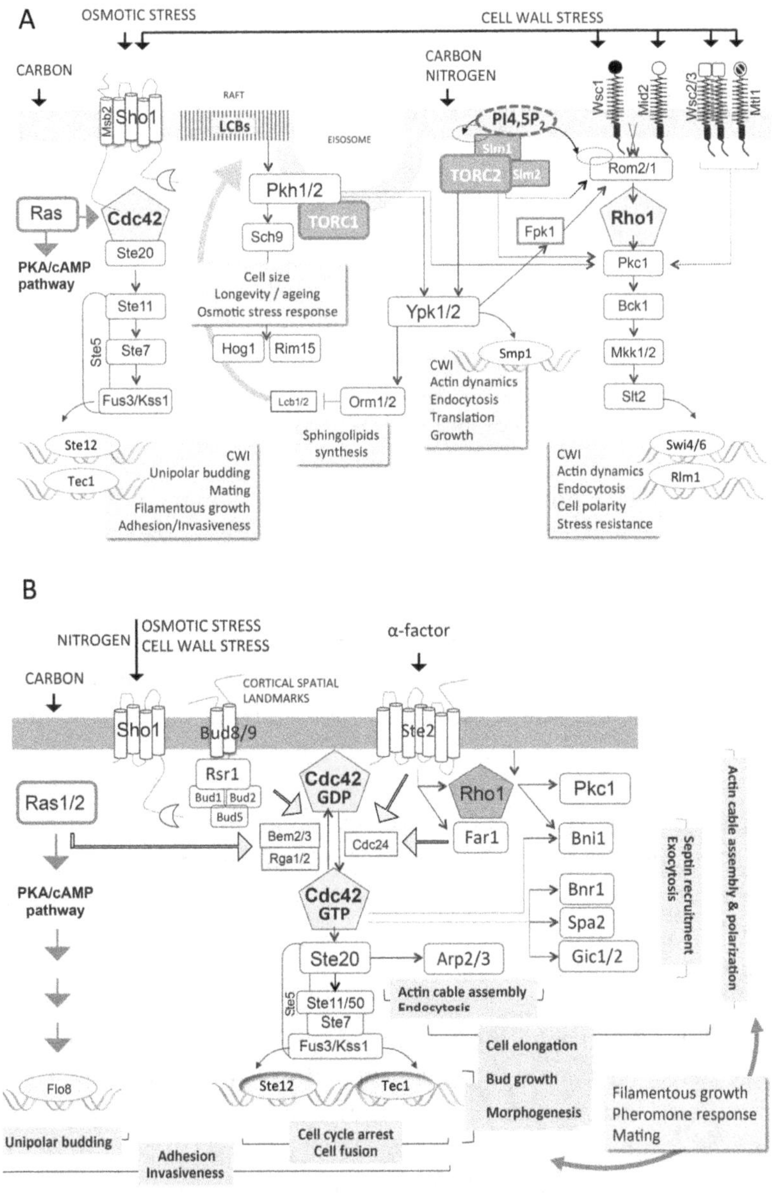

Figure 1. Major players implicated in the phenotypes associated with the deletion of *GUP1* in *S. cerevisiae* and *C. albicans* (grey boxes). Plain arrows indicate established interactions; dashed arrows indicate possible interactions. Pentagons refer to nodal proteins exerting multiple signalling, affecting positively or negatively many other proteins. (**A**) The Sho1 downward cascade belongs to the HOG (High Osmolarity Glycerol) pathway, and the Rom/Rho cascade constitutes the Cell Wall Integrity—CWI/PKC pathway. Crosstalk between HOG, PKC, the complex lipids and TORC1/2 signalling pathways relevant for *GUP1* associated phenotypes is shown; (**B**) Centrality of Cdc42 phosphorylation (pink boxes) in the control of pathways ultimately promoting differentiation, polarity, growth and reproduction. The central role of one protein shared by many pathways is designated in the text as *star*-type pathway.

Bearing in mind that the deletion of *GUP1* causes defects in cell wall [73], membrane [37,55] and endocytosis [59], the cause of resistance to rapamycin could be the inability of the drug to enter the cell. Rapamycin is practically insoluble in water and poorly soluble in ethanol or glycerol [85]. This, together with its large molecular mass (≈900 Da), suggests an endocytic entrance. Additionally, *gup1*Δ is also resistant to the calcineurin inhibitor FK506 [65]. In mammalian cells, this immunosuppressant drug binds to FKBP12 inhibiting the Ca^{2+} signalling. Actually, FKBP12 bridges FK506, calcineurin and IP3 (inositol 3-phosphate) receptor, promoting the dephosphorylation of the latter, thereby modulating Ca^{2+} flux and signalling [86]. In yeast, FK506 affects the Ca^{2+}-mediated response to high temperature, high NaCl-osmotic stress and α-factor pheromone [87]. High temperatures and high sodium, as discussed above, cause severe phenotypes in *gup1*Δ [19,28], while α-factor secretion is higher in *gup1*Δ [59].

The regulatory roles of phosphoinositides in yeast are not as well understood as in higher eukaryotes [88]. Still, their synthesis and turnover pathways are genetically characterised and the effects of some of the enzymes on major cellular processes have been recognised. At least seven unique second-messenger molecules are generated that regulate growth, differentiation, cytoskeletal rearrangements, membrane trafficking and protein vacuole sorting [88]. The phosphoinositide signalling operates in yeast as a *star-type* pathway, i.e., by cross-talking with numerous other pathways controlling their associated transcription, thus promoting the orchestrated cellular response to numerous stimuli. Major effectors associated with phosphoinosides in yeast belong to TORC2 and PKC pathways.

Yeast TOR pathways respond to carbon and nitrogen availability [89]. In silico analysis of the promoter regions of both *GUP1* and *GUP2* unveiled a series of putative regulatory sequences, from which the presence of repeated consensus sequences for Gcn4, a major transcription factor that responds to amino acid starvation and nitrogen in the dependence of TORC1, stands out [90]. Accordingly, wild-type strain and *gcn4*Δ mutant expressing lacZ reporter constructions for *GUP1* and *GUP2* promoter regions showed that β-galactosidase activity depended on Gcn4 for full expression of the *GUP1* but not *GUP2* in synthetic medium. If the medium is supplemented with 0.5% ammonium sulphate and 15 mM glycerol, the expression of Gup1 in *gcn4*Δ reverts to the values of wild-type strain [73]. The *gup1*Δ mutant dual dependence on nitrogen source and glycerol suggests an intricate position of the Gup1 protein in the midst of crossroads of N and C regulatory pathways. Moreover, the operationality of PKC signalling observed in the *gup1*Δ mutant suggests that the TORC2 pathway is not affected by the function of Gup1 protein. However, as discussed below, many of the cellular functions controlled by TORC2 are defective in the absence of *GUP1*. Besides wall damage-associated phenotypes, the most striking of these is the altered polarity during division exhibited by *gup1*Δ mutant [57,58].

The *gup1*Δ was found to display profoundly aberrant random bud site selection [57,58,76]. This implies a large defect on the proper establishment of cytoskeleton polarity [91,92], which is known to depend on Rho1/Pkc1 signalling [78]. It is as well dependent on other pathways associated with the multiple roles of the signalling node Cdc42 [93] (Figure 1B), including the Sho1 branch of the HOG pathway [94]. Cdc42 has the same fundamental role in mammalian cells (Figure 1B). This intricate crosstalk implicates the coordinated control of actin cytoskeleton remodelling for directional growth (elongation), which is mandatory for cell division through budding or mating. However, more importantly, underlies the morphogenic switch into filamentation that mediates invasion and adhesion [6,95].

3.3. High Osmolarity Glycerol (HOG) Pathway

In spite of the *GUP1* deletion, the CWI/PKC pathway is working properly at the level of Slt2 phosphorylation when stimulated by hypotonic shock [28]. Nevertheless, the wall of the mutant is severely affected. Two other pathways contribute to the wall integrity and remodelling: the Sho1 branch of the HOG pathway, and the Long-Chain Base (LCB)/sphingolipids YPK pathway (Figure 1A). HOG stands for High Osmolarity Glycerol, a name derived from control of the production and accumulation of the yeast osmolyte glycerol in response to high osmotic stress (reviewed by [51]). The HOG pathway controls many other processes such as polarity, adhesion and invasiveness, filamentation (i.e., differentiation), mating and cell wall biogenesis (reviewed by [96]). Importantly, the HOG pathway is functionally equivalent to the mammalian Wnt/p38 stress responsive MAPK pathway (Figure 2) [97,98]. Null mutants from HOG pathway have been successfully complemented by murine [99] or human p38 and associated kinases [100,101]. The p38 pathway is a non-canonical branch of Wnt MAPK, as are the JNK and ERK5 branches. Nevertheless, it has been shown to crosstalk with the Wnt canonical ERK1/2/β-catenin branch [102]. Actually, the four Wnt branches (ERK1/2, p38, JNK and ERK5) are all partially redundant in the control of growth/proliferation, i.e., cell cycle progression, differentiation, adhesion, migration and apoptosis in response to growth factors, cytokines and stress (reviewed by [103,104]). The HOG pathway Sho1 sensor functions in an exceptionally complex manner [105]. It binds to effectors that deliver different responses: Msb2 binds to Cdc42, activating the Ste11/Ste7/Fuss3/Kss1 filamentation/mating and cell wall biogenesis pathway, and Hkr1 or Opy2 that activate the HOG pathway (Figures 1 and 2) [105,106].

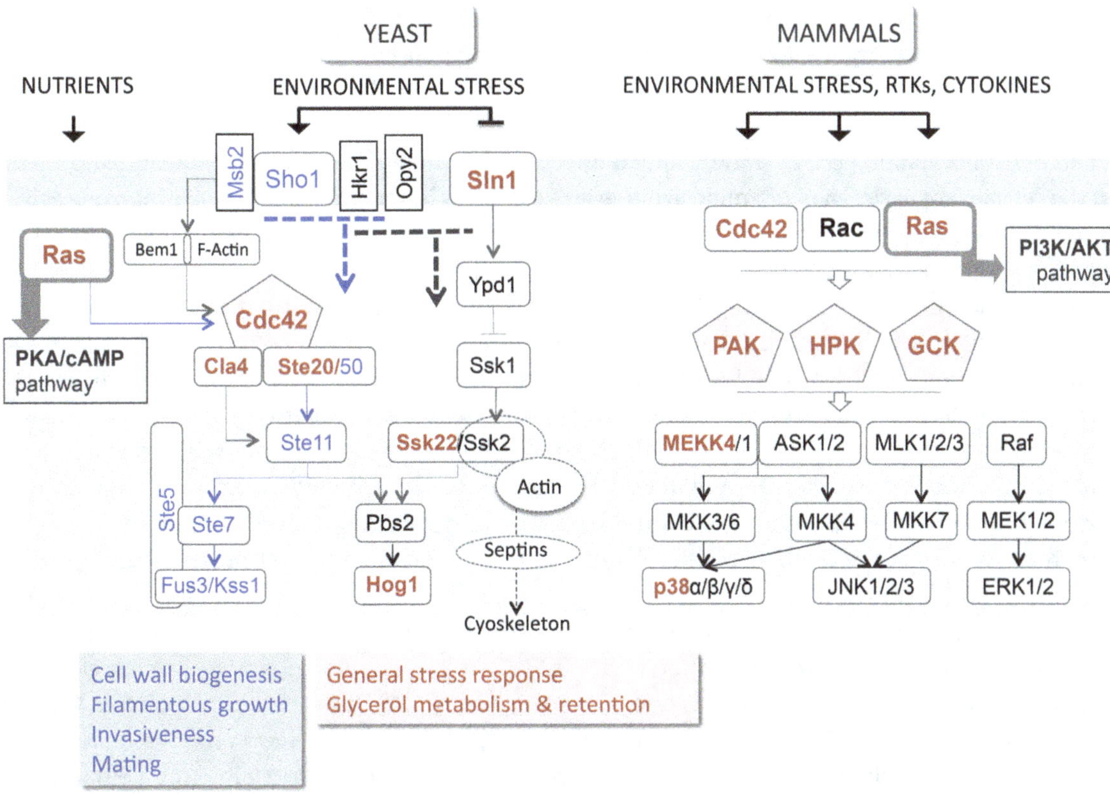

Figure 2. Parallelism between MAPK signalling cascades in yeast and in mammalian cells. In bold: proteins that functionally complement yeast mutants. In orange: proteins that are identical in yeast and mammalian pathways. In blue: proteins involved in the yeast phenotypes affected by *GUP1* deletion. Pentagons refer to nodal proteins exerting multiple signalling, affecting positive or negatively many other proteins (*star-type* pathway). Required nomenclature disambiguation: PAK: P21 Activated Kinase; HPK: Histidine Protein Kinase (Sln1 is also a histidine-aspartate phosphotransferase/kinase); GCK: Germinal Center Kinase; MEKK4 = MTK1.

The strains without *GUP1* are unable to grow under high osmotic stress [19,28]. This could indicate a malfunction of the HOG pathway and a concomitant deficient production and/or accumulation of glycerol. Hyperosmotic stress also stimulates the CWI/PKC pathway [107], a fact that is not surprising since the adaptation to high osmotic stress implicates changes in cell volume and turgor that demand remodelling of the cell wall. The simultaneous activation of HOG and CWI/PKC pathways can also result from other stimuli, such as high temperature [108] and the action of zymolyase, an enzyme cocktail with β1,3-glucanase activity [109,110]. *gup1*Δ is extremely sensitive to both of these stimuli [28]. The effect of zymolyase is particularly severe, namely in comparison with lyticase, another wall disrupting enzyme [28]. This severity could be the consequence of the increased percentage of β1,3-glucans in the mutant cell wall. The malfunction of HOG rather than CWI/PKC pathway was also inferred from the pronounced phenotype of *gup1Δ* mutant in the presence of Calcofluor White [28] compared to Congo Red, which caused a negligible defect [111]. While some insults to the cell wall like Congo Red only trigger CWI/PKC [81,112], others trigger both HOG and CWI/PKC pathways [107–110]. In spite of the intricate interplay between both pathways (reviewed by [107]) the phenotypes caused by *GUP1* deletion associated with osmoregulation [19] and the cell wall [28] are consistent with a relation of Gup1 with both pathways.

3.4. Plasma Membrane Composition and Associated Signalling

Gup1 is implicated in several membrane and lipids associated phenotypes [37,55,60,113]. The signalling pathway associated with complex lipids, the LCB/sphingolipid YPK pathway (partially depicted in Figure 1A), is not very well studied in yeasts, and no parallel has been established with the mammalian counterpart. The activation of Ypk1/2 is dependent on phosphorylation by Pkh1/2 kinases [114,115] or by TORC2 [115,116]. Pkh1/2 responds to LCBs and is implicated in membrane trafficking events. These include endocytosis through Pkc1-associated actin remodelling [114], protein phosphorylation [117], the trafficking of glycosylphosphatidylinositol (GPI)-anchored proteins, and the cell surface expression of amino acid permeases [118]. In mammalian cells, sphingosine acts as an inhibitor of PKC, or as activator of casein kinase II, Pkcζ, Pak1 (p21-activated kinase) and PDK1 (phosphoinositide-dependent kinase 1) (reviewed by [119]).

The lipid membrane composition and rafts content have implications in membrane fluidity/rigidity. In this regard, membrane fluidity homeostasis has been suggested to be under the control of a novel pathway that involves Rho1 effector [75]. According to these authors, the overlap between the CWI pathway with the membrane fluidity homeostasis pathway includes at least six components: Rho1, Pkc1, Bck1, Mkk1, Mkk2, Slt2 and Swi6. This is comparable to the overlap between mating response, pseudohyphal growth and the HOG pathways [107]. Membrane fluidity depends ultimately on the amount of complex lipids. Accordingly, the depletion of sphingolipids elicits the activation of TORC2 through Ypk1/2 and Orm1/2 that in turn stimulate their synthesis (Figure 1A) [120].

Down-regulation of TORC2-Ypk1 signalling allows cell survival under high osmotic stress independently of HOG pathway activation [121]. TORC2-Ypk1 signalling operates through closing of the glycerol channel Fps1, by blocking its phosphorylation by Ypk1. This is crucial to maintain the Fps1 channel open in the absence of stress [121]. This TORC2-Ypk1-mediated control occurs independently of the control of the channel by Hog1 [122]. Moreover, the enzyme responsible for deviating the glycolytic flux towards glycerol production Gpd1 (cytosolic glycerol 3 phosphate dehydrogenase) is also a target for Ypk1/2 [122], while its mitochondrial isogene Gpd2, from the glycerol NAD(H) shuttle, is under the direct control of glucose and is activated by phosphorylation via the AMP-activated kinase Snf1 [122]. Moreover, glycerol consumption and active uptake through Stl1 permease [26] is severely affected by the absence of *PKC1* [123]. This strengthens the notion that glycerol is a major driver in maintaining the signalling competent state of cells, as suggested before [54], namely in the crossroads of responses to nutrients availability and stress through TORC2, sphingolipids, CWI/PKC and HOG pathways. The defects caused by *GUP1* deletion appear in the centre of this puzzle, suggesting a pivotal role in signalling for the Gup1 protein.

LCB/sphingolipids form Detergent-Resistant Microdomains (DRM) in the plasma membrane, also known as lipid rafts due to their particular molecular/physicochemical properties. These properties derive from the predominantly saturated hydrocarbon tails of the sphingolipids and the high concentration of sterols: cholesterol in animal cells (e.g., [124]) and ergosterol in yeasts [125,126]. In yeast, as in mammalian cells, lipid rafts participate in protein sorting (contributing to the functions of transporters, pumps, sensors, receptors or G-proteins), dynamically including or excluding proteins [127–129]. Importantly, the components of lipid rafts are also actively involved in signalling and in stabilization of actin cytoskeleton–membrane interactions and therefore polarity [128,129].

In yeast, many proteins are known to functionally localise in lipid rafts forming a complex patchwork [128], which biogenesis is still poorly known. The first proteins found in yeast membrane microdomains were the H^+-ATPase pump Pma1 [130] and the amino acid permease Can1 [131]. Presently, almost 300 proteins have been found to differentially localise into lipid rafts [128], including many permeases involved in ion homeostasis, water, glycerol or drugs transport as well as sugar or nitrogen transport/sensing, cell wall metabolism, stress response, signalling, flocculation, bud site selection and eisosome formation. Some are GPI-anchored like the β1,3-glucanosyltransferase Gas1 [125], while others are prenylated, or farnesylated and palmitoylated like Ras2 [132], but most just have transmembrane domains (see [128] for a comprehensive list). Different types of membrane lipid microdomains were identified and designated identified on the basis of the proteins they harbour: Membrane Compartment occupied by Pma1 (MCP), Can1 (MCC) [131] or TORC2 (MCT) [133].

The plasma membrane localisation of Gup1 [19,41] and Gup2 [45] has not been associated with raft residence [128]. Still, a patchy distribution of the labelled Gup1 protein was observed by electron microscopy [41]. Importantly, *gup1Δ* is affected in lipid rafts integrity and assembly [37]. This was evidenced by the even, not punctate, distribution of sterols in the mutant plasma membrane, as well as the 40% lower amounts of DRM recovered [37]. Additionally, Pma1 (H^+ pump known to reside in lipid rafts, specifically MCPs [128]) is absent from *gup1Δ* DRMs, and the amounts of Gas1 (the most abundant yeast GPI-anchored protein) are residual, suggesting that rafts were unselectively affected by *GUP1* deletion [37]. Accordingly, Gup1 interferes in sterol and sphingolipids synthesis, as evidenced by respectively increased and decreased resistance of the mutant to inhibitors of specific biosynthetic steps [37].

The yeast cell is a facultative anaerobic organism that can efficiently endure environments with virtually inexistent oxygen. In these conditions, the yeast cell does not synthesise sterols. Instead, it must take up sterols from the environment using the Pdr11 ATP-binding cassette (ABC) multidrug transporter [134], another permease known to locate in lipid rafts [128]. *GUP1* was found to be crucial for survival in anaerobiosis [63]. In *gup1Δ*, not only is the biosynthesis of sterols affected, but also the uptake of sterols is impaired, underlying its inability to grow in anaerobic conditions [60].

Yeast lipid rafts assembly, unlike mammalian counterparts, takes place in the ER, only subsequently stepping into the Golgi [125]. Moreover, MCPs and MCCs are likely to segregate differently, since Pma1 is only incorporated into the rafts outside the ER [130]. ER resident proteins targeted to sphingolipids/ergosterol-poor membranes, such as the vacuole membrane, are not involved in rafts assembling [129]. Therefore, the defect in rafts assembly generated by *GUP1* deletion must occur in the ER (one of Gup1 localisations) and may induce the segregation of specific membrane/protein assemblages towards their correct location. Another group of results concurs for a clear implication of the Gup1 protein in the secretory/endocytic pathways. In yeast, the vacuole performs the role of mammalian lysosomes. Proteins are addressed to the vacuole for degradation by the same pathway as carboxypeptidase Y (CPY), through the late endosome (reviewed by [135]). *GUP1* was identified as one of the genes required for efficient sorting of CPY to the vacuole, as well as for the maintenance of this organelle regular size and morphology [59]. Its deletion also causes the excessive secretion of α-factor [59], the ER-resident protein Pdi1 [28], and the GPI-anchored β1,3-glucanosyltransferase Gas1 [37].

The deletion of *GUP1*, besides affecting sterols and sphingolipids synthesis and impairing sterols uptake, also causes changes in the regular concentrations of major lipid types: a high increase in diacyl- and triacylglycerol (Figure 3A) and a decrease in phospholipids [55]. Fatty acids synthesis, on the other hand, remains normal [37]. However, Gup1 is not recognised as an enzyme from lipid biosynthesis pathways. In this regard, Bosson and co-workers [113] proposed the involvement of Gup1 in the GPI synthesis/remodelling process, which occurs at the level of the ER in *S. cerevisiae* [136,137]. The anchors of mature GPI-anchored proteins of *S. cerevisiae* contain either ceramide or diacylglycerol with a C26 fatty acid in the *sn*2 position [138,139]. Gup1 was proposed to be the enzyme that adds C26 fatty acids to the *sn*2 position of lyso-phosphatidylinositol (lyso-PI)-containing GPI anchors (Figure 3A,B) [113]. Accordingly, the *GUP1* deleted mutant, in contrast to the wild-type strain, was shown to accumulate lyso-PI and to contain phosphatidylinositol tails with C16 and C18 fatty acids instead of C26, all found in much smaller amounts compared to wild type (Figure 3B) [113]. The direct involvement of Gup1 as an enzyme in this process is nevertheless controversial. The wild-type GPI-anchor types can still be found in the mutant [113]. Concomitantly, it was suggested that Cwh43 was able to catalyse a bypass to the Gup1-dependent step in inositol phosphoceramide (IPC) *tail* maturation (Figure 3B) [140,141], using indifferently PI or Lyso-PI as substrates (Figure 3B). Moreover, the metabolic step associated with Gup1 is preceded by the transformation of PI into lyso-PI by Per1 (Figure 3B) [139]. In *gup1Δ* mutants, the lyso-PI harbours different fatty acid chains, suggesting that the Per1-dependent step is affected by the deletion of *GUP1* as well [113]. Altogether, it is prudent to restrain from naming Gup1 as a mere GPI remodellase, since this classification does not take into account the multiple phenotypes associated with the protein, neither does it consider the localisation of Gup1 in the plasma membrane and its association with the mitochondria. In view of the complexity associated with Gup1 protein so far, the effects observed in GPI anchor synthesis could be indirect. GPI-anchor remodelling, if confirmed to be a direct function of Gup1 as an enzyme, will be among the other functions. The enzymes that catalyse the last steps of PI(4,5)P2 synthesis have associated phenotypes common to the *gup1Δ* mutant (Figure 4) (Saccharomyces Genome Database). This could provide the rationale for considering the possibility of Gup1 acting upstream of these enzymes, at the level of PI production (Figure 3). A change in PI availability would provide a reason for the changes observed at the level of the ceramide and diacylglycerol types (Figure 3B). The inosite metabolism was though never studied in this mutant.

3.5. Cell Death

Apoptosis is the most common process of Programmed Cell Death (PCD) in eukaryotes. It is vital for the fast elimination of useless or injured cells and for the differential development of tissues and organs. The existence of PCD processes in yeasts is presently unarguable [142]. As mentioned above, multicellular aggregates of microbial cells, like colonies or biofilms, are spatially organised and require the specialization of cells differentially localised to ensure supply of nutrients and water to the whole cell ensemble [143]. PCD occurs in a differentiated way within these aggregates [4]. The underlying rationale is that older cells of the colony centre that have already multiplied extensively, dying can contribute to the survival of adjacent cells with limited access to nutrients and water [4].

Several types of stimuli, both endogenous and exogenous, are known to cause yeast cells to enter a PCD process. H_2O_2 and acetic acid are major exogenous triggers commonly used to induce apoptosis in yeast (reviewed by [142]). However, several additional agents have been reported to induce apoptotic phenotypes, namely, salt stress, ethanol, heavy metals, UV radiation and high temperatures. Endogenous triggers on the other hand, include defects in N-glycosylation, chromatid cohesion, mRNA stability and ubiquitination (reviewed by [142]). Moreover, DNA damage (resulting from oxygen metabolism and ROS generation) and replication failure can stimulate the activation of yeast cell death programmes. Apoptosis may also play a role in ageing, replicatively [144] or chronologically [145,146].

Gup1 is required for several cellular processes that are related to apoptosis, namely the above discussed rafts integrity and stability [37], lipid metabolism [55], including GPI anchor correct remodelling [113], mitochondrial [60] and vacuole [59,147] functions and actin dynamics [57,58]. Accordingly, the deletion of *GUP1* induces the hypersensitivity of yeast cells to two known apoptosis inducing conditions [148], chronological life span (CLS) [149] and acetic acid (reviewed by [150]). Two apparently contradictory works suggest that the mutation decreases [148], or extends [147] CLS. In this last work, replicative life span (RLS) is also described to increase in the absence of *GUP1*. The main difference between both studies resides in the amino acid availability in the yeast growth medium, which was four times higher in the study by Li et al. [147] than in the one by Tulha et al. [148]. Considering the above-mentioned suggestions that Gup1 might be implicated in ammonium and TOR signalling, which is controlled by nitrogen/amino acid abundance [79], the mutant will likely respond differently to apoptosis induction in either cultivation condition.

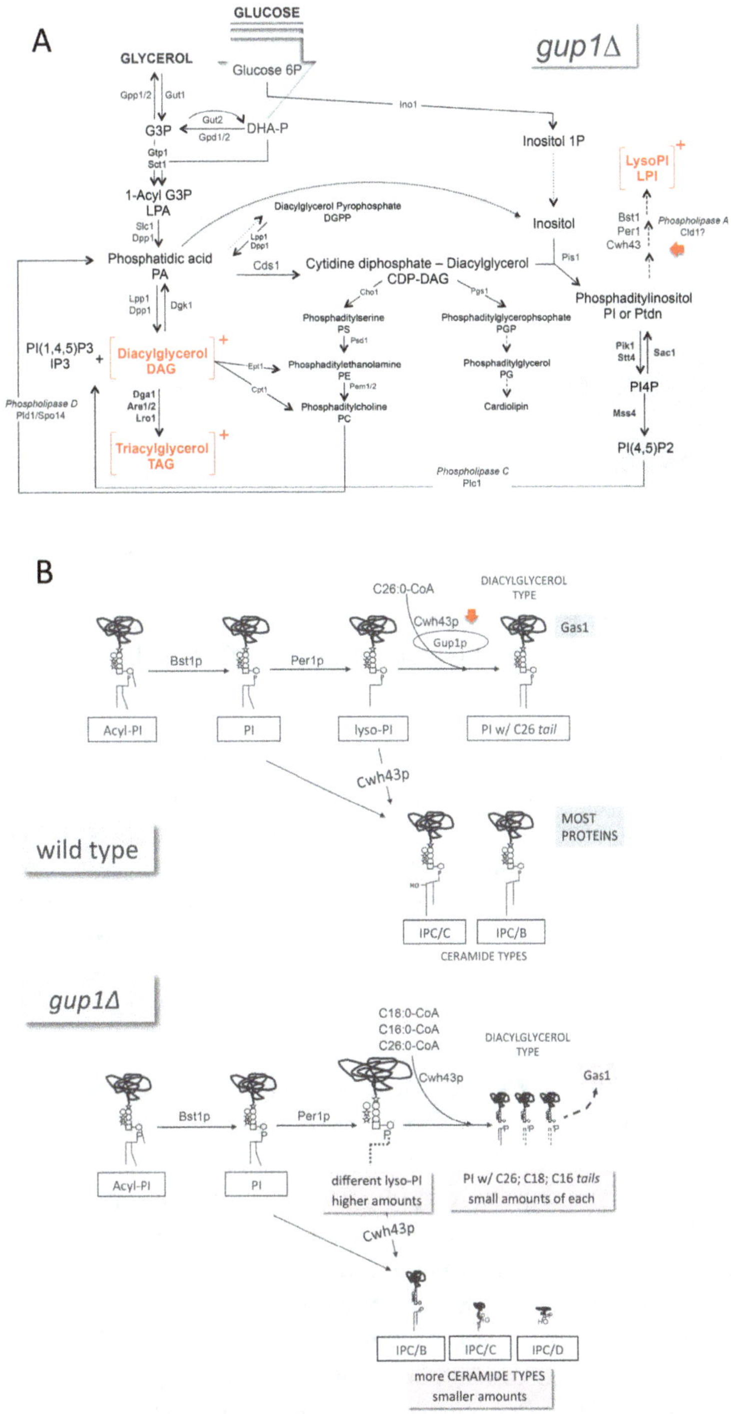

Figure 3. (**A**) Key lipid synthesis pathways from *S. cerevisiae*. In red are shown the types of lipids which amounts are affected by the *GUP1* deletion. Steps in GPI anchor synthesis occurring in *S. cerevisiae* wild-type strain (**B**) and in *gup1*Δ mutant (C). Gas1 is an example of a GPI anchored protein that is found excessively liberated into the medium in the mutant cultures. Red arrows indicate the step in which Bosson et al. (2006) suggest that Gup1 acts as an acyltransferase. GPI anchor backbone is composed of mannose (O), glucosamine (☆) and ethanolamine-P (□). Dark thick scrawls represent peptide chains. PI: phosphatidylinositol; IPC: inositol phosphoceramide. Types B, C and D were defined from bands obtained using TLC [136].

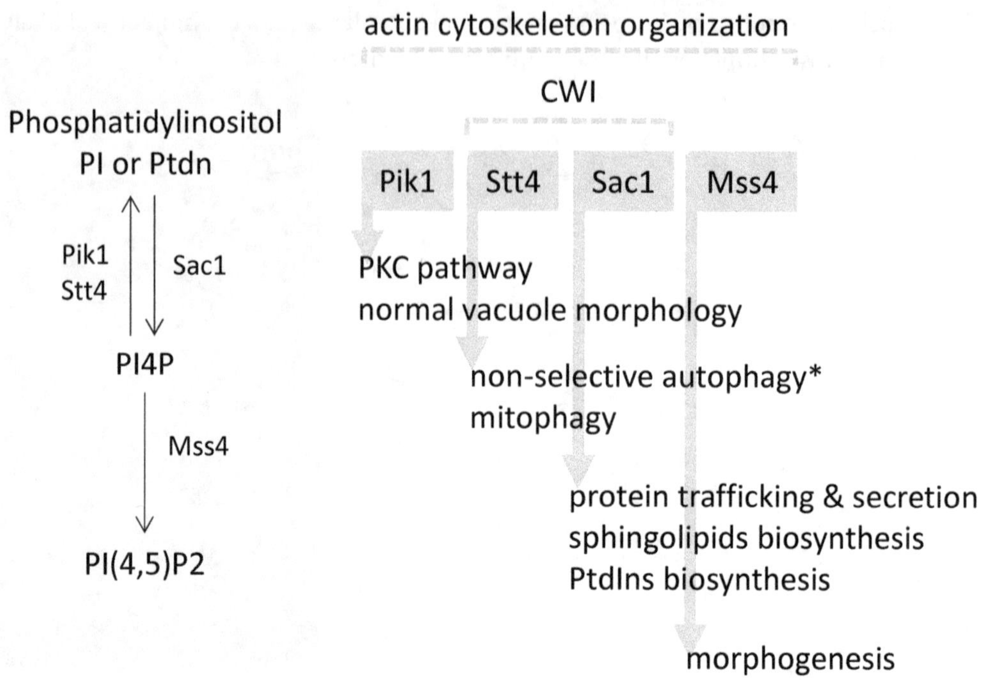

Figure 4. Phenotypes associated to the enzymes known to catalyse the steps from PI to PI(4,5)P2 in *S. cerevisiae*. These phenotypes are also associated with the *GUP1* deletion, with the exception of mitophagy. * Preliminary results [45].

On the other hand, in the presence of lethal concentrations of acetic acid, *gup1*Δ unlike the wild type undergoes a necrotic-like cell death process [148]. This is inferred from the absence of typical apoptotic features: (a) maintenance of the membrane integrity, (b) phosphatidylserine externalisation, (c) depolarisation of mitochondrial membrane, and (d) chromatin condensation. Moreover, preliminary assays on autophagy inducing conditions, i.e., nitrogen starvation, show that the deletion of *GUP1* causes a clear decrease of autophagic activity as determined by GFP-Atg8 (autophagy-related 8) degradation assay [45,151].

3.6. Differentiation and Morphology

The Hedgehog pathway has well-established roles in the development and maintenance of high eukaryote differentiation. In yeasts, changes in morphology and differentiation occur at two very distinct levels. Individual cells can differentiate by changing shape, shifting from yeasts to very elongated polarised cells of true or pseudo hyphae. Colonies can display multiple 3D shapes [152] that relate with the ability to invade and adhere. At this level, *Candida* species, namely *C. albicans*, have been more studied than *S. cerevisiae*, mainly due to their pathogenicity.

C. albicans is a commensal constituent of normal human microflora that acts as an opportunistic pathogen causing infections such as denture stomatitis, thrush and urinary tract infections, but can also provoke more severe systemic infections, in particular in immunocompromised individuals [153,154]. Among the underlying virulence mechanisms, the most well studied so far is probably the morphological switch from budding yeast to filamentous fungus, which, together with cell surface expression of adhesins and invasins (reviewed by [153]), allows colonisation of host tissues and their aggressive invasion, eventually overcoming the endothelial barrier [155]. The infection-associated yeast fast reproduction promotes formation of biofilms, not only on human and other organism tissues, but also on inert surfaces from clinical devices [154,156], constituting a major threat for hospitalised patients, in particular if immunocompromised.

C. albicans has only one GUP orthologue: ca*GUP1*. caGup1 has been implicated in *C. albicans* virulence [29]. Accordingly, the *cagup1*Δ null mutant [157] was strongly affected in the ability to develop hyphae, to adhere, invade and form biofilm, and its colonies exhibited aberrant morphology/differentiation patterns [29]. Interestingly, *M. musculus* HHATL cDNA was able to complement the *cagup1Δ* null mutant morphogenic defects associated with colony invasive growth (Figure 5) [29]. Preliminary data suggests that the hyphae development defects of the mutant can be reverted although not completely by *S. cerevisiae GUP1*, as well as the human HHATL and *D. melanogaster* RASP(HHAT) [158]. This partial reversal corresponds to a notorious cell elongation into pseudohyphae instead of true hyphae. The amino acid sequences of *C. albicans* and *M. musculus* Gup proteins share only 20% similarity, whereas *S. cerevisiae* and *C. albicans GUP1* sequences share 58%. Moreover, as mentioned above, the two key MBOAT amino acid residues differ in yeast Gup1/2 and mammalian HHAT(L). It remains to be seen in the future how far the high eukaryotes HHAT(L) ability goes in complementing yeast phenotypes, namely using alternative expression strategies and if the substitutions in the MBOAT residues play a role.

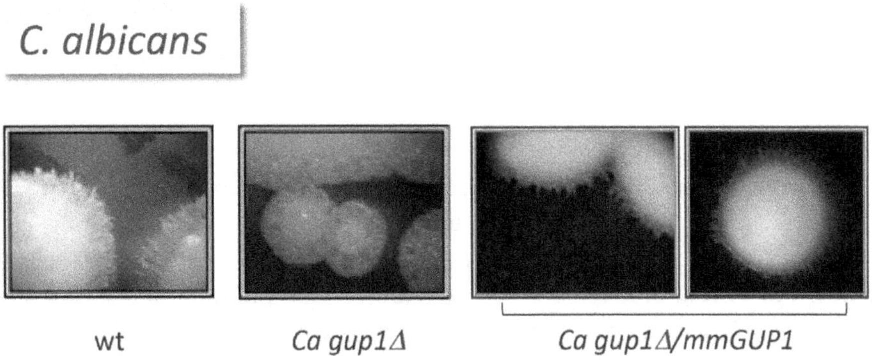

Figure 5. *Mus musculus* HHATL complementation of *C. albicans* colonies hyphae defect caused by the double deletion of *GUP1* alleles [29,156]. Pictures of colony borders were taken after 3 days of incubation in spider medium at 37 °C.

Similarly to *S. cerevisiae*, deletion of *GUP1* in *C. albicans* modifies ergosterol distribution at the level of plasma membrane [29]. Among various lipids of yeast membranes, ergosterol has been used as target of the most common antifungals like polyenes and azoles [159,160]. In *S. cerevisiae*, deletion of *GUP1* causes high resistance to ergosterol synthesis inhibitors including azoles [37]. Likewise, *C. albicans cagup1Δ* null mutant displayed increased resistance to fluconazole, ketoconazole and clotrimazole [29].

Another important phenotype of the *cagup1Δ* null mutant is the delayed ability to form biofilm [29]. The formation of biofilms, as the formation of colonies, derives from extensive colonisation of the environment, generating multicellular structures that allow regular feeding of all cells, comparably to a proto-tissue. Traditionally, *C. albicans* or other closely related yeasts have been extensively used to study biofilm formation. *S. cerevisiae* can be used to mimic biofilm by promoting its growth onto large communities such as giant colonies or mats [2,161–163]. Analogously to the tissues from higher eukaryotes, these multicellular aggregates depend on the production and secretion of extracellular matrix (ECM) [163]. In *S. cerevisiae*, this was shown to be made of two different polysaccharides [10] and to harbour a large and abundant proteome composed of 684 well-identified proteins belonging to very different functional classes [11]. The deletion of *GUP1* in *S. cerevisiae* provokes a profound difference in the composition of the secretome, yielding a sludgy-textured ECM that lost one of the polysaccharides present in wild-type ECM and harbours approximately 15% less proteins, 26% of which are not found in the wild-type ECM [10,11]. Additionally, DIGE analysis identified as well common proteins though present in statistically different amounts [11]. Among the missing proteins, there are some key regulators of metabolic pathways, namely Fbp1, Pyc2 and Pdc5/6 from glycolysis/gluconeogenesis/fermentation; important effectors in cytoskeleton organization like Tub2, Ent2/3, Rvs161 or VPS; stress response and secretory pathway proteins like Vma6/7, Vph1 or Ctt1 and Ecm4/38; and proteins involved in ubiquitination and sumoylation [11]. Moreover, the selective presence in the wild-type ECM of alpha-saccharides remodelling enzymes, such as Glc3 1,4-α-glucan branching enzyme, Mnn2 α-1,2-mannosyltransferase and Sga1 α-glucoamylase, suggest the presence of α-based polysaccharides in opposition to the cell wall beta sugars backbone. These seem to be important for ECM texture as inferred from the absence of these enzymes in *gup1Δ* mutant ECM [11]. On the other hand, the mutant ECM comprises a great number of proteins involved in cell wall remodelling that were not found in the wild-type ECM, including the homologous GPI-anchored putative mannosidases Dcw1 and Dfg1, required for cell wall biosynthesis, Utr2, Kre6 and Krt2 proteins involved in the biosynthesis of β-glucans, Pir1 and Pir2 proteins involved in stabilization of the cell wall, and GPI-anchored protease Yps1 required for cell wall growth and maintenance. Clearly *GUP1* considerably affects the secretion of both glycosides and proteins into the yeast ECM.

4. Proteins Interacting with Yeast Gup1 and Gup2

Although much work is still necessary to fully identify the full interactome of Gup1/2 proteins in yeast, these have putative physical partners, predicted by whole genome screenings, or ascertained in particular single-gene studies in the case of Gup1 (Table 3). Gup2 partners include interesting proteins (Table 3), like Ste2, the receptor for α-factor, a yeast pheromone that initiates signalling response to mating by interacting both with the pheromone and a heterotrimeric G-protein [164]. On the other hand, Gup1 co-localises with Mep2 [165] and Por1 [45] and immunoprecipitates with Por1 and Pil1 [45]. Pil1 is a major constituent of eisosomes, and Por1 is the mitochondrial VDAC (voltage-dependent anion channel). The nature of these two last interactions remains unknown for the time being. Biochemically, Por1 interacts with Sfk1, a suppressor of the calcineurin inhibitor immunosuppressant drug FK506, by which it modulates the mitochondrial ionic balance [166]. *gup1Δ* is resistant to FK506 [65], suggesting that Gup1 might be necessary for the interaction of Por1 with Sfk1, which would therefore be impaired from inhibiting FK506 in the mutant.

Mep2 belongs to a highly conserved Mep/Amt/Rh superfamily that also includes the Rh blood-group human antigens [167]. There are three well-characterised Mep-type ammonium transporters in *S. cerevisiae*, Mep1, 2 and 3 [168]. *S. cerevisiae* Mep2 and its positive regulator Npr1 kinase [169,170], and *C. albicans* Mep1 and Mep2 [171], are indispensable for differentiation, i.e., for true or pseudo-hyphae formation respectively. Mep2 has accordingly been associated with TORC1 signalling, implicating in CLS and ageing [172]. Moreover, Mep2 promotes activation of the PKA pathway by ammonium on nitrogen-starved cells [173]. *GUP1* is required for proper distribution of Mep2 over the plasma membrane, while another protein that also putatively interacts with Gup1, Vma4 (Table 3), is required for Mep2 expression [165]. In *gup1Δ*, Mep2 is localised in patches instead of homogenously over the entire plasma membrane, while both transport and signalling associated to Mep2 function are increased [165]. Although these authors suggest that the effect of Gup1 on Mep2 might be indirect in view of the multiple phenotypes associated with *GUP1* gene involving plasma membrane, rafts and wall, it seems more likely that Gup1 directly negatively regulates Mep2. This would justify the higher Mep2-mediated transport and signalling observed in the strain deleted in *GUP1* (with Mep2 uninhibited). However, Mep2 is released naturally onto the wild-type yeast ECM [11], suggesting that the picture is much more complex.

Table 3. Gup1 and Gup2 ascertained and predicted physical partners.

	Partner	Assay	Type of Study	Key Reference	Assigned Function
Gup1	Fet3	PCA	HTP	[174]	Iron-O_2-oxidoreductase; multicopper oxidase that oxidizes ferrous (Fe^{2+}) to ferric iron (Fe^{3+}) for subsequent cellular uptake by transmembrane permease Ftr1.
	Frk1	BA	HTP	[175]	Protein kinase of unknown cellular role; interacts with rRNA transcription and ribosome biogenesis factors, and the long chain fatty acyl-CoA synthetase Faa3p.
	Hek2	AC	HTP	[176]	RNA-binding protein involved in asymmetric localization of ASH1 mRNA; represses translation of ASH1 mRNA, an effect reversed by Yck1-dependent phosphorylation; regulates telomere position effect and length.
	Mep2	2H/Co-L	Single study	[165]	NH4+ permease; regulation of pseudohyphal growth; expression regulated by nitrogen catabolite repression (NCR).
	Msc7	PCA	HTP	[177]	Cytoplasmic protein of unknown function.
	Nab2	AC	HTP	[178]	Nuclear poly(A)-binding protein; required for nuclear mRNA export and tail length control.
	Pil1	Co-IP	Single study	[45]	Eisosome core component; detected in phosphorylated state in mitochondria; phosphorylated upon Pkc1 hyperactivation in a Slt2p MAPK-dependent fashion.
	Por1	Co-IP/Co-L	Single study	[45]	Mitochondrial voltage-dependent anion channel (VDAC); required for maintenance of mitochondrial osmotic stability and membrane permeability; couples the glutathione pools of the intermembrane space and the cytosol.
	Sat4	BA	HTP	[175]	Ser/Thr protein kinase involved in salt tolerance; functions in regulation of Trk1-Trk2 potassium transporter.
	Vtc4	PCA	HTP	[174]	Vacuolar membrane polyP polymerase; subunit of the vacuolar transporter chaperone (VTC) complex; regulates membrane trafficking; role in non-autophagic vacuolar fusion.
	YHL042W	PCA	HTP	[174]	Putative protein of unknown function.
Gup2	Aqy1	PCA	HTP	[174]	Spore-specific aquaporin.
	Aus1	PCA	HTP	[179]	Plasma membrane sterol transporter of the ATP-binding cassette family; required, along with Pdr11, for uptake of exogenous sterols and their incorporation into the plasma membrane.
	Hsp30	PCA	HTP	[174]	Negative regulator of the H(+)-ATPase Pma1; stress-responsive; induced by heat shock, ethanol, weak organic acid, glucose limitation, and entry into stationary phase.
	Ifa38	PCA	HTP	[174]	Microsomal beta-keto-reductase; mutants exhibit reduced VLCFA synthesis, accumulate high levels of dihydrosphingosine, phytosphingosine and medium-chain ceramides.
	Nam7	AC	HTP	[180]	ATP-dependent RNA helicase.
	Pdr10	PCA	HTP	[179]	ATP-binding cassette (ABC) transporter; multidrug transporter involved in the pleiotropic drug resistance network; regulated by Pdr1p and Pdr3p.
	Pho88	PCA	HTP	[174]	Probable membrane protein involved in phosphate transport; role in the maturation of secretory proteins.
	Sss1	PCA	HTP	[174]	Subunit of the Sec61 translocation complex (Sec61-Sss1-Sbh1); this complex forms a channel for passage of secretory proteins through the ER membrane.
	Ste2	PCA	HTP	[174]	Receptor for α-factor pheromone; interacts with both pheromone and a heterotrimeric G-protein to initiate the signaling response that leads to mating.

HTP: high throughput; PCA: protein complementation assay; AC: affinity capture (RNA); BA: biochemical activity; 2H: two-hybrid; Co-IP: co-immunoprecipitation; Co-L: co-localization.

5. Gup1/2 Homologues from High Eukaryotes

5.1. Hedgehog Pathway

Yeast Gup1 and its close homologue Gup2 share significant sequence homology with proteins belonging to higher eukaryotes, from *Caenorhabditis elegans*, to *Drosophila melanogaster*, to mammals. All of those proteins, identical to Gup1/2 from yeast [23], were identified as members of the membrane bound *O*-acyltransferase (MBOAT) family based on their sequence homology. They all share two highly conserved regions of amino acids: one hydrophobic region containing an invariant histidine and a region composed of moderately hydrophobic amino acids that surround a well-conserved asparagine/aspartic acid residue [23]. The only exception found so far is the mammalian homologue of Gup1 HHATL (Hedgehog acyltransferase-like), in which the homologous regions are maintained, but the conserved histidine residue is replaced with a leucine. Thanks to the high degree of conservation of these two regions, at least 11 members of the MBOAT family have been identified so far in the human genome [38].

MBOAT family proteins also share a common topology. They all have 8–12 TMDs and localise in the ER or Golgi [38,181,182]. HHAT was first predicted to have eight TMDs [40], but this topology seems inconsistent when compared to other members of the MBOAT family. Specifically, the MBOAT signature amino acid residues would be located in the cytoplasm, which would be a feature unique to HHAT among all the other MBOAT members [42]. Recently, the topology of human HHAT has been further investigated using different prediction software, and the results obtained by two independent research groups showed that HHAT contains 10 TMDs and two re-entrant loops. The invariant His is located in the lumenal side of the endoplasmic reticulum, whereas the Asp residue is on the opposite side [42,183]. Conceivably, HHATL should have a similar structure, due to the high sequence similarity.

Most members of the MBOAT family catalyse the transfer of long chain fatty acids to hydroxyl groups of other lipophilic molecules [184–186], or the formation of wax esters [187]. They are categorised into three subgroups based on their biochemical reaction, due to the variety of processes in which they are involved [38]. Mammalian HHAT and its fly counterpart, together with Porcupine and Ghrelin *O*-acyltransferase (GOAT), belong to the group involved in the acylation of secreted proteins [38,182]. In particular, HHAT is the protein responsible for the N-palmitoylation of the Hedgehog morphogens [41,42] and Spitz (also known as Torpedo or DER), a fly protein homologous to TGF-alpha, that either acts as a secreted or as a cell-bound EGFR ligand [42]. *M. musculus* and human HHATL, as mentioned above, have the invariant His necessary for the palmitoylation activity of HHAT replaced by a Leu [40,42]. It is, therefore, expected to not act as an acyltransferase, although it was found to co-localise with both HHAT and Sonic in the ER [21]. It was shown that HHATL interacts directly with Shh decreasing the efficiency of Shh palmitoylation [22], being proposed as a negative regulator of the HHAT-driven N-palmitoylation of Shh and of the Hh pathway. However, the precise mechanism of action has yet to be described.

In mammals as well as in *Drosophila*, HHAT promotes Hedgehog secreted morphogens N-palmitoylation [20–22,40]. Hh morphogens are diffusible proteins involved in long distance cell-cell communication, playing fundamental roles in embryonic development, patterning and cellular differentiation [188–191], as well as in adult tissue homeostasis and cancer [192]. Proper signalling is achieved by a specific gradient of Hh morphogens concentration from the cells that produce it to those destined to bind them [169,193–195]. Hh post-translational lipophilic modifications have a key role in maintaining the protein concentration gradient, influencing the long range signalling [196,197], which depends on the formation of diffusible multimeric forms [198–200]. On the other hand, secretion of Hh morphogens might operate through several mechanisms. One is the formation of specific fat body-derived diffusible lipoprotein particles [196,201], which in *Drosophila* largely depends on lipid rafts [202–204]. Another mechanism is an exosome-mediated pathway [205,206]. Exosomes derive from the endocytic compartment where vesicles are packed [207]. Yeasts also produce and secrete exosomes [208,209], which proteome was characterised in *C. albicans* [210,211]. Still there is no idea regarding their biological roles.

In mammalian cells, Hedgehog responsive cells activate transcription through Gli1. This occurs in the Hh responding cells by the triangulation Hh ligand-Ptch1-Smo. Gli1 can be activated by Smo-independent mechanisms, increasing Hh signalling even in the absence of the Hh ligand. This non-canonical activation of Gli1 was described in association with oesophageal adenocarcinoma [212]. In that case, Gli1 is controlled by a TNFα-derived TORC1/S6K1 pathway. TORC1 promotes the phosphorylation of Gli1 (at S^{384}) by S6K1, freeing it from SuFu and easing the entrance to the nucleus. Additionally, TORC2 promotes the stabilisation of Gli1 by keeping it phosphorylated (at S^{230}) through the action of AKT [212,213]. AKT commands a *star*-type pathway and demands intracellular signalling by phosphoinosides, from which the designation of PI3K-PKB/AKT pathway (reviewed by [214]). Other possible cross-talks with Hedgehog have been suggested to operate at the level of Gli1 activation: MAPK/ERK [215] and KRAS [216].

5.2. HHAT(L) Proteins Expression and Associated Pathologies

Palmitoylation at the N-terminal domain of Hedgehog proteins is necessary for the proper Hh signalling at both long and short range [217]. It was shown in vitro that palmitoylated forms of Shh are 40–160-fold more active compared to unmodified Shh [218]. Palmitoylated Sonic Hh is required for the formation of multimeric complexes that are fundamental for extracellular migration [199]. Therefore, the proper functionality of HHAT is fundamental for maintenance of the correct Hh function. This should stand for the Hh pathway function, namely during embryogenesis. Accordingly, HHAT loss of function by a single G287V missense mutation resulted in serious sex development defects [219]. In general, Hh signalling malfunction, causes severe and often life-threatening developmental disorders such as holoprosencephaly, craniofacial defects and skeletal malformations [31,220–222].

In human tissues, the HHAT(L) picture appears somewhat more complex than predicted [22]. The HHATL gene is expressed in very few tissues and in different amounts: heart > skeletal muscle > brain [34,35,223]. Actually, the gene maps to a locus of dilated cardiomyopathy (CMD1E/OMIM 601154), a disorder characterised by ventricular dilation and impaired systolic function, resulting in congestive heart failure and arrhythmia, eventually causing premature death that also harbours another gene encoding a cardiac muscle Na^+ channel (SCN5A/HH1). Despite this association, the absence of HHATL disease-specific mutations in a cohort of 48 patients [34], leaves no clear indication that HHATL might be responsible for the disease.

It has been estimated that 25% of all human tumours required Hh signalling to maintain tumour cell viability [167]. Corrupted Hh signalling, namely through over-activation, underlies many types of cancer, including small and non-small cell lung cancer, squamous cell and basal cell carcinomas and myeloid leukaemias [224–228]. The importance of Hh signal palmitoylation on this regard has recently been demonstrated. The knockdown of HHAT caused in vitro reduced proliferation and invasiveness of human pancreatic ductal adenocarcinoma [229]. Therefore, it is conceivable that blocking the Hedgehog pathway using HHAT inhibitors may block tumour progression [230,231]. Moreover, HHATL transfected cells presented lower capability of cell proliferation, invasion and tumorigenicity, supporting the notion that the protein might function as a tumour suppressor [36]. This function is compatible with the negative regulation of HHAT-mediated Hh palmitoylation described for HHATL [22]. The reason why this gene is only expressed in few tissues appears intriguing. Other works reported the expression of KIAA1173 (another *GUP1*/HHATL alias—Table 1) in normal skin [232] and nasopharyngeal mucosa [36], and its extreme down-regulation in cells of skin squamous cell carcinoma [232]. Still, the KIAA1173 protein is N-172 amino acids shorter than the cardiac isoform above-mentioned [34], suggesting that the HHATL ORF might undergo extensive RNA or protein splicing and perform different roles in different tissues.

The HHAT gene has also been associated with type 2 diabetes [233]. Its changed expression was found in patients undergoing bariatric surgery [233], further suggesting a correlation with glucose homeostasis and therefore obesity. *Diabetes mellitus* and obesity are closely related [234], through deficient glycerol consumption by glycerol kinase [235] and deficient tissue regulation of glycerol transport by aquaglyceroporins [236]. This once again raises the notion that in yeasts Gup1 implicates in glycerol transport, consumption and production for osmoregulation purposes [19,28] and that it apparently connects nutrient regulation to morphogenic-related Hedgehog-like phenotypes, through a signalling entanglement between the CWI/Pkc1, TORC2/YPK TORC1/Pkh1/2 and HOG/Sho1 pathways.

6. Similarity Analysis of the GUP/HHAT(L) Amino Acid Sequences

Gup proteins from both yeast and higher eukaryotes belong to the MBOAT superfamily of membrane-bound *O*-acyltransferases, although only the mammalian HHA*T* has been assigned a function as an enzyme performing the palmitoylation of the Sonic Hedgehog ligand [21]. All eukaryotes have at least one such gene/protein. Therefore, the databases present an extremely large number of sequences designated as Gup or HHAT(L). This generalised presence does not necessarily mean that the amino acid sequence is highly conserved. Multiple sequence alignment similarity matrices were built using Clustal Omega [237], based on the accession numbers available at UniProt database and confronting them with NCBI available sequences (Supplementary File S1). These were grouped in a simple manner: yeasts, filamentous fungi, mammals, fish, teleostomi, reptiles, birds, amphibian, insects, nematodes, testudines and arthopoda (Supplementary File S2). A few isolated organisms presenting GUP/HHAT(L) sequences were grouped as *others* in Supplementary File S2. Gup1/HHATL is absent from nematodes and insects, while Gup2/HHAT is absent from reptiles, amphibian, testudines, teleostomi and *Zea mays*, the only plant where a Gup amino acid sequence was so far identified. Within each matrix, the percentage of similarity varied between two extreme values (shadowed in File S2). Notably, Gup1/HHATL sequences from nematodes are the most diverse. Generally, Gup1/HHATL sequences are less diverse than the Gup2/HHAT, which is perceptible in the respective cladograms (Figure S1A,B) built using Mega 7 (Molecular Evolutionary Genetics Analysis) [238]. Still, the difference in similarity between the Gup1/HHATL and Gup2/HHAT sequences is very small. As an example, we present in Table 4 the similarity matrix between the better studied Gup proteins: *S. cerevisiae* Gup1 and Gup2, *C. albicans* Gup1, *M. musculus* HHATL and HHAT, *H. sapiens* HHATL and HHAT and *D. melanogaster* RASP/HHAT. The sequence similarity presented is very high. *M. musculus* HHAT shares 20% similarity with yeast Gup2 and 19.3% with yeast Gup1, while HHATL shares 21.9% similarity with yeast Gup2 and 20.7% with yeast Gup1. The frontier between proteins considered HHAT and HHATL is very narrow. Moreover, some sequences actually appear to be too different from the yeast Gup proteins to even be considered as such (Supplementary File S2).

Table 4. Similarity matrix between the most studied GUP proteins.

	C. albicans **SC5314 Gup1**	*S. cerevisiae* **S288c Gup1**	*S. cerevisiae* **S288c Gup2**	*D. melanogaster* **Gup2/RASP**	*M. musculus* **Gup1/HHATL**	*H. sapiens* **Gup1/HHATL**	*M. musculus* **Gup2/HHAT**	*H. sapiens* **Gup2/HHAT**
C. albicans SC5314 Gup1	100.00	57.64	47.02	16.63	19.15	18.45	18.49	18.77
S. cerevisiae S288c Gup1	57.64	100.00	56.61	17.84	20.70	20.45	19.30	19.80
S. cerevisiae S288c Gup2	47.02	56.61	100.00	17.80	20.92	20.20	20.00	21.07
D. melanogaster Gup2/RASP	16.63	17.84	17.80	100.00	21.62	21.14	25.94	26.58
M. musculus Gup1/HHATL	19.15	20.70	20.92	21.62	100.00	90.07	28.16	27.92
H. sapiens Gup1/HHATL	18.45	20.45	20.20	21.14	90.07	100.00	28.33	27.86
M. musculus Gup2/HHAT	18.49	19.30	20.00	25.94	28.16	28.33	100.00	82.91
H. sapiens Gup2/HHAT	18.77	19.80	21.07	26.58	27.92	27.86	82.91	100.00

7. Conclusions

Paracrine signalling has not yet been recognised as occurring in microbial communities. However, the notion that microbes multicellular aggregates can be regarded as tissues, suggested for the first time by Shapiro in 1988 (reviewed by [239]), predicts that each individual cell does not live exclusively in response to environmental stimuli, as happens in planktonic life, but must involve cell–cell communication. It remains unclear whether this may happen through a diffusible chemical, like ammonia [12,240,241] or the quorum-sensing chemicals [13,14], or through a peptide signal like in higher Eukaryotes. The large number and diverse proteome from the yeast ECM [11,210,211] hampers the identification of a putative peptide signal. Additionally, the also high number of proteins identified as being differently expressed in the *GUP1*-deleted mutant ECM compared to the wt strain, does not allow the suggestion of a candidate [11]. Nevertheless, the fact that all Eukaryotes present a Gup/HHAT(L) protein suggests a conserved mechanism in which these proteins may participate. One should, however, not discard the hypothesis that proteins from the same group might perform very different roles in different organisms. The plethora of phenotypes from the yeast *gup1Δ* mutant, and the multiple localisation of the Gup1 protein, actually favour this as having multiple roles.

The most prominent Gup1 and Gup2 localisation is the plasma membrane [19,41,45]. In tmammalian or fly cells, HHAT is almost exclusively located in the ER, while the localisation HHATL is still scarcely documented. Data suggest that yeast Gup1 is, or locates at, a hub between CWI/Pkc1, TORC2/YPK, TORC1/Pkh1/2 and HOG/Sho1 pathways, controlling response to nutrients and stress and morphogenic-related Hedgehog-like phenotypes. Therefore, it is not surprising that *GUP1* deletion-associated phenotypes cover a vast number of pathways controlling basic processes of yeast life (summarised in Figure 6), namely associated with growth and development and that the deletion of the gene does not, in spite of its central role, become lethal. Altogether only fitness appears to be decreased in most of the environmental conditions studied (temperature, pH, stress and so on).

Importantly, most of the phenotypes caused in yeast by the deletion of *GUP1* or *GUP2* are associated with the *GUP1* deletion, in opposition of mammalian and fly cells in which case the information available almost exclusively regards Gup2/HHAT (Figure 6). This does not mean that Gup2 has no function in yeasts. In mammals, the Gup2/HHAT is the enzyme that performs the palmitoylation of the Hh signal [21], while Gup1/HHATL acts as an inhibitor of that palmitoylation [21] through an as yet unknown mechanism. If that would be the case for yeasts, Gup2 could be performing a role that is inhibited by Gup1 and, therefore, all of the phenotypes associated with the deletion of *GUP1* would actually derive from the uninhibited function of Gup2.

A last comment goes to nomenclature. It is important that Gup proteins are not further taken as glycerol transporters. This mistaken allusion has equally appeared in the literature dedicated to yeasts or mammalian cells. Due to the poor understanding of these proteins, the Gup1 from *S. cerevisiae* has even been cloned and heterologously expressed as if it were the actual glycerol permease (e.g., [27]). Moreover, careful naming would be recommended in the fields of research using high eukaryotes, in view of the extreme similarity between HHAT and HHATL and between these and the original *S. cerevisiae* Gup1 and Gup2 proteins. In order to avoid the misguidance generated by the multiple designations (Table 1) we propose a single notification. Based on the examples of the Ras and TOR proteins, for which the name used in the higher eukaryotes and the numeration of yeasts were combined, we propose that HHAT1 and HHAT2 are consistently adopted hereon to be designated Gup1/HHATL and Gup2/HHAT, regardless of the biological system under study.

A lot is still required to unveil the true nature and biological function of the Gup proteins. The results so far contribute to stress that studies based on either yeasts or higher eukaryotes are equally important and complement themselves. In particular, in the near future, the search for the molecular partners/targets of the Gup proteins in yeast will be important, along with an understanding of why Gup1/HHATL is only present and important in a few adult human tissues, the physiology of which does not straightforwardly relate to the Hedgehog pathway.

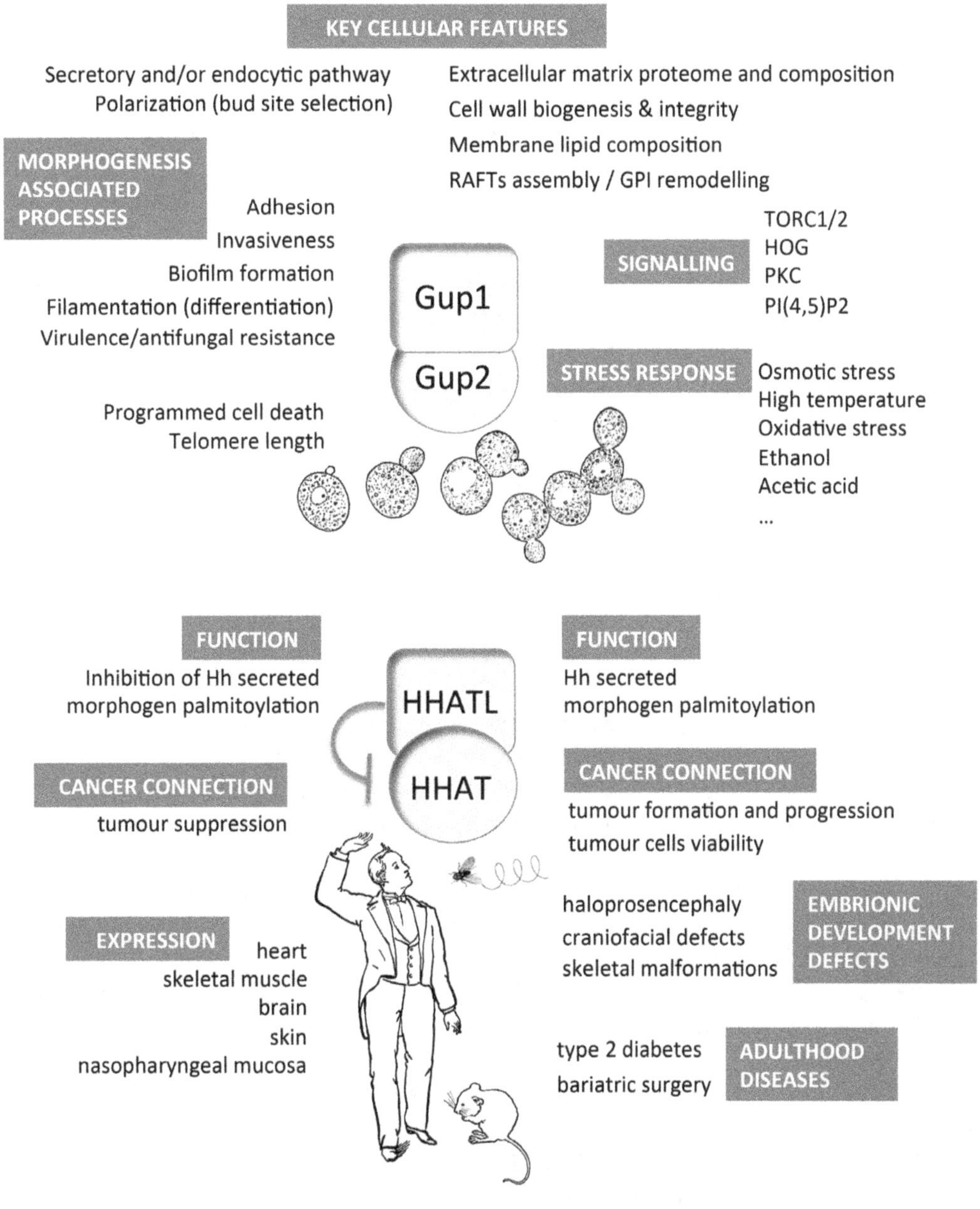

Figure 6. Phenotypes and biological processes associated to the Gup proteins in yeasts (*S. cerevisiae* and *C. albicans*) and in the higher eukaryotes *M. musculus*, *H. sapiens* and *D. melanogaster*. Clipart was downloaded from the free access supplier http://etc.usf.edu/clipart/home/.

Acknowledgments: This work was supported by the strategic programme UID/BIA/04050/2013 (POCI-01-0145-FEDER-007569) funded by national funds through the FCT I.P. and by the ERDF through the COMPETE2020—Programa Operacional Competitividade e Internacionalização (POCI); and by Marie Curie Initial Training Network GLYCOPHARM (PITN-GA-2012-317297). Giulia Cazzanelli is PhD grantee from GLYCOPHARM. Joana Tulha is PhD grantee SFRH/BD/76025/2011 from FCT (Fundação para a Ciência e Tecnologia) from the Portuguese Ministry of Science, Technology and Higher Education. Ricardo Franco-Duarte is post-doc grantee UMINHO/BPD/54/2015. We thank Fábio Faria-Oliveira presently a post-doc at the University of Göteborg, Sweden, for critical reading of the manuscript, and Pedro Ferraz, PhD student from our group at CBMA for all the text corrections and kind availability and support.

Author Contributions: Cândida Lucas, Giulia Cazzanelli and Joana Tulha wrote the paper. Célia Ferreira read, corrected and commented the manuscript. Ricado Franco-Duarte performed all the amino acids sequences analyses.

Conflicts of Interest: The authors declare no conflict of interest.

Nomenclature

According to the rules from Yeast scientific nomenclature, genes, proteins and mutants are presented in, capitals/italic (ex.: *GUP1*), title case (ex.: Gup1), and lower case/italics (ex.: *gup1*Δ), respectively. In the case of other model organisms, proteins and genes are presented as in the corresponding literature.

References

1. Verduyckt, M.; Vignaud, H.; Bynens, T.; van den Brande, J.; Franssens, V.; Cullin, C.; Winderickx, J. Yeast as a model for Alzheimer's disease: Latest studies and advanced strategies. *Meth. Mol. Biol.* **2016**, *1303*, 197–215.
2. Brückner, S.; Mösch, H.U. Choosing the right lifestyle: Adhesion and development in *Saccharomyces cerevisiae. FEMS Microbiol. Rev.* **2012**, *36*, 25–58. [CrossRef] [PubMed]
3. Čáp, M.; Štepánek, L.; Harant, K.; Váchová, L.; Palková, Z. Cell differentiation within a yeast colony: Metabolic and regulatory parallels with a tumor-affected organism. *Mol. Cell* **2012**, *46*, 436–448. [CrossRef] [PubMed]
4. Váchová, L.; Palková, Z. Physiological regulation of yeast cell death in multicellular colonies is triggered by ammonia. *J. Cell Biol.* **2005**, *169*, 711–717. [CrossRef] [PubMed]
5. Váchová, L.; Palková, Z. Caspases in yeast apoptosis-like death: Facts and artefacts. *FEMS Yeast Res.* **2007**, *7*, 12–21. [CrossRef] [PubMed]
6. Turrà, D.; Nordzieke, D.; Vitale, S.; El Ghalid, M.; di Pietro, A. Hyphal chemotropism in fungal pathogenicity. *Semin. Cell Dev. Biol.* **2016**, *57*, 69–75. [CrossRef] [PubMed]
7. St'ovíček, V.; Váchová, L.; Kuthan, M.; Palková, Z. General factors important for the formation of structured biofilm-like yeast colonies. *Fungal Genet. Biol.* **2010**, *47*, 1012–1022. [CrossRef] [PubMed]
8. Váchová, L.; Kucerová, H.; Devaux, F.; Ulehlová, M.; Palková, Z. Metabolic diversification of cells during the development of yeast colonies. *Environ. Microbiol.* **2009**, *11*, 494–504. [CrossRef] [PubMed]
9. Zupan, J.; Raspor, P. Invasive growth of *Saccharomyces cerevisiae* depends on environmental triggers: A quantitative model. *Yeast* **2010**, *27*, 217–228. [PubMed]
10. Faria-Oliveira, F.; Carvalho, J.; Belmiro, C.L.; Ramalho, G.; Pavão, M.; Lucas, C.; Ferreira, C. Elemental biochemical analysis of the polysaccharides in the extracellular matrix of the yeast *Saccharomyces cerevisiae. J. Basic Microbiol.* **2015**, *55*, 685–694. [CrossRef] [PubMed]
11. Faria-Oliveira, F.; Carvalho, J.; Ferreira, C.; Hernáez, M.L.; Gil, C.; Lucas, C. Quantitative differential proteomics of yeast extracellular matrix: There is more to it than meets the eye. *BMC Microbiol.* **2015**, *15*, 271. [CrossRef] [PubMed]
12. Palková, Z.; Váchová, L. Ammonia signaling in yeast colony formation. *Int. Rev. Cytol.* **2003**, *225*, 229–272. [PubMed]
13. Chen, H.; Fujita, M.; Feng, Q.; Clardy, J.; Fink, G.R. Tyrosol is a quorum sensing molecule in *Candida albicans. Proc. Nat. Acad. Sci. USA* **2004**, *101*, 5048–5052. [CrossRef] [PubMed]
14. Sprague, G.F.; Winans, S.C. Eukaryotes learn how to count: Quorum sensing by yeast. *Genes Dev.* **2006**, *20*, 1045–1049. [CrossRef] [PubMed]
15. Guerrero, I.; Kornberg, T.B. Hedgehog and its circuitous journey from producing to target cells. *Semin. Cell Dev. Biol.* **2014**, *33*, 52–62. [CrossRef] [PubMed]
16. Fukuda, T.; Nagai, Y.; Ohya, Y. Molecular mechanism of VDE-initiated intein homing in yeast nuclear genome. *Adv. Biophys.* **2004**, *38*, 215–232. [CrossRef]
17. Chong, S.; Shao, Y.; Paulus, H.; Benner, J.; Perler, F.B.; Xu, M.Q. Protein splicing involving the *Saccharomyces cerevisiae* VMA intein. The steps in the splicing pathway, side reactions leading to protein cleavage, and establishment of an in vitro splicing system. *J. Biol. Chem.* **1997**, *271*, 22159–22168.
18. McGuire, C.; Cotter, K.; Stransky, L.; Forgac, M. Regulation of V-ATPase assembly and function of V-ATPases in tumor cell invasiveness. *Biochim. Biophys. Acta* **2016**, *1857*, 1213–1218. [CrossRef] [PubMed]
19. Hölst, B.; Lunde, C.; Lages, F.; Oliveira, R.; Lucas, C.; Kielland-Brandt, M.C. *GUP1* and its close homologue *GUP2*, encoding multimembrane-spanning proteins involved in active glycerol uptake in *Saccharomyces cerevisiae. Mol. Microbiol.* **2000**, *37*, 108–124. [CrossRef] [PubMed]
20. Chamoun, Z.; Mann, R.K.; Nellen, D.; von Kessler, D.P.; Bellotto, M.; Beachy, P.A.; Basler, K. Skinny Hedgehog, an acyltransferase required for palmitoylation and activity of the Hedgehog signal. *Science* **2001**, *293*, 2080–2084. [CrossRef] [PubMed]

21. Buglino, J.A.; Resh, M.D. Hhat is a palmitoylacyltransferase with specificity for N-palmitoylation of Sonic Hedgehog. *J. Biol. Chem.* **2008**, *283*, 22076–22088. [CrossRef] [PubMed]
22. Abe, Y.; Kita, Y.; Niikura, T. Mammalian Gup1, a homolog of *Saccharomyces cerevisiae* glycerol uptake/transporter 1, acts as a negative regulator for N-terminal palmitoylation of Sonic hedgehog. *FEBS J.* **2008**, *275*, 318–331. [CrossRef] [PubMed]
23. Hofmann, K. A superfamily of membrane-bound *O*-acyltransferases with implications for Wnt signaling. *Trends Biochem. Sci.* **2000**, *26*, 111–112. [CrossRef]
24. Neves, L.; Oliveira, R.; Lucas, C. Yeast orthologues associated with glycerol transport and metabolism. *FEMS Yeast Res.* **2004**, *5*, 51–62. [CrossRef] [PubMed]
25. Neves, L.; Lages, F.; Lucas, C. New insights on glycerol transport in *Saccharomyces cerevisiae*. *FEBS Lett.* **2004**, *565*, 160–162. [CrossRef] [PubMed]
26. Ferreira, C.; van Voorst, F.; Martins, A.; Neves, L.; Oliveira, R.; Kielland-Brandt, M.; Lucas, C.; Brandt, A. A member of the sugar transporter family, Stl1p is the glycerol/H^+ symporter in *Saccharomyces cerevisiae*. *Mol. Biol. Cell* **2005**, *16*, 2068–2076. [CrossRef] [PubMed]
27. Yu, K.O.; Kim, S.W.; Han, S.O. Engineering of glycerol utilization pathway for ethanol production by *Saccharomyces cerevisiae*. *Bioresour. Technol.* **2010**, *101*, 4157–4161. [CrossRef] [PubMed]
28. Ferreira, C.; Silva, S.; van Voorst, F.; Aguiar, C.; Kielland-Brandt, M.C.; Brandt, A.; Lucas, C. Absence of Gup1p in *Saccharomyces cerevisiae* results in defective cell wall composition, assembly, stability and morphology. *FEMS Yeast Res.* **2006**, *6*, 1027–1038. [CrossRef] [PubMed]
29. Ferreira, C.; Silva, S.; Faria-Oliveira, F.; Pinho, E.; Henriques, M.; Lucas, C. *Candida albicans* virulence and drug-resistance requires the *O*-acyltransferase Gup1p. *BMC Microbiol.* **2010**, *10*, 238. [CrossRef] [PubMed]
30. Micchelli, C.A.; The, I.; Selva, E.; Mogila, V.; Perrimon, N. Rasp, a putative transmembrane acyltransferase, is required for Hedgehog signaling. *Development* **2002**, *129*, 843–851. [PubMed]
31. Lee, J.D.; Treisman, J.E. Sightless has homology to transmembrane acyltransferases and is required to generate active Hedgehog protein. *Curr. Biol.* **2001**, *11*, 1147–1152. [CrossRef]
32. Amanai, K.; Jiang, J. Distinct roles of Central Missing and Dispatched in sending the Hedgehog signal. *Development* **2001**, *128*, 5119–5127. [PubMed]
33. NCBI—National Center for Biotechnological Information. Available online: http://www.ncbi.nlm.nih.gov/ (assessed on 28 October 2016).
34. Soejima, H.; Kawamoto, S.; Akai, J.; Miyoshi, O.; Arai, Y.; Morohka, T.; Matsuo, S.; Niikawa, N.; Kimura, A.; Okubo, K.; et al. Isolation of novel heart-specific genes using the Body Map database. *Genomics* **2001**, *74*, 115–120. [CrossRef] [PubMed]
35. Zhang, S.Q.; Peng, H.; Jiang, H.Y.; Hu, H.; Zhang, J.H.; Yao, K.T.; Zhao, T. Cloning of human KIAA1173 gene and biological characterization of transfected 6–10B cells. *Di Yi Jun Yi Da Xue Xue Bao* **2005**, *25*, 1216–1220. [PubMed]
36. Zhang, S.Q.; Peng, H.; Song, L.Y.; Li, X.M.; Jiang, H.Y.; Yao, K.T.; Zhao, T. Detection of KIAA1173 gene expression in nasopharyngeal carcinoma tissues and cell lines on tissue microarray. *Ai Zheng* **2005**, *24*, 1322–1326. [PubMed]
37. Ferreira, C.; Lucas, C. The yeast *O*-acyltransferase Gup1p interferes in lipid metabolism with direct consequences on the sphingolipid-sterol-ordered domains integrity/assembly. *Biochim. Biophys. Acta* **2008**, *1778*, 2648–2653. [CrossRef] [PubMed]
38. Chang, C.C.Y.; Sun, J.; Chang, T.-Y. Membrane-bound *O*-acyltransferases (MBOATs). *Front. Biol.* **2011**, *6*, 177–182.
39. Bleve, G.; di Sansebastiano, G.P.; Grieco, F. Over-expression of functional *Saccharomyces cerevisiae GUP1*, induces proliferation of intracellular membranes containing ER and Golgi resident proteins. *Biochim. Biophys. Acta* **2011**, *1808*, 733–744. [CrossRef] [PubMed]
40. Buglino, J.A.; Resh, M.D. Identification of conserved regions and residues within Hedgehog acyltransferase critical for palmitoylation of Sonic Hedgehog. *PLoS ONE* **2010**, *5*, e11195. [CrossRef] [PubMed]
41. Bleve, G.; Zacheo, G.; Cappello, M.S.; Dellaglio, F.; Grieco, F. Subcellular localization and functional expression of the glycerol uptake protein 1 (*GUP1*) of *Saccharomyces cerevisiae* tagged with green fluorescent protein. *Biochem. J.* **2005**, *390*, 145–155. [CrossRef] [PubMed]

42. Konitsiotis, A.D.; Jovanović, B.; Ciepla, P.; Spitaler, M.; Lanyon-Hogg, T.; Tate, E.W.; Magee, A.I. Topological analysis of Hedgehog acyltransferase, a multipalmitoylated transmembrane protein. *J. Biol. Chem.* **2015**, *290*, 3293–3307. [CrossRef] [PubMed]
43. Transmembrane Helix Prediction. Available online: http://www.cbs.dtu.dk/services/TMHMM/ (assessed on 15 July 2016).
44. CSS-Palm. Prediction of Palmitoylation Site. Available online: http://csspalm.biocuckoo.org (assessed on 15 July 2016).
45. Tulha, J.; Lucas, C.; CBMA-Centre of Molecular and Environmental Biology, University of Minho, Braga, Portugal. Unpublished work, 2016.
46. YEASTRACT, Yeast Search for Transcriptional Regulators and Consensus Tracking. Available online: http://www.yeastract.com/ (assessed on 15 July 2016).
47. Oliveira, R.; Lucas, C. Expression studies of *GUP1* and *GUP2*, genes involved in glycerol active transport in *Saccharomyces cerevisiae*, using semi-quantitative RT-PCR. *Curr. Genet.* **2004**, *46*, 140–146. [CrossRef] [PubMed]
48. Posas, F.; Chambers, J.R.; Heyman, J.A.; Hoeffler, J.P.; de Nadal, E.; Ariño, J. The transcriptional response of yeast to saline stress. *J. Biol. Chem.* **2000**, *275*, 17249–17255. [CrossRef] [PubMed]
49. Rep, M.; Krantz, M.; Thevelein, J.M.; Hohmann, S. The transcriptional response of *Saccharomyces cerevisiae* to osmotic shock. Hot1p and Msn2p/Msn4p are required for the induction of subsets of high osmolarity glycerol pathway-dependent genes. *J. Biol. Chem.* **2000**, *275*, 8290–8300. [CrossRef] [PubMed]
50. Yale, J.; Bohnert, H.J. Transcript expression in *Saccharomyces cerevisiae* at high salinity. *J. Biol. Chem.* **2001**, *276*, 15996–16007. [CrossRef] [PubMed]
51. Hohmann, S. Osmotic stress signaling and osmoadaptation in yeasts. *Microbiol. Mol. Biol. Rev.* **2002**, *66*, 300–372. [CrossRef] [PubMed]
52. Henry, S.A.; Kohlwein, S.D.; Carman, G.M. Metabolism and regulation of glycerolipids in the yeast *Saccharomyces cerevisiae*. *Genetics* **2012**, *190*, 317–349. [CrossRef] [PubMed]
53. Larsson, C.; Påhlman, I.-L.; Ansell, R.; Rigoulet, M.; Adler, L.; Gustafsson, L. The importance of the glycerol 3-phosphate shuttle during aerobic growth of *Saccharomyces cerevisiae*. *Yeast* **1988**, *14*, 347–357. [CrossRef]
54. Siderius, M.; van Wuytswinkel, O.; Reijenga, K.A.; Kelders, M.; Mager, W.H. The control of intracellular glycerol in *Saccharomyces cerevisiae* influences osmotic stress response and resistance to increased temperature. *Mol. Microbiol.* **2000**, *36*, 1381–1390. [CrossRef] [PubMed]
55. Oelkers, P.; Tinkelenberg, A.; Erdeniz, N.; Cromley, D.; Billheimer, T.; Sturley, S.L. A lecithin cholesterol acyltransferase-like gene mediates diacylglycerol esterification in yeast. *J. Biol. Chem.* **2000**, *275*, 15609–15612. [CrossRef] [PubMed]
56. Dardalhon, M.; Lin, W.; Nicolas, A.; Averbeck, D. Specific transcriptional responses induced by 8-methoxypsoralen and UVA in yeast. *FEMS Yeast Res.* **2007**, *7*, 866–878. [CrossRef] [PubMed]
57. Ni, L.; Snyder, M. A genomic study of the bipolar bud site selection pattern in *Saccharomyces cerevisiae*. *Mol. Biol. Cell* **2001**, *12*, 2147–2170. [CrossRef] [PubMed]
58. Casamayor, A.; Snyder, M. Bud-site selection and cell polarity in budding yeast. *Curr. Opin. Microbiol.* **2002**, *5*, 179–186. [CrossRef]
59. Bonangelino, C.J.; Chavez, E.M.; Bonifacino, J.S. Genomic screen for vacuolar protein sorting genes in *Saccharomyces cerevisiae*. *Mol. Biol. Cell* **2002**, *13*, 2486–2501. [CrossRef] [PubMed]
60. Reiner, S.; Micolod, D.; Zellnig, G.; Schneiter, R. A genomewide screen reveals a role of mitochondria in anaerobic uptake of sterols in yeast. *Mol. Biol. Cell* **2006**, *17*, 90–103. [CrossRef] [PubMed]
61. Askree, S.H.; Yehuda, T.; Smolikov, S.; Gurevich, R.; Hawk, J.; Coker, C.; Krauskopf, A.; Kupiec, M.; McEachern, M.J. A genome-wide screen for *Saccharomyces cerevisiae* deletion mutants that affect telomere length. *Proc. Natl. Acad. Sci. USA* **2004**, *101*, 8658–8663. [CrossRef] [PubMed]
62. Dos Santos, S.C.; Sá-Correia, I. Genome-wide identification of genes required for yeast growth under imatinib stress: Vacuolar H^+-ATPase function is an important target of this anticancer drug. *OMICS* **2009**, *13*, 185–198. [CrossRef] [PubMed]
63. Liao, C.; Hu, B.; Arno, M.J.; Panaretou, B. Genomic screening in vivo reveals the role played by vacuolar H^+ ATPase and cytosolic acidification in sensitivity to DNA-damaging agents such as cisplatin. *Mol. Pharmacol.* **2007**, *71*, 2416–2425. [CrossRef] [PubMed]

64. Tucker, C.L.; Fields, S. Quantitative genome-wide analysis of yeast deletion strain sensitivities to oxidative and chemical stress. *Comp. Funct. Genom.* **2004**, *5*, 216–224. [CrossRef] [PubMed]
65. Viladevall, L.; Serrano, R.; Ruiz, A.; Domenech, G.; Giraldo, J.; Barceló, A.; Ariño, J. Characterization of the calcium-mediated response to alkaline stress in *Saccharomyces cerevisiae*. *J. Biol. Chem.* **2004**, *279*, 43614–43624. [CrossRef] [PubMed]
66. Darvishi, E.; Omidi, M.; Bushehri, A.A.; Golshani, A.; Smith, M.L. Thymol antifungal mode of action involves telomerase inhibition. *Med. Mycol.* **2013**, *51*, 826–834. [CrossRef] [PubMed]
67. Zhao, J.; Lin, W.; Ma, X.; Lu, Q.; Ma, X.; Bian, G.; Jiang, L. The protein kinase Hal5p is the high-copy suppressor of lithium-sensitive mutations of genes involved in the sporulation and meiosis as well as the ergosterol biosynthesis in *Saccharomyces cerevisiae*. *Genomics* **2010**, *95*, 290–298. [CrossRef] [PubMed]
68. Zhou, X.; Arita, A.; Ellen, T.P.; Liu, X.; Bai, J.; Rooney, J.P.; Kurtz, A.D.; Klein, C.B.; Dai, W.; Begley, T.J.; et al. A genome-wide screen in *Saccharomyces cerevisiae* reveals pathways affected by arsenic toxicity. *Genomics* **2009**, *94*, 294–307. [CrossRef] [PubMed]
69. Rossouw, D.; Bauer, F.F. Comparing the transcriptomes of wine yeast strains: Toward understanding the interaction between environment and transcriptome during fermentation. *Appl. Microbiol. Biotechnol.* **2009**, *84*, 937–954. [CrossRef] [PubMed]
70. Ma, M.; Liu, L.Z. Quantitative transcription dynamic analysis reveals candidate genes and key regulators for ethanol tolerance in *Saccharomyces cerevisiae*. *BMC Microbiol.* **2010**, *10*, 169. [CrossRef] [PubMed]
71. Lesage, G.; Shapiro, J.; Specht, C.A.; Sdicu, A.M.; Ménard, P.; Hussein, S.; Tong, A.H.; Boone, C.; Bussey, H. An interactional network of genes involved in chitin synthesis in *Saccharomyces cerevisiae*. *BMC Genet.* **2005**, *6*, 8. [CrossRef] [PubMed]
72. De Nobel, H.; Ruiz, C.; Martin, H.; Morris, W.; Brul, S.; Molina, M.; Klis, F.M. Cell wall perturbation in yeast results in dual phosphorylation of the Slt2/Mpk1 MAP kinase and in an Slt2-mediated increase in *FKS2*-lacZ expression, glucanase resistance and thermotolerance. *Microbiology* **2000**, *46*, 2121–2132. [CrossRef] [PubMed]
73. Ferreira, C. Identification and characterisation of the glycerol/H^+ symporter in *Saccharomyces cerevisiaie* and the involvement of related genes in the cell wall integrity. Ph.D. Thesis, University of Minho, Braga, Portugal, September 2005. Available online: http://repositorium.sdum.uminho.pt/ (assessed on 28 October 2016).
74. Levin, D.E. Regulation of cell wall biogenesis in *Saccharomyces cerevisiae*: The cell wall integrity signaling pathway. *Genetics* **2011**, *189*, 1145–1175. [CrossRef] [PubMed]
75. Lockshon, D.; Olsen, C.P.; Brett, C.L.; Chertov, A.; Merz, A.J.; Lorenz, D.A.; van Gilst, M.R.; Kennedy, B.K. Rho signaling participates in membrane fluidity homeostasis. *PLoS ONE* **2012**, *7*, e45049. [CrossRef] [PubMed]
76. Audhya, A.; Loewith, R.; Parsons, A.B.; Gao, L.; Tabuchi, M.; Zhou, H.; Boone, C.; Hall, M.N.; Emr, S.D. Genome-wide lethality screen identifies new PI4,5P2 effectors that regulate the actin cytoskeleton. *EMBO J.* **2004**, *23*, 3747–3757. [CrossRef] [PubMed]
77. Yan, G.; Lai, Y.; Jiang, Y. The TOR complex 1 is a direct target of Rho1 GTPase. *Mol. Cell* **2012**, *45*, 743–753. [CrossRef] [PubMed]
78. Perez, P.; Rincón, S.A. Rho GTPases: Regulation of cell polarity and growth in yeasts. *Biochem. J.* **2010**, *426*, 243–253. [CrossRef] [PubMed]
79. Loewith, R.; Hall, M.N. Target of Rapamycin (TOR) in nutrient signaling and growth control. *Genetics* **2011**, *189*, 1177–1201. [CrossRef] [PubMed]
80. Jacinto, E.; Lorberg, A. TOR regulation of AGC kinases in yeast and mammals. *Biochem. J.* **2008**, *410*, 19–37. [CrossRef] [PubMed]
81. Kuranda, K.; Leberre, V.; Sokol, S.; Palamarczyk, G.; François, J. Investigating the caffeine effects in the yeast *Saccharomyces cerevisiae* brings new insights into the connection between TOR, PKC and Ras/cAMP signalling pathways. *Mol. Microbiol.* **2006**, *61*, 1147–1166. [CrossRef] [PubMed]
82. Hardwick, J.S.; Kuruvilla, F.G.; Tong, J.K.; Shamji, A.F.; Schreiber, S.L. Rapamycin-modulated transcription defines the subset of nutrient-sensitive signaling pathways directly controlled by the Tor proteins. *Proc. Natl. Acad. Sci. USA* **1999**, *96*, 14866–14870. [CrossRef] [PubMed]
83. Lorenz, M.C.; Heitman, J. TOR mutations confer rapamycin resistance by preventing interaction with FKBP12-rapamycin. *J. Biol. Chem.* **1995**, *270*, 27531–27537. [PubMed]

84. Kunz, J.; Loeschmann, A.; Deuter-Reinhard, M.; Hall, M.N. *FAP1*, a homologue of human transcription factor NF-X1, competes with rapamycin for binding to FKBP12 in yeast. *Mol. Microbiol.* **2000**, *37*, 1480–9143. [CrossRef] [PubMed]
85. Simamora, P.; Alvarez, J.M.; Yalkowsky, S.H. Solubilization of rapamycin. *Int. J. Pharm.* **2001**, *213*, 25–29. [CrossRef]
86. Cameron, A.M.; Nucifora, F.C., Jr.; Fung, E.T.; Livingston, D.J.; Aldape, R.A.; Ross, C.A.; Snyder, S.H. FKBP12 binds the inositol 1,4,5-trisphosphate receptor at leucine-proline (1400–1401) and anchors calcineurin to this FK506-like domain. *J. Biol. Chem.* **1997**, *272*, 27582–27588. [CrossRef] [PubMed]
87. Yoshimoto, H.; Saltsman, K.; Gasch, A.P.; Li, H.X.; Ogawa, N.; Botstein, D.; Brown, P.O.; Cyert, M.S. Genome-wide analysis of gene expression regulated by the Calcineurin/Crz1p signaling pathway in *Saccharomyces cerevisiae*. *J. Biol. Chem.* **2002**, *277*, 31079–31088. [CrossRef] [PubMed]
88. Odorizzi, G.; Babst, M.; Emr, S.D. Phosphoinositide signaling and the regulation of membrane trafficking in yeast. *Trends Biol. Sci.* **2000**, *25*, 229–235. [CrossRef]
89. Staschke, K.A.; Dey, S.; Zaborske, J.M.; Palam, L.R.; McClintick, J.N.; Pan, T.; Edenberg, H.J.; Wek, R.C. Integration of general amino acid control and target of rapamycin (TOR) regulatory pathways in nitrogen assimilation in yeast. *J. Biol. Chem.* **2010**, *285*, 16893–16911. [CrossRef] [PubMed]
90. Valenzuela, L.; Aranda, C.; González, A. TOR modulates *GCN4*-dependent expression of genes turned on by nitrogen limitation. *J. Bacteriol.* **2001**, *183*, 2331–2334. [CrossRef] [PubMed]
91. Schuldiner, M.; Collins, S.R.; Thompson, N.J.; Denic, V.; Bhamidipati, A.; Punna, T.; Ihmels, J.; Andrews, B.; Boone, C.; Greenblatt, J.F.; et al. Exploration of the function and organization of the yeast early secretory pathway through an epistatic miniarray profile. *Cell* **2005**, *123*, 507–519. [CrossRef] [PubMed]
92. Burston, H.E.; Maldonado-Báez, L.; Davey, M.; Montpetit, B.; Schluter, C.; Wendland, B.; Conibear, E. Regulators of yeast endocytosis identified by systematic quantitative analysis. *J. Cell Biol.* **2009**, *185*, 1097–1110. [CrossRef] [PubMed]
93. Martin, S.G. Spontaneous cell polarization: Feedback control of Cdc42 GTPase breaks cellular symmetry. *Bioessays* **2015**, *37*, 1193–1201. [CrossRef] [PubMed]
94. Raitt, D.C.; Posas, F.; Saito, H. Yeast Cdc42 GTPase and Ste20 PAK-like kinase regulate Sho1-dependent activation of the Hog1 MAPK pathway. *EMBO J.* **2000**, *19*, 4623–4631. [CrossRef] [PubMed]
95. Styles, E.; Youn, J.Y.; Mattiazzi Usaj, M.; Andrews, B. Functional genomics in the study of yeast cell polarity: Moving in the right direction. *Philos. Trans. R. Soc. B* **2016**, *368*, 20130118. [CrossRef] [PubMed]
96. Hohmann, S. An integrated view on a eukaryotic osmoregulation system. *Curr. Genet.* **2015**, *61*, 373–382. [CrossRef] [PubMed]
97. Gordon, M.D.; Nusse, R. Wnt signaling: Multiple pathways, multiple receptors, and multiple transcription factors. *J. Biol. Chem.* **2006**, *281*, 22429–22433. [CrossRef] [PubMed]
98. Cuadrado, A.; Nebreda, A.R. Mechanisms and functions of p38 MAPK signalling. *Biochem. J.* **2010**, *429*, 403–417. [CrossRef] [PubMed]
99. Han, J.; Lee, J.D.; Bibbs, L.; Ulevitch, R.J. A MAP kinase targeted by endotoxin and hyperosmolarity in mammalian cells. *Science* **1994**, *265*, 808–811. [CrossRef] [PubMed]
100. Kumar, S.; McLaughlin, M.M.; McDonnell, P.C.; Lee, J.C.; Livi, G.P.; Young, P.R. Human mitogen-activated protein kinase CSBP1, but not CSBP2, complements a *hog1* deletion in yeast. *J. Biol. Chem.* **1995**, *270*, 29043–29046. [PubMed]
101. Takekawa, M.; Posas, F.; Saito, H. A human homolog of the yeast Ssk2/Ssk22 MAP kinase kinase kinases, MTK1, mediates stress-induced activation of the p38 and JNK pathways. *EMBO J.* **1997**, *16*, 4973–4982. [CrossRef] [PubMed]
102. Bikkavilli, R.K.; Feigin, M.E.; Malbon, C.C. p38 mitogen-activated protein kinase regulates canonical Wnt–β-catenin signaling by inactivation of GSK3β. *J. Cell Sci.* **2008**, *121*, 3598–3607. [CrossRef] [PubMed]
103. Gallo, K.A.; Johnson, G.L. Mixed-lineage kinase control of JNK and p38 MAPK pathways. *Nat. Rev.* **2002**, *3*, 663–672. [CrossRef] [PubMed]
104. The Wnt Homepage. Available online: http://web.stanford.edu/group/nusselab/cgi-bin/wnt/ (assessed on 8 August 2016).
105. Tatebayashi, H.; Yamamoto, K.; Nagoya, M.; Takayama, T.; Nishimura, A.; Sakurai, M.; Momma, T.; Saito, H. Osmosensing and scaffolding functions of the oligomeric four-transmembrane domain osmosensor Sho1. *Nat. Commun.* **2015**, *6*, 6975. [CrossRef] [PubMed]

106. Nishimura, A.; Yamamoto, K.; Oyama, M.; Kozuka-Hata, H.; Saito, H.; Tatebayashi, K. Scaffold protein Ahk1, which associates with Hkr1, Sho1, Ste11, and Pbs2, inhibits cross talk signaling from the Hkr1 osmosensor to the Kss1 mitogen-activated protein kinase. *Mol. Cell Biol.* **2016**, *36*, 1109–1123. [CrossRef] [PubMed]
107. Rodríguez-Peña, J.M.; García, R.; Nombela, C.; Arroyo, J. The high-osmolarity glycerol (HOG) and cell wall integrity (CWI) signalling pathways interplay: A yeast dialogue between MAPK routes. *Yeast* **2010**, *27*, 495–502. [CrossRef] [PubMed]
108. Winkler, A.; Arkind, C.; Mattison, C.P.; Burkholder, A.; Knoche, K.; Ota, I. Heat stress activates the yeast high-osmolarity glycerol mitogen-activated protein kinase pathway, and protein tyrosine phosphatases are essential under heat stress. *Eukaryot. Cell* **2002**, *1*, 163–173. [CrossRef] [PubMed]
109. Alonso-Monge, R.; Real, E.; Wojda, I.; Bebelman, J.P.; Mager, W.H.; Siderius, M. Hyperosmotic stress response and regulation of cell wall integrity in *Saccharomyces cerevisiae* share common functional aspects. *Mol. Microbiol.* **2001**, *41*, 717–730. [CrossRef] [PubMed]
110. Bermejo, C.; Rodríguez, E.; García, R.; Rodríguez-Peña, J.M.; de la Concepción, M.L.R.; Rivas, C.; Arias, P.; Nombela, C.; Posas, F.; Arroyo, J. The sequential activation of the yeast HOG and *SLT2* pathways is required for cell survival to cell wall stress. *Mol. Biol. Cell* **2008**, *19*, 1113–1124. [CrossRef] [PubMed]
111. Faria-Oliveira, F. First molecular and biochemical characterization of the extracelular matrix of *Saccharomyces cerevisiae*. Ph.D. Thesis, University of Minho, Braga, Portugal, September 2013. Available online: http://repositorium.sdum.uminho.pt/ (assessed on 28 October 2016).
112. García, R.; Bermejo, C.; Grau, C.; Pérez, R.; Rodríguez-Peña, J.M.; Francois, J.; Nombela, C.; Arroyo, J. The global transcriptional response to transient cell wall damage in Saccharomyces cerevisiae and its regulation by the cell integrity signaling pathway. *J. Biol. Chem.* **2004**, *279*, 15183–15195. [CrossRef] [PubMed]
113. Bosson, R.; Jaquenoud, M.; Conzelmann, A. *GUP1* of *Saccharomyces cerevisiae* encodes an O-acyltransferase involved in remodeling of the GPI anchor. *Mol. Biol. Cell* **2006**, *17*, 2736–2645. [CrossRef] [PubMed]
114. Friant, S.; Lombardi, R.; Schmelze, T.; Hall, M.N.; Riezman, H. Sphingolipid base signalling via PKh kinases is required for endocytosis in yeast. *EMBO J.* **2001**, *20*, 6783–6792. [CrossRef] [PubMed]
115. Roelants, F.M.; Torrance, P.D.; Bezman, N.; Thorner, J. Pkh1 and Pkh2 differentially phosphorylate and activate Ypk1 and Ykr2 and define protein kinase modules required for maintenance of cell wall integrity. *Mol. Biol. Cell* **2002**, *13*, 3005–3028. [CrossRef] [PubMed]
116. Roelants, F.M.; Torrance, P.D.; Thorner, J. Differential roles of PDK1- and PDK2-phosphorylation sites in the yeast AGC kinases Ypk1, Pck1 and Sch9. *Nat. Rev. Microbiol.* **2004**, *150*, 3289–3304.
117. Friant, S.; Zanolari, B.; Riezman, H. Increased protein kinase or decreased PP2A activity bypasses sphigoid base requirement in endocytosis. *EMBO J.* **2000**, *19*, 2834–2844. [CrossRef] [PubMed]
118. Dupré, S.; Haguenauer-Tsapis, R. Raft partitioning of the yeast uracil permease during trafficking along the endocytic pathway. *Traffic* **2003**, *4*, 83–96. [CrossRef] [PubMed]
119. Neubauer, H.A.; Pitson, S.M. Roles, regulation and inhibitors of sphingosine kinase 2. *FEBS J.* **2013**, *280*, 5317–5336. [CrossRef] [PubMed]
120. Roelants, F.M.; Breslow, D.K.; Muir, A.; Weissman, J.S.; Thornera, J. Protein kinase Ypk1 phosphorylates regulatory proteins Orm1 and Orm2 to control sphingolipid homeostasis in *Saccharomyces cerevisiae. Proc. Natl. Acad. Sci. USA* **2011**, *108*, 19222–19227. [CrossRef] [PubMed]
121. Muir, A.; Roelants, F.M.; Timmons, G.; Leskoske, K.L.; Thorner, J. Down-regulation of TORC2-Ypk1 signaling promotes MAPK-independent survival under hyperosmotic stress. *eLife* **2015**, *4*, e09336. [CrossRef] [PubMed]
122. Lee, J.; Reiter, W.; Dohnal, I.; Gregori, C.; Beese-Sims, S.; Kuchler, K.; Ammerer, G.; Levin, D.E. *MAPK Hog1 closes the S. cerevisiae* glycerol channel Fps1 by phosphorylating and displacing its positive regulators. *Genes Dev.* **2013**, *27*, 2590–2601. [CrossRef] [PubMed]
123. Gomes, K.N.; Freitas, S.M.; Pais, T.M.; Fietto, J.L.; Totola, A.H.; Arantes, R.M.; Martins, A.; Lucas, C.; Schuller, D.; Casal, M.; et al. Deficiency of Pkc1 activity affects glycerol metabolism in *Saccharomyces cerevisiae. FEMS Yeast Res.* **2005**, *5*, 767–776. [CrossRef] [PubMed]
124. Van Meer, G. Lipid traffic in animal cells. *Annu. Rev. Cell Biol.* **1989**, *5*, 247–275. [CrossRef] [PubMed]
125. Bagnat, M.; Keranen, S.; Shevchenko, A.; Simons, K. Lipid rafts function in biosynthetic delivery of proteins to the cell surface in yeast. *Proc. Natl. Acad. Sci. USA* **2000**, *97*, 3254–3259. [CrossRef] [PubMed]
126. Jacobson, K.; Mouritsen, O.G.; Anderson, R.G.W. Lipid rafts: At a crossroad between cell biology and physics. *Nat. Cell Biol.* **2007**, *9*, 7–14. [CrossRef] [PubMed]

127. Mollinedo, F. Lipid raft involvement in yeast cell growth and death. *Front. Oncol.* **2012**, *2*, 140. [CrossRef] [PubMed]

128. Spira, F.; Mueller, N.S.; Beck, G.; von Olshausen, P.; Beig, J.; Wedlich-Söldner, R. Patchwork organization of the yeast plasma membrane into numerous coexisting domains. *Nat. Cell Biol.* **2012**, *14*, 640–648. [CrossRef] [PubMed]

129. Bagnat, M.; Simons, K. Lipid rafts in protein sorting and cell polarity in budding yeast *Saccharomyces cerevisiae*. *Biol. Chem.* **2002**, *383*, 1475–1480. [CrossRef] [PubMed]

130. Bagnat, M.; Chang, A.; Simons, K. Plasma membrane proton ATPase Pma1p requires raft association for surface delivery in yeast. *Mol. Biol. Cell* **2001**, *12*, 4129–4138. [CrossRef] [PubMed]

131. Malinska, K.; Malinsky, J.; Opekarova, M.; Tanner, W. Visualization of protein compartmentation within the plasma membrane of living yeast cells. *Mol. Biol. Cell* **2003**, *14*, 4427–4436. [CrossRef] [PubMed]

132. Kuroda, Y.; Suzuki, N.; Kataoka, T. The effect of posttranslational modifications on the interaction of Ras2 with adenylyl cyclase. *Science* **1993**, *259*, 683–686. [CrossRef] [PubMed]

133. Berchtold, D.; Walther, T. TORC2 Plasma membrane localization is essential for cell viability and restricted to a distinct domain. *Mol. Biol. Cell* **2009**, *20*, 1565–1575. [CrossRef] [PubMed]

134. Li, Y.; Prinz, W.A. ATP-binding cassette (ABC) transporters mediate nonvesicular, raft-modulated sterol movement from the plasma membrane to the endoplasmic reticulum. *J. Biol. Chem.* **2004**, *279*, 45226–45234. [CrossRef] [PubMed]

135. Odorizzi, G.; Cowles, C.R.; Emr, S.D. The AP-3 complex: A coat of many colours. *Trends Cell Biol.* **1998**, *8*, 282–288. [CrossRef]

136. Bosson, R.; Guillas, I.; Vionnet, C.; Roubaty, C.; Conzelmann, A. Incorporation of ceramides into *Saccharomyces cerevisiae* glycosylphosphatidylinositol-anchored proteins can be monitored *in vitro*. *Euk. Cell* **2009**, *8*, 306–314. [CrossRef] [PubMed]

137. Fujita, M.; Watanabe, R.; Jaensch, N.; Romanova-Michaelides, M.; Satoh, T.; Kato, M.; Riezman, H.; Yamaguchi, Y.; Maeda, Y.; Kinoshita, T. Sorting of GPI-anchored proteins into ER exit sites by p24 proteins is dependent on remodeled GPI. *J. Cell Biol.* **2011**, *194*, 61–75. [CrossRef] [PubMed]

138. Sipos, G.; Reggiori, F.; Vionnet, C.; Conzelmann, A. Alternative lipid remodelling pathways for glycosylphosphatidylinositol membrane anchors in *Saccharomyces cerevisiae*. *EMBO J.* **1997**, *16*, 3494–3505. [CrossRef] [PubMed]

139. Fujita, M.; Umemura, M.; Yoko-o, T.; Jigami, Y. *PER1* is required for GPI-phospholipase A2 activity and involved in lipid remodeling of GPI-anchored proteins. *Mol. Biol. Cell* **2006**, *17*, 5253–5264. [CrossRef] [PubMed]

140. Umemura, M.; Fujita, M.; Yoko-o, T.; Fukamizu, A.; Jigami, Y. *Saccharomyces cerevisiae CWH43* is involved in the remodelling of the lipid moiety of GPI anchors to ceramides. *Mol. Biol. Cell* **2007**, *18*, 4304–4316. [CrossRef] [PubMed]

141. Yoko-o, T.; Ichikawa, D.; Miyagishi, Y.; Kato, A.; Umemura, M.; Takase, K.; Ra, M.; Ikeda, K.; Taguchi, R.; Jigami, Y. Determination and physiological roles of the glycosylphosphatidylinositol lipid remodelling pathway in yeast. *Mol. Microbiol.* **2013**, *88*, 140–155. [CrossRef] [PubMed]

142. Carmona-Gutierrez, D.; Eisenberg, T.; Büttner, S.; Meisinger, C.; Kroemer, G.; Madeo, F. Apoptosis in yeast: Triggers, pathways, subroutines. *Cell Death Differ.* **2010**, *17*, 763–773. [CrossRef] [PubMed]

143. Palková, Z.; Váchová, L. Yeast cell differentiation: Lessons from pathogenic and non-pathogenic yeasts. *Semin. Cell Dev. Biol.* **2016**, *57*, 110–119. [CrossRef] [PubMed]

144. Laun, P.; Pichova, A.; Madeo, F.; Fuchs, J.; Ellinger, A.; Kohlwein, S.; Dawes, I.; Fröhlich, K.U.; Breitenbach, M. Aged mother cells of *Saccharomyces cerevisiae* show markers of oxidative stress and apoptosis. *Mol. Microbiol.* **2001**, *39*, 1166–1173. [CrossRef] [PubMed]

145. Fabrizio, P.; Pletcher, S.D.; Minois, N.; Vaupel, J.W.; Longo, V.D. Chronological aging-independent replicative life span regulation by Msn2/Msn4 and Sod2 in *Saccharomyces cerevisiae*. *FEBS Lett.* **2004**, *557*, 136–142. [CrossRef]

146. Herker, E.; Jungwirth, H.; Lehmann, K.A.; Maldener, C.; Fröhlich, K.U.; Wissing, S.; Büttner, S.; Fehr, M.; Sigrist, S.; Madeo, F. Chronological aging leads to apoptosis in yeast. *J. Cell Biol.* **2004**, *164*, 501–507. [CrossRef] [PubMed]

147. Li, B.; Skinner, C.; Castello, P.R.; Kato, M.; Easlon, E.; Xie, L.; Li, T.; Lu, S.P.; Wang, C.; Tsang, F.; et al. Identification of potential calorie restriction-mimicking yeast mutants with increased mitochondrial respiratory chain and nitric oxide levels. *J. Aging Res.* **2011**, *2011*, 673185. [CrossRef] [PubMed]
148. Tulha, J.; Faria-Oliveira, F.; Costa, C.; Lucas, C.; Ferreira, C. *Saccharomyces cerevisiae* Gup1p is required for a programmed cell death process. *BMC Microbiol.* **2012**, *12*, 80. [CrossRef] [PubMed]
149. Fabrizio, P.; Longo, V.D. Chronological aging-induced apoptosis in yeast. *Biochim. Biophys. Acta* **2008**, *1783*, 1280–1285. [CrossRef] [PubMed]
150. Guaragnella, N.; Antonacci, L.; Passarella, S.; Marra, E.; Giannattasio, S. Achievements and perspectives in yeast acetic acid-induced programmed cell death pathways. *Biochem. Soc. Trans.* **2011**, *39*, 1538–1543. [CrossRef] [PubMed]
151. Shintani, T.; Klionsky, D.J. Autophagy in health and disease: A double-edged sword. *Science* **2004**, *306*, 990–995. [CrossRef] [PubMed]
152. Granek, J.A.; Magwene, P.M. Environmental and genetic determinants of colony morphology in yeast. *PLoS Genet.* **2010**, *6*, e1000823. [CrossRef] [PubMed]
153. Mayer, F.L.; Wilson, D.; Hube, B. *Candida albicans* pathogenicity mechanisms. *Virulence* **2013**, *4*, 119–128. [CrossRef] [PubMed]
154. Eggimann, P.; Pittet, D. Candida colonization index and subsequent infection in critically ill surgical patients: 20 years later. *Intensive Care Med.* **2014**, *40*, 1429–1448. [CrossRef] [PubMed]
155. Cheng, S.C.; Joosten, L.A.; Kullberg, B.J.; Netea, M.G. Interplay between *Candida albicans* and the mammalian innate host defense. *Infect. Immun.* **2012**, *80*, 1304–1313. [CrossRef] [PubMed]
156. Nobile, C.J.; Johnson, A.D. *Candida albicans* biofilms and human disease. *Annu. Rev. Microbiol.* **2015**, *69*, 71–92. [CrossRef] [PubMed]
157. Kayingo, G.; Martins, A.; Andrie, R.; Neves, L.; Lucas, C.; Wong, B. A permease encoded by *STL1* is required for active glycerol uptake by *Candida albicans*. *Microbiology* **2009**, *155*, 1547–1557. [CrossRef] [PubMed]
158. Armada, R. Human, mouse, fly and yeast *GUP1* orthologues in *Candida albicans*. Master's Thesis, University of Minho, Braga, Portugal, September 2011. Available online: http://repositorium.sdum.uminho.pt/ (assessed on 28 October 2016).
159. Pemán, J.; Cantón, E.; Espinel-Ingroff, A. Antifungal drug resistance mechanisms. *Expert Rev. Anti Infect. Ther.* **2009**, *7*, 453–460. [CrossRef] [PubMed]
160. Sanglard, D.; Ischer, F.; Parkinson, T.; Falconer, D.; Bille, J. *Candida albicans* mutations in the ergosterol biosynthetic pathway and resistance to several antifungal agentes. *Antimicrob. Agents Chemother.* **2003**, *47*, 2040–2412. [CrossRef]
161. Reynolds, T.B.; Fink, G.R. Bakers' yeast, a model for fungal biofilm formation. *Science* **2001**, *291*, 878–881. [CrossRef] [PubMed]
162. Bojsen, R.K.; Andersen, K.S.; Regenberg, B. *Saccharomyces cerevisiae*—A model to uncover molecular mechanisms for yeast biofilm biology. *FEMS Immunol. Med. Microbiol.* **2012**, *65*, 169–182. [CrossRef] [PubMed]
163. Faria-Oliveira, F.; Carvalho, J.; Belmiro, C.L.; Martinez-Gomariz, M.; Hernaez, M.L.; Pavão, M.; Gil, C.; Lucas, C.; Ferreira, C. Methodologies to generate, extract, purify and fractionate yeast ECM for analytical use in proteomics and glycomics. *BMC Microbiol.* **2014**, *14*, 244. [CrossRef] [PubMed]
164. Alvaro, C.G.; O'Donnell, A.F.; Prosser, D.C.; Augustine, A.A.; Goldman, A.; Brodsky, J.L.; Cyert, M.S.; Wendland, B.; Thorner, J. Specific α-arrestins negatively regulate *Saccharomyces cerevisiae* pheromone response by down-modulating the G-protein-coupled receptor Ste2. *Mol. Cell Biol.* **2014**, *34*, 2660–2681. [CrossRef] [PubMed]
165. Van Zeebroeck, G.; Kimpe, M.; Vandormael, P.; Thevelein, J.M. A split-ubiquitin two-hybrid screen for proteins physically interacting with the yeast amino acid transceptor Gap1 and ammonium transceptor Mep2. *PLoS ONE* **2011**, *6*, e24275. [CrossRef] [PubMed]
166. Butcher, R.A.; Schreiber, S.L. A small molecule suppressor of FK506 that targets the mitochondria and modulates ionic balance in *Saccharomyces cerevisiae*. *Chem. Biol.* **2003**, *10*, 521–531. [CrossRef]
167. Andrade, S.L.; Einsle, O. The Amt/Mep/Rh family of ammonium transport proteins. *Mol. Membr. Biol.* **2007**, *24*, 357–365. [CrossRef] [PubMed]
168. Marini, A.-M.; Soussi-Boudekou, S.; Vissers, S.; André, B. A family of ammonium transporters in *Saccharomyces cerevisiae*. *Mol. Cell Biol.* **1997**, *17*, 4282–4293. [CrossRef] [PubMed]

169. Boeckstaens, M.; André, B.; Marini, A.M. The yeast ammonium transport protein Mep2 and its postivie regulator, the Npr1 kinase, play an importante role in normal and pesudohyphal growth on various nitrogen media through retrieval of excreted ammonium. *Mol. Microbiol.* **2007**, *64*, 534–546. [CrossRef] [PubMed]
170. Boeckstaens, M.; André, B.; Marini, A.M. Distinct transport mechanisms in yeast ammonium transport/sensor proteins of the Mep/Amt/Rh family and impact on filamentation. *J. Biol. Chem.* **2008**, *283*, 21362–21370. [CrossRef] [PubMed]
171. Neuhäuser, B.; Dunkel, N.; Satheesh, S.V.; Morschhäuser, J. Role of the Npr1 kinase in ammonium transport and signaling by the ammonium permease Mep2 in *Candida albicans*. *Euk. Cell* **2011**, *10*, 332–342. [CrossRef] [PubMed]
172. Powers, R.W., III; Kaeberlein, M.; Caldwell, S.D.; Kennedy, B.K.; Fields, S. Extension of chronological life span in yeast by decreased TOR pathway signaling. *Genes Dev.* **2006**, *20*, 174–184. [CrossRef] [PubMed]
173. Thevelein, J.M.; Geladé, R.; Holsbeeks, I.; Lagatie, O.; Popova, Y.; Rolland, F.; Stolz, F.; van de Velde, S.; van Dijck, P.; Vandormael, P.; et al. Nutrient sensing systems for rapid activation of the protein kinase A pathway in yeast. *Biochem. Soc. Trans.* **2005**, *33*, 253–256. [CrossRef] [PubMed]
174. Miller, J.P.; Lo, R.S.; Ben-Hur, A.; Desmarais, C.; Stagljar, I.; Noble, W.S.; Fields, S. Large-scale identification of yeast integral membrane protein interactions. *Proc. Natl. Acad. Sci. USA* **2005**, *102*, 12123–12128. [CrossRef] [PubMed]
175. Ptacek, J.; Devgan, G.; Michaud, G.; Zhu, H.; Zhu, X.; Fasolo, J.; Guo, H.; Jona, G.; Breitkreutz, A.; Sopko, R.; et al. Global analysis of protein phosphorylation in yeast. *Nature* **2005**, *438*, 679–684. [CrossRef] [PubMed]
176. Hasegawa, Y.; Irie, K.; Gerber, A.P. Distinct roles for Khd1p in the localization and expression of bud-localized mRNAs in yeast. *RNA* **2008**, *14*, 2333–2347. [CrossRef] [PubMed]
177. Schlecht, U.; Miranda, M.; Suresh, S.; Davis, R.W.; St Onge, R.P. Multiplex assay for condition-dependent changes in protein-protein interactions. *Proc. Natl. Acad. Sci. USA* **2012**, *109*, 9213–9218. [CrossRef] [PubMed]
178. Batisse, J.; Batisse, C.; Budd, A.; Boettcher, B.; Hurt, E. Purification of nuclear poly(A)-binding protein Nab2 reveals association with the yeast transcriptome and a messenger ribonucleoprotein core structure. *J. Biol. Chem.* **2009**, *284*, 34911–34917. [CrossRef] [PubMed]
179. Snider, J.; Hanif, A.; Lee, M.E.; Jin, K.; Yu, A.R.; Graham, C.; Chuk, M.; Damjanovic, D.; Wierzbicka, M.; Tang, P.; et al. Mapping the functional yeast ABC transporter interactome. *Nat. Chem. Biol.* **2013**, *9*, 565–572. [CrossRef] [PubMed]
180. Johansson, M.J.; He, F.; Spatrick, P.; Li, C.; Jacobson, A. Association of yeast Upf1p with direct substrates of the NMD pathway. *Proc. Natl. Acad. Sci. USA* **2007**, *104*, 20872–20877. [CrossRef] [PubMed]
181. Chang, S.C.; Magee, A.I. Acyltransferases for secreted signalling proteins. *Mol. Membr. Biol.* **2009**, *26*, 104–113. [CrossRef] [PubMed]
182. Masumoto, N.; Lanyon-Hogg, T.; Rodgers, U.R.; Konitsiotis, A.D.; Magee, A.I.; Tate, E.W. Membrane bound *O*-acyltransferases and their inhibitors. *Biochem. Soc. Trans.* **2015**, *43*, 246–252. [CrossRef] [PubMed]
183. Matevossian, A.; Resh, M.D. Membrane topology of hedgehog acyltransferase. *J. Biol. Chem.* **2014**, *290*, 2235–2243. [CrossRef] [PubMed]
184. Cases, S.; Smith, S.J.; Zheng, Y.W.; Myers, H.M.; Lear, S.R.; Sande, E.; Novak, S.; Collins, C.; Welch, C.B.; Lusis, A.J.; et al. Identification of a gene encoding an acyl CoA:diacylglycerol acyltransferase, a key enzyme in triacylglycerol synthesis. *Proc. Natl. Acad. Sci. USA* **1998**, *95*, 13018–13023. [CrossRef] [PubMed]
185. Chang, C.; Lee, C.Y.; Chang, E.T.; Cruz, J.C.; Levesque, M.C.; Chang, T.Y. Recombinant acyl-CoA:cholesterol acyltransferase-1 (ACAT-1) purified to essential homogeneity utilizes cholesterol in mixed micelles or in vesicles in a highly cooperative manner. *J. Biol. Chem.* **1998**, *273*, 35132–35141. [CrossRef] [PubMed]
186. Stahl, U.; Stalberg, K.; Stymne, S.; Ronne, H. A family of eukaryotic lysophospholipid acyltransferases with broad specificity. *FEBS Lett.* **2008**, *582*, 305–309. [CrossRef] [PubMed]
187. Yen, C.L.; Monetti, M.; Burri, B.J.; Farese, R.V., Jr. The triacylglycerol synthesis enzyme DGAT1 also catalyzes the synthesis of diacylglycerols, waxes, and retinyl esters. *J. Lipid Res.* **2005**, *46*, 1502–1511. [CrossRef] [PubMed]
188. Echelard, Y.; Epstein, D.J.; St-Jacques, B.; Shen, L.; Mohler, J.; McMahon, J.A.; McMahon, A.P. Sonic hedgehog, a member of a family of putative signaling molecules, is implicated in the regulation of CNS polarity. *Cell* **1993**, *75*, 1417–1430. [CrossRef]

189. Ma, C.; Zhou, Y.; Beachy, P.A.; Moses, K. The segment polarity gene hedgehog is required for progression of the morphogenetic furrow in the developing Drosophila eye. *Cell* **1993**, *75*, 927–938. [CrossRef]
190. Heemskerk, J.; DiNardo, S. Drosophila hedgehog acts as a morphogen in cellular patterning. *Cell* **1994**, *76*, 449–460. [CrossRef]
191. Ericson, J.; Muhr, J.; Placzek, M.; Lints, T.; Jessell, T.M.; Edlund, T. Sonic hedgehog induces the differentiation of ventral forebrain neurons: A common signal for ventral patterning within the neural tube. *Cell* **1995**, *81*, 747–756. [CrossRef]
192. Nybakken, K.; Perrimon, N. Hedgehog signal transduction: Recent findings. *Curr. Opin. Genet. Dev.* **2002**, *12*, 503–511. [CrossRef]
193. Gritli-Linde, A.; Lewis, P.; McMahon, A.P.; Linde, A. The whereabouts of a morphogen: Direct evidence for short- and graded long-range activity of hedgehog signaling peptides. *Dev. Biol.* **2001**, *236*, 364–386. [CrossRef] [PubMed]
194. Stamataki, D.; Ulloa, F.; Tsoni, S.V.; Mynett, A.; Briscoe, J. A gradient of Gli activity mediates graded Sonic Hedgehog signaling in the neural tube. *Genes Dev.* **2005**, *19*, 626–641. [CrossRef] [PubMed]
195. Fuccillo, M.; Joyner, A.L.; Fishell, G. Morphogen to mitogen: The multiple roles of hedgehog signalling in vertebrate neural development. *Nat. Rev. Neurosci.* **2006**, *7*, 772–783. [CrossRef] [PubMed]
196. Panaková, D.; Sprong, H.; Marois, E.; Thiele, C.; Eaton, S. Lipoprotein particles are required for Hedgehog and Wingless signalling. *Nature* **2005**, *435*, 58–65. [CrossRef] [PubMed]
197. Callejo, A.; Culi, J.; Guerrero, I. Patched, the receptor of Hedgehog, is a lipoprotein receptor. *Proc. Natl. Acad. Sci. USA* **2008**, *105*, 912–917. [CrossRef] [PubMed]
198. Zeng, X.; Goetz, J.A.; Suber, L.M.; Scott, W.J.; Schreiner, C.M.; Robbins, D.J. A freely diffusible form of Sonic hedgehog mediates long-range signalling. *Nature* **2001**, *411*, 716–720. [CrossRef] [PubMed]
199. Chen, M.H.; Li, Y.J.; Kawakami, T.; Xu, M.; Chuang, P.T. Palmitoylation is required for the production of a soluble multimeric Hedgehog protein complex and long-range signaling in vertebrates. *Genes Dev.* **2004**, *18*, 641–659. [CrossRef] [PubMed]
200. Goetz, J.A.; Singh, S.; Suber, L.M.; Kull, F.J.; Robbins, D.J. A highly conserved amino-terminal region of sonic hedgehog is required for the formation of its freely diffusible multimeric form. *J. Biol. Chem.* **2006**, *281*, 4087–4093. [CrossRef] [PubMed]
201. Steinhauer, J.; Treisman, J.E. Lipid-modified morphogens: Functions of fats. *Curr. Opin. Genet. Dev.* **2009**, *19*, 308–314. [CrossRef] [PubMed]
202. Burke, R.; Nellen, D.; Bellotto, M.; Hafen, E.; Senti, K.A.; Dickson, B.J.; Basler, K. Dispatched, a novel sterol-sensing domain protein dedicated to the release of cholesterol-modified hedgehog from signaling cells. *Cell* **1999**, *99*, 803–815. [CrossRef]
203. Katanaev, V.L.; Solis, G.P.; Hausmann, G.; Buestorf, S.; Katanayeva, N.; Schrock, Y.; Stuermer, C.A.; Basler, K. Reggie-1/Flotillin-2 promotes secretion of the long-range signalling forms of Wingless and Hedgehog in Drosophila. *EMBO J.* **2008**, *27*, 509–521. [CrossRef] [PubMed]
204. Palm, W.; Swierczynska, M.M.; Kumari, V.; Ehrhart-Bornstein, M.; Bornstein, S.R.; Eaton, S. Secretion and signaling activities of lipoprotein-associated hedgehog and non-sterol-modified hedgehog in flies and mammals. *PLoS Biol.* **2013**, *11*, e1001505. [CrossRef] [PubMed]
205. Liegeois, S.; Benedetto, A.; Garnier, J.M.; Schwab, Y.; Labouesse, M. The V0-ATPase mediates apical secretion of exosomes containing Hedgehog-related proteins in *Caenorhabditis elegans*. *J. Cell Biol.* **2006**, *173*, 949–961. [CrossRef] [PubMed]
206. Vyas, N.; Walvekar, A.; Tate, D.; Lakshmanan, V.; Bansal, D.; Lo Cicero, A.; Raposo, G.; Palakodeti, D.; Dhawan, J. Vertebrate Hedgehog is secreted on two types of extracellular vesicles with different signaling properties. *Nat. Sci. Rep.* **2014**, *4*, 7357. [CrossRef] [PubMed]
207. Raposo, G.; Stoorvogel, W. Extracellular vesicles: Exosomes, microvesicles, and friends. *J. Cell Biol.* **2013**, *200*, 373–383. [CrossRef] [PubMed]
208. Oliveira, D.L.; Nakayasu, E.S.; Joffe, L.S.; Guimarães, A.J.; Sobreira, T.J.P.; Nosanchuk, J.D.; Cordero, R.J.B.; Frases, S.; Casadevall, A.; Almeida, I.C.; et al. Biogenesis of extracellular vesicles in yeast. Many questions and few answers. *Commun. Integr. Biol.* **2010**, *3*, 533–535. [CrossRef] [PubMed]

209. Oliveira, D.L.; Nakayasu, E.S.; Joffe, L.S.; Guimarães, A.J.; Sobreira, T.J.P.; Nosanchuk, J.D.; Cordero, R.J.B.; Frases, S.; Casadevall, A.; Almeida, I.C.; et al. Characterization of yeast extracellular vesicles: Evidence for the participation of different pathways of cellular traffic in vesicle biogenesis. *PLoS ONE* **2010**, *5*, e11113. [CrossRef] [PubMed]
210. Gil-Bona, A.; Llama-Palacios, A.; Parra, C.M.; Vivanco, F.; Nombela, C.; Monteoliva, L.; Gil, C. Proteomics unravels extracellular vesicles as carriers of classical cytoplasmic proteins in *Candida albicans*. *J. Prot. Res.* **2015**, *14*, 142–153. [CrossRef] [PubMed]
211. Gil-Bona, A.; Monteoliva, L.; Gil, C. Global proteomic profiling of the secretome of *Candida albicans ecm33* cell wall mutant reveals the involvement of Ecm33 in Sap2 secretion. *J. Prot. Res.* **2015**, *14*, 4270–4281. [CrossRef] [PubMed]
212. Wang, Y.; Ding, Q.; Yen, C.J.; Xia, W.; Izzo, J.G.; Lang, J.Y.; Li, C.W.; Hsu, J.L.; Miller, S.A.; Wang, X.; et al. The crosstalk of mTOR/S6K1 and Hedgehog pathways. *Cancer Cell* **2012**, *21*, 374–387. [CrossRef] [PubMed]
213. Stecca, B.; Mas, C.; Clement, V.; Zbinden, M.; Correa, R.; Piguet, V.; Beermann, F.; Ruiz, I.; Altaba, A. Melanomas require Hedgehog-GLI signaling regulated by interactions between GLI1 and the RAS-MEK/AKT pathways. *Proc. Natl. Acad. Sci. USA* **2007**, *104*, 5895–5900. [CrossRef] [PubMed]
214. Hemmings, B.A.; Restuccia, D.F. PI3K-PKB/AKT pathway. *Cold Spring Harb. Perspect. Biol.* **2012**, *4*, a011189. [CrossRef] [PubMed]
215. Seto, M.; Ohta, M.; Asaoka, Y.; Ikenoue, T.; Tada, M.; Miyabayashi, K.; Mohri, D.; Tanaka, Y.; Ijichi, H.; Tateishi, K.; et al. Regulation of the hedgehog signaling by the mitogen-activated protein kinase cascade in gastric cancer. *Mol. Carcinog.* **2009**, *48*, 703–712. [CrossRef] [PubMed]
216. Nolan-Stevaux, O.; Lau, J.; Truitt, M.L.; Chu, G.C.; Hebrok, M.; Fernández-Zapico, M.E.; Hanahan, D. GLI1 is regulated through Smoothened-independent mechanisms in neoplastic pancreatic ducts and mediates PDAC cell survival and transformation. *Genes Dev.* **2009**, *23*, 24–36. [CrossRef] [PubMed]
217. Lee, J.D.; Kraus, P.; Gaiano, N.; Nery, S.; Kohtz, J.; Fishell, G.; Loomis, C.A.; Treisman, J.E. An acylatable residue of Hedgehog is differentially required in Drosophila and mouse limb development. *Dev. Biol.* **2001**, *233*, 122–136. [CrossRef] [PubMed]
218. Pepinsky, R.B.; Zeng, C.; Wen, D.; Rayhorn, P.; Baker, D.P.; Williams, K.P.; Bixler, S.A.; Ambrose, C.M.; Garber, E.A.; Miatkowski, K.; et al. Identification of a palmitic acid-modified form of human Sonic hedgehog. *J. Biol. Chem.* **1998**, *273*, 14037–14045. [CrossRef] [PubMed]
219. Callier, P.; Calvel, P.; Matevossian, A.; Makrythanasis, P.; Bernard, P.; Kurosaka, H.; Vannier, A.; Thauvin-Robinet, C.; Borel, C.; Mazaud-Guittot, S.; et al. Loss of function mutation in the palmitoyltransferase HHAT leads to syndromic 46,XY disorder of sex development by impeding Hedgehog protein palmitoylation and signaling. *PLoS Genet.* **2014**, *10*, e1004340. [CrossRef] [PubMed]
220. Cooper, M.K.; Porter, J.A.; Young, K.E.; Beachy, P.A. Teratogen-mediated inhibition of target tissue response to Shh signaling. *Science* **1998**, *280*, 1603–1607. [CrossRef] [PubMed]
221. Hahn, H.; Christiansen, J.; Wicking, C.; Zaphiropoulos, P.G.; Chidambaram, A.; Gerrard, B.; Vorechovsky, I.; Bale, A.E.; Toftgard, R.; Dean, M.; et al. A mammalian patched homolog is expressed in target tissues of sonic hedgehog and maps to a region associated with developmental abnormalities. *J. Biol. Chem.* **1996**, *271*, 12125–12128. [CrossRef] [PubMed]
222. Incardona, J.P.; Roelink, H. The role of cholesterol in Shh signaling and teratogen-induced holoprosencephaly. *Cell Mol. Life Sci.* **2000**, *57*, 1709–1719. [CrossRef] [PubMed]
223. The Human Protein Atlas. Available online: http://www.proteinatlas.org/ (assessed on 17 August 2016).
224. Karhadkar, S.S.; Bova, G.S.; Abdallah, N.; Dhara, S.; Gardner, D.; Maitra, A.; Isaacs, J.T.; Berman, D.M.; Beachy, P.A. Hedgehog signalling in prostate regeneration, neoplasia and metastasis. *Nature* **2004**, *431*, 707–712. [CrossRef] [PubMed]
225. Feldmann, G.; Dhara, S.; Fendrich, V.; Bedja, D.; Beaty, R.; Mullendore, M.; Karikari, C.; Alvarez, H.; Iacobuzio-Donahue, C.; Jimeno, A.; et al. Blockade of hedgehog signaling inhibits pancreatic cancer invasion and metastases: A new paradigm for combination therapy in solid cancers. *Cancer Res.* **2007**, *67*, 2187–2196. [CrossRef] [PubMed]
226. Yauch, R.L.; Gould, S.E.; Scales, S.J.; Tang, T.; Tian, H.; Ahn, C.P.; Marshall, D.; Fu, L.; Januario, T.; Kallop, D.; et al. A paracrine requirement for hedgehog signalling in cancer. *Nature* **2008**, *455*, 406–410. [CrossRef] [PubMed]

227. Tian, H.; Callahan, C.A.; DuPree, K.J.; Darbonne, W.C.; Ahn, C.P.; Scales, S.J.; de Sauvage, F.J. Hedgehog signaling is restricted to the stromal compartment during pancreatic carcinogenesis. *Proc. Natl. Acad. Sci. USA* **2009**, *106*, 4254–4259. [CrossRef] [PubMed]
228. Jacob, L.; Lum, L. Deconstructing the Hedgehog pathway in development and disease. *Science* **2007**, *318*, 66–68. [CrossRef] [PubMed]
229. Konitsiotis, A.D.; Chang, S.C.; Jovanović, B.; Ciepla, P.; Masumoto, N.; Palmer, C.P.; Tate, E.W.; Couchman, J.R.; Magee, A.I. Attenuation of hedgehog acyltransferase-catalyzed Sonic Hedgehog palmitoylation causes reduced signaling, proliferation and invasiveness of human carcinoma cells. *PLoS ONE* **2014**, *9*, e89899. [CrossRef] [PubMed]
230. Matevossian, A.; Resh, M.D. Hedgehog acyltransferase as a target in estrogen receptor positive, *HER2* amplified, and tamoxifen resistant breast cancer cells. *Mol. Cancer* **2015**, *14*, 72. [CrossRef] [PubMed]
231. Petrova, E.; Matevossian, A.; Resh, M.D. Hedgehog acyltransferase as a target in pancreatic ductal adenocarcinoma. *Oncogene* **2015**, *34*, 263–268. [CrossRef] [PubMed]
232. Zhang, S.Q.; Tian, X.; Luo, Y.W.; Zhong, D.Q.; Huang, Z.M.; Zhang, X.B. Expression, clinical and pathological significance of KIAA1173 gene in skin squamous cell carcinoma. *Zhonghua Yi Xue Za Zhi* **2010**, *90*, 1243–1246. [PubMed]
233. Berisha, S.Z.; Serre, D.; Schauer, P.; Kashyap, S.R.; Smith, J.D. Changes in whole blood gene expression in obese subjects with type 2 diabetes following bariatric surgery: A pilot study. *PLoS ONE* **2011**, *6*, e16729. [CrossRef] [PubMed]
234. Al-Goblan, A.; Al-Alfi, M.; Khan, M.Z. Mechanism linking diabetes *mellitus* and obesity. *Diab. Metabol. Synd. Obes. Targ. Ther.* **2014**, *7*, 587–591. [CrossRef] [PubMed]
235. Rahib, L.; MacLennan, N.K.; Horvath, S.; Liao, J.C.; Dipple, K.M. Glycerol kinase deficiency alters expression of genes involved in lipid metabolism, carbohydrate metabolism, and insulin signaling. *Eur. J. Hum. Genet.* **2007**, *15*, 646–657. [CrossRef] [PubMed]
236. Maeda, N.; Funahashi, T.; Shimomura, I. Metabolic impact of adipose and hepatic glycerol channels aquaporin 7 and aquaporin 9. *Nat. Clin. Pract. Endocrinol. Metab.* **2008**, *4*, 627–634. [CrossRef] [PubMed]
237. Sievers, F.; Wilm, A.; Dineen, D.; Gibson, T.J.; Karplus, K.; Li, W.; Lopez, R.; McWilliam, H.; Remmert, M.; Söding, J.; et al. Fast, scalable generation of high-quality protein multiple sequence alignments using Clustal Omega. *Mol. Syst. Biol.* **2011**, *7*, 539. Available online: www.ebi.ac.uk/Tools/msa/clustalo/ (assessed on 8 July 2016). [CrossRef] [PubMed]
238. Kumar, S.; Stecher, G.; Tamura, K. MEGA7: Molecular Evolutionary Genetics Analysis version 7.0 for bigger datasets. *Mol. Biol. Evol.* **2016**, *33*, 1870–1874. [CrossRef] [PubMed]
239. Shapiro, J.A. Thinking about bacterial populations as multicellular organisms. *Annu. Rev. Microbiol.* **1998**, *52*, 81–104. [CrossRef] [PubMed]
240. Palková, Z.; Janderová, B.; Gabriel, J.; Zikánová, B.; Pospísek, M.; Forstová, J. Ammonia mediates communication between yeast colonies. *Nature* **1997**, *390*, 532–536. [CrossRef] [PubMed]
241. Palková, Z.; Forstová, J. Yeast colonies synchronise their growth and development. *J. Cell Sci.* **2000**, *113*, 1923–1928. [PubMed]

10

Asxl2$^{-/-}$ Mice Exhibit De Novo Cardiomyocyte Production during Adulthood

Rachel Brunner [1], Hsiao-Lei Lai [1,2], Zane Deliu [1], Elan Melman [1,3], David L. Geenen [4,5] and Q. Tian Wang [1,6,*]

1 Department of Biological Sciences, University of Illinois at Chicago, Chicago, IL 60607, USA; rbrunn2@uic.edu (R.B.); hsiaoleilai@gmail.com (H.-L.L.); zdeliu2@uic.edu (Z.D.); emelma2@illinois.edu (E.M.)
2 PTM Biolabs Inc., Chicago, IL 60612, USA
3 The School of Molecular and Cellular Biology, University of Illinois Urbana-Champaign, Champaign, IL 61801, USA
4 Department of Physiology and Biophysics, University of Illinois at Chicago, Chicago, IL 60612, USA; geenend@gvsu.edu
5 Physician Assistant Studies, Grand Valley State University, Grand Rapids, MI 49503, USA
6 Congressionally Directed Medical Research Programs, Frederick, MD 21702, USA
* Correspondence: quntianw@live.com

Academic Editor: Simon J. Conway

Abstract: Heart attacks affect more than seven million people worldwide each year. A heart attack, or myocardial infarction, may result in the death of a billion cardiomyocytes within hours. The adult mammalian heart does not have an effective mechanism to replace lost cardiomyocytes. Instead, lost muscle is replaced with scar tissue, which decreases blood pumping ability and leads to heart failure over time. Here, we report that the loss of the chromatin factor ASXL2 results in spontaneous proliferation and cardiogenic differentiation of a subset of interstitial non-cardiomyocytes. The adult *Asxl2*$^{-/-}$ heart displays spontaneous overgrowth without cardiomyocyte hypertrophy. Thymidine analog labeling and Ki67 staining of 12-week-old hearts revealed 3- and 5-fold increases of proliferation rate for vimentin^{+} non-cardiomyocytes in *Asxl2*$^{-/-}$ over age- and sex-matched wildtype controls, respectively. Approximately 10% of proliferating non-cardiomyocytes in the *Asxl2*$^{-/-}$ heart express the cardiogenic marker NKX2-5, a frequency that is ~7-fold higher than that observed in the wildtype. EdU lineage tracing experiments showed that ~6% of pulsed-labeled non-cardiomyocytes in *Asxl2*$^{-/-}$ hearts differentiate into mature cardiomyocytes after a four-week chase, a phenomenon not observed for similarly pulse-chased wildtype controls. Taken together, these data indicate de novo cardiomyocyte production in the *Asxl2*$^{-/-}$ heart due to activation of a population of proliferative cardiogenic non-cardiomyocytes. Our study suggests the existence of an epigenetic barrier to cardiogenicity in the adult heart and raises the intriguing possibility of unlocking regenerative potential via transient modulation of epigenetic activity.

Keywords: heart; cardiogenic; cardiomyocyte; proliferation; regeneration; chromatin factor; epigenetic

1. Introduction

Heart disease is the leading cause of death in developed countries. Many forms of heart disease result in a loss of functional muscle. Unfortunately, the heart's natural ability to generate new muscle is severely limited. The development of effective therapies for cardiac regeneration will benefit from a thorough understanding of endogenous cardiogenic mechanisms and how they are regulated.

Very limited cardiomyocyte production occurs in the adult heart and past research has reported two mechanisms by which this occurs. The most indisputable mechanism is renewal via cardiomyocyte proliferation [1–9]. Genetic fate-mapping studies suggest that this is the dominant mechanism for cardiomyocyte production in the adult mammalian heart [5,7,8,10]. However, the majority of studies estimate the rate of cardiomyocyte proliferation to be very low during adulthood, at or below ~1% per year [1,2,4–9]. It is possible to stimulate proliferation by manipulating the activity of certain genes [11–24].

The second mechanism of cardiomyocyte production in the adult heart involves the differentiation of resident cardiac cells with progenitor activity [25]. For example, c-kit$^+$ cardiac stem cells (CSCs) have been reported to produce cardiomyocytes in the adult heart [26]. However, the estimated efficiency of this process is drastically different between different studies [27–30]. The adult epicardial cells have also been reported to produce cardiomyocytes after thymosin β4 treatment in the normal or infarcted adult heart [31–33], though an effort to duplicate this result was unsuccessful [34]. Mechanisms for cardiomyocyte production in the adult heart have been identified, however, a better understanding of their regulation will help pave the way to develop therapies to replace lost muscle tissue.

Aside from cardiomyocyte proliferation and cardiogenic differentiation from CSCs, cardiomyocytes can be derived from several populations of cells isolated from the adult heart, such as Sca1$^+$ cells [35], side population (SP) cells [36], Isl1$^+$c-kit$^+$ cells [37], and cardiac colony-forming units-fibroblast (cCFU-Fs) [38]. It is unclear whether these cells are normally cardiogenic in vivo or if cardiogenicity was induced by in vitro culturing.

Finally, cardiac fibroblasts (CFs) can be reprogrammed into cardiomyocytes both in vitro and in vivo through the transduction of a cocktail of three transcription factors, GATA4, MEF2C, and TBX5 [39–41]. This mechanism does not appear to be active in the normal adult heart, as genetic tracing of the CF lineage using Periostin (Postn)-Cre did not find sign of spontaneous trans-differentiation of CFs into cardiomyocytes [41,42].

Here we report de novo cardiomyocyte production at a significant rate in adult mice carrying a mutation in *Asxl2*, a chromatin-associated factor. The hearts of *Asxl2*$^{-/-}$ mice exhibit significant growth between two and four months after birth. This growth is not due to cardiomyocyte hypertrophy. Rather, we present data that the adult *Asxl2*$^{-/-}$ heart harbors a population of proliferative interstitial cells that undergo spontaneous cardiogenic differentiation.

We have previously shown that ASXL2 is an important regulator of histone H3 methylation and H2A deubiquitination [43,44]. The implication of ASXL2 in the regulation of cardiomyocyte production during adult life raises new therapeutic possibilities for cardiac repair and regeneration.

2. Materials and Methods

2.1. Animals

Asxl2 mutant mice were generated by utilizing a gene-trapped embryonic stem cell line from the Gene-Trap Consortium (http://www.genetrap.org/) [43]. The *Asxl2*$^-$ allele yields an mRNA with the first 19 amino acids of ASXL2 followed by the gene trap cassette. The resultant fusion protein contains none of the conserved domains of ASXL2.

The *Asxl2*$^-$ allele is currently in two inbred genetic backgrounds, C57BL/6J and 129/Sv. Viable homozygous mutants (*Asxl2*$^{-/-}$) are not recovered in either inbred genetic background, but are recovered when heterozygous C57BL/6J mice are mated to heterozygous 129/Sv mice. Experiments presented here are from such F1 progeny.

All animal studies were performed in accordance with the University of Illinois at Chicago Institutional Animal Care and Use Committee (IACUC) and Animal Care Committee policies (ACC 13-117).

2.2. Assessment of Heart Growth

Transthoracic echocardiography and left ventricular (LV) mass calculation were performed as previously described [45,46]. LVMI is LV mass (mg) expressed relative to body mass (g).

Direct measurement of heart weight was performed with freshly dissected hearts that had been trimmed of excess large vessel tissue and gently blotted. Heart mass index is the heart mass (mg) expressed relative to body mass (g).

Quantitative morphometric analysis of total ventricular muscle volume was performed using a previously described method for calculating scar volume [5], with some modifications. Paraffin embedded hearts from wildtype and *Asxl2*$^{-/-}$ animals at 8- and 16-week of age were cross-sectioned (5 μm). Whole hematoxylin and eosin stained sections at 125 μm intervals were imaged at 10× on a Zeiss Observer.Z1 with ZenPro software using the tiling feature. The area covered by the section was calculated using ImageJ, then multiplied by the interval of 125 μm, and these volumes were totaled to give the ventricle muscle volume per heart. To normalize for differences in body mass, ventricle muscle volume (mm^3) is expressed relative to body mass (g).

2.3. Morphometric Analysis of Isolated Cardiomyocytes

Adult cardiomyocytes were isolated as previously described [47]. Glutaraldehyde fixed isolated cardiomyocytes from 8- and 16-week wildtype and *Asxl2*$^{-/-}$ hearts were analyzed by Martin Gerdes' Lab (New York Institute of Technology) to determine the following: cell volume (Coulter Counter/Channelyzer), cell length and cell profile area (microscopically by image analysis), cross-sectional area (calculated from cell volume/cell length), and nucleation status (Figure 2 and Supplemental Figure S3).

Cardiomyocytes were also isolated from hearts of wildtype and *Asxl2*$^{-/-}$ that were pulse-labeled with EdU at 12-weeks, and chased 4-weeks (Figure 6). Isolated cardiomyocytes were fixed in 4% paraformaldehyde for 10 min on ice with frequent agitation to prevent aggregation of the cardiomyocytes, permeabilized with 0.5% Tritonx-100 in PBS for 10 min, and then labeled for EdU incorporation (Click-iT® EdU Imaging Kit, Invitrogen, Waltham, MA, USA, C10337), per the manufacturer's instructions. Cardiomyocytes were then co-labeled with mouse-α-cTnT (Thermo Scientific, Waltham, MA, USA, MS-295), labeled with biotin-conjugated anti-mouse IgG (Vector, M.O.M Kit, BMK-2202), and subsequently labeled with streptavidin conjugated to AlexaFluor 594 (Jackson ImmunoResearch, 016-580-084). The cardiomyocytes were re-suspended in mounting medium (Vectashield with DAPI (4′,6-diamidino-2-phenylindole), Vector Laboratories, H-1200).

Imaging was performed at the Northwestern University Center for Advanced Microscopy (generously supported by CCSG P30 CA060553 awarded to the Robert H Laurie Comprehensive Cancer Center). Cardiomyocytes were visualized using a Zeiss upright AXIO microscope at 10× magnification. The Tissue Gnostics system (Vienna, Austria) and Tissue FAXs software were utilized to image entire slides, with the individual tiles (defined as one 10× image field) being exported for analysis. Each tile was manually analyzed for cardiomyocytes with EdU$^+$ nuclei. To estimate the total number of cardiomyocytes, they were manually counted in every tenth tile. These numbers were then averaged per slide, and subsequently multiplied by ten to give the estimated total number of cardiomyocytes per slide. From these data, the percentage of EdU$^+$ labeled cardiomyocytes among all analyzed isolated cardiomyocytes from chased hearts was calculated.

To analyze the morphology and nucleation status of EdU$^+$ cardiomyocytes, the 20× objective and ZenPro software was used to acquire images of EdU$^+$ cardiomyocytes. All imaged cardiomyocytes were in a longitudinal orientation, had smooth membranes, and sarcomeres were clearly evident. A total of 70 EdU$^+$ cardiomyocytes (from two *Asxl2*$^{-/-}$ hearts) were analyzed. The nucleation status (mono- or bi-nucleated; EdU$^+$ cardiomyocytes with more than two nuclei were not observed) was recorded for each EdU$^+$ cardiomyocyte. Additionally, the length of each EdU$^+$ cardiomyocyte was measured using Pixel Stick (Plum Amazing). From this analysis, EdU$^+$ cardiomyocytes were placed into one of three categories: (1) mononuclear and shorter than the shortest cardiomyocytes observed from four-month-old wildtype hearts; (2) mononuclear and within range of lengths of cardiomyocytes from four-month-old wildtype hearts; or (3) binuclear and within range of lengths of cardiomyocytes from four-month-old wildtype hearts.

2.4. Thymidine Analog Pulse and Pulse-Chase Assays

Mice were assayed for proliferation at two time-points, week 8 and week 12. Mice were injected with a thymidine analog, either 100 mg/kg 5-bromo-2-deoxyuridine (BrdU, Sigma, St. Louis, MO, USA, B5002) or 50 mg/kg 5-Ethynyl-2′-deoxyuridine (EdU, Invitrogen, A10044), once daily for three consecutive days. Hearts were harvested approximately 6 h post the final injection.

Paraffin sections were labeled for EdU or BrdU, imaged, and analyzed for proliferation (for details see methods section on "Immunofluorescence" below). $BrdU^+$ or EdU^+ nuclei were counted manually and total nuclei per image was found using Cell Profiler software [48]. The proliferation index is expressed as the percentage of $BrdU^+$ nuclei of total nuclei.

To determine the fate of cells that proliferated at 12-week, mice were injected with EdU at 12 weeks of age (as above), and hearts were harvested 4 weeks after the final EdU injection, allowing time for the newly proliferated to differentiate.

2.5. Preparation of Histological Sections

Prior to harvest, mice were injected with heparin, then euthanized with carbon dioxide, followed by cervical dislocation. Hearts were extracted and submerged in 1 M potassium chloride (in phosphate buffered saline). Hearts were then trimmed to remove lung and large vessel tissues, gently blotted dry, and weighed.

Hearts to be paraffin-embedded were fixed in cold 4% paraformaldehyde overnight. Hearts were then washed in PBS and a small amount of tissue was cut from the dorsal and ventral sides of the heart to expose the left and right ventricles to facilitate infiltration (this step was omitted for hearts used for quantitative morphometric ventricle volume analysis). Hearts were dehydrated through a graded series of isopropanol (Sigma, St. Louis, MO, USA, 534021-4L) in 20 min/10 mL washes as follows: 50%, 75%, 95%, and 100% twice. Isopropanol was dissolved in dH_2O and samples were shaken gently on a horizontal rotator in six-well plates. Hearts were cleared with mineral oil (Sigma, M8410) as follows, 25%, 50%, 75% (mixed with isopropanol), and three changes of 100% mineral oil (using the same conditions as dehydration steps). The hearts were then infiltrated with paraffin (McCormick Scientific, St. Louis, MO, USA, 39502004) as follows: 25%, 50%, 75% (dissolved in mineral oil, 20 min, 59 °C), and followed by two changes in 100% paraffin. Samples were then positioned in molds and allowed to cool to room temperature. Five-micron sections were cut, dried vertically overnight, and heat-fixed to the slides the following day for 2 h at 42 °C.

For frozen sections, freshly dissected, and trimmed hearts were placed in optimum cutting temperature (O.C.T.) medium (TissueTek, 4583) and snap-frozen in a hexane/dry ice bath. Five-micron sections were cut on a cryotome, mounted on charged slides, and stored at −20 °C until use.

2.6. Immunofluorescence

For immunofluorescent analysis of paraffin sections, slides were de-paraffinized in xylene and rehydrated through a graded series of ethanol. Antigen unmasking was performed with 10 mM Tris, 1 mM EDTA, 0.05% Tween-20, pH 9.0 at 92 °C for 20 min. Sections were then permeabilized with 0.2% TritonX-100 (in PBS) for 10 min. Blocking was done using 5% normal serum (in PBS) of the species the secondary antibody was raised in. EdU incorporation was detected using a Click-iT® EdU Imaging Kit (Invitrogen, C10337), per manufacturer's instructions.

Sections were incubated in primary antibody overnight at 4 °C in a humidified chamber. Primary antibodies used on paraffin sections include rat-α-BrdU (1:100, Abcam, Cambridge, MA, USA, ab6326), mouse-α-cTnT (1:100, Thermo Scientific, MS-295), rabbit-α-Ki67 (1:400, Abcam, ab15580), rabbit-α-vimentin (1:100, Abcam, ab92547), rabbit-α-CD31 (1:40, Lifespan BioSciences, Seattle, WA, USA, LS-B1932), rabbit-α-α-smooth muscle actin (1:800, Sigma, A2547), rabbit-α-NKX2-5 (1:100, Abcam, ab22611), rabbit-α-MEF2C (1:100, Abcam, ab64644), rabbit-α-GATA4 (1:100, Santa Cruz, Dallas, TX, USA, sc-9053), and rabbit-α-connexin43 (1:200, Santa Cruz, sc-6560). Cell membranes were labeled with wheat germ agglutinin (WGA) conjugates, either WGA conjugated to fluorescein isothiocyanate (FITC) (20 μL/mL, Vector, FL-1021) or WGA-CF™ 594 conjugate (1:200, Biotium, Fremont, CA, USA, 29023-1). All secondary antibodies were from Jackson ImmunoResearch and used at a concentration of 1:200. All nuclei were labeled either with DAPI (Vectashield + DAPI mounting media, Vector Laboratories, Burlingame, CA, USA, H-1200) or Hoechst (1:2000, Invitrogen). If Hoechst was used, slides were mounted with Vectashield (Vector Laboratories, H-1000).

For immunofluorescence on frozen sections, sections were fixed with methanol. After fixation, sections were incubated in primary antibody overnight at 4 °C in a humidified chamber. Primary antibodies used on frozen sections include rabbit-α-C-Kit (Abcam, ab5506) and rat-α-PDGFRα (BD Pharmingen, 558774).

Sections were imaged using a Zeiss Axiovert 200 M or a Zeiss Observer.Z1 microscope using the 20× objective. Twenty-five images were taken per frontal section as follows: 10 images of the left ventricular free wall (five images each of epicardial side and endocardial side), 10 images of the septum (five images each of the left and right ventricular sides) and five images of the right ventricular free wall.

3. Results

3.1. Overgrowth of the Adult *Asxl2*$^{-/-}$ Heart without Cardiomyocyte Hypertrophy

Our previously published results suggested that *Asxl2*$^{-/-}$ animals have enlarged hearts [43]. To further investigate this phenotype, we took three approaches to evaluate the growth of *Asxl2*$^{-/-}$ and wildtype hearts: echocardiographic measurement of heart dimensions in live animals, weight measurement of freshly dissected hearts, and quantitative morphometric analysis of paraffin embedded hearts. All three approaches showed that from approximately eight weeks onwards, the rate of cardiac growth in *Asxl2*$^{-/-}$ mice outpaced the rate of body weight growth (Figure 1, Supplemental Figures S1 and S2); in contrast, the rate of cardiac growth in wildtype mice either parallels or is slower than that of body weight growth depending on the means of measurement (Supplemental Figure S1). By 12 weeks, the hearts of *Asxl2*$^{-/-}$ animals are disproportionally larger than those of wildtype controls.

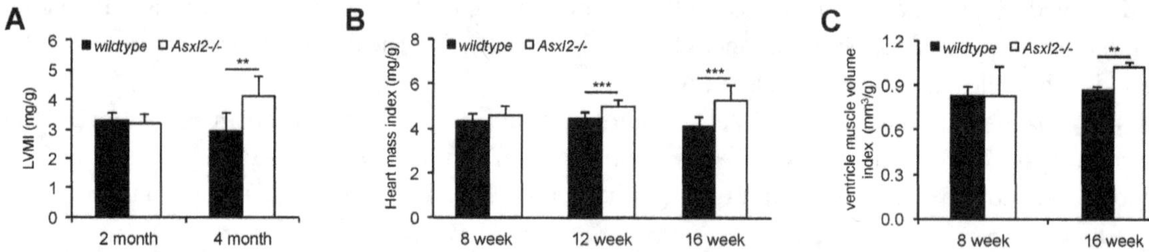

Figure 1. Overgrowth of the adult *Asxl2*$^{-/-}$ heart. (**A**) Left ventricular mass index found by echocardiography at 2-month (wildtype, $n = 4$; *Asxl2*$^{-/-}$, $n = 4$) and 4-month (wildtype, $n = 7$; *Asxl2*$^{-/-}$, $n = 7$) of age, normalized to body mass; (**B**) Mass of freshly dissected hearts normalized to body mass at 8-week (wildtype, $n = 6$; *Asxl2*$^{-/-}$, $n = 6$), 12-week (wildtype, $n = 16$; *Asxl2*$^{-/-}$, $n = 11$), and 16-week (wildtype, $n = 12$; *Asxl2*$^{-/-}$, $n = 9$) of age; (**C**) Quantitative morphometric analysis of ventricle muscle volume normalized to body mass at 8-week (wildtype, $n = 3$; *Asxl2*$^{-/-}$, $n = 3$) and 16-week (wildtype, $n = 3$; *Asxl2*$^{-/-}$, $n = 3$) of age. Bars indicate standard deviation. *p*-Values (Student's *t*-test): ** <0.01; *** <0.005.

Growth of the normal adult heart is predominantly the result of growth in size of existing cardiomyocytes [49,50]. We isolated cardiomyocytes from *Asxl2*$^{-/-}$ and wildtype hearts and assessed nucleation, cell dimensions, and volume (using the high-throughput Coulter Counter/Channelyzer method [51]). *Asxl2*$^{-/-}$ cardiomyocytes are not larger than wildtype nor is there a higher percentage of binucleated cardiomyocytes at either 8- or 16-week of age (Figure 2 and Supplemental Figure S3).

In fact, while not statistically significant, *Asxl2*$^{-/-}$ cardiomyocytes appear to be somewhat smaller than wildtype (Figure 2). These observations are consistent with a previous imaging-based measurement showing that *Asxl2*$^{-/-}$ cardiomyocytes are not hypertrophic at six months of age [45]. We conclude that the overgrowth of the *Asxl2*$^{-/-}$ heart observed after approximately eight weeks cannot be accounted for by cardiomyocyte hypertrophy.

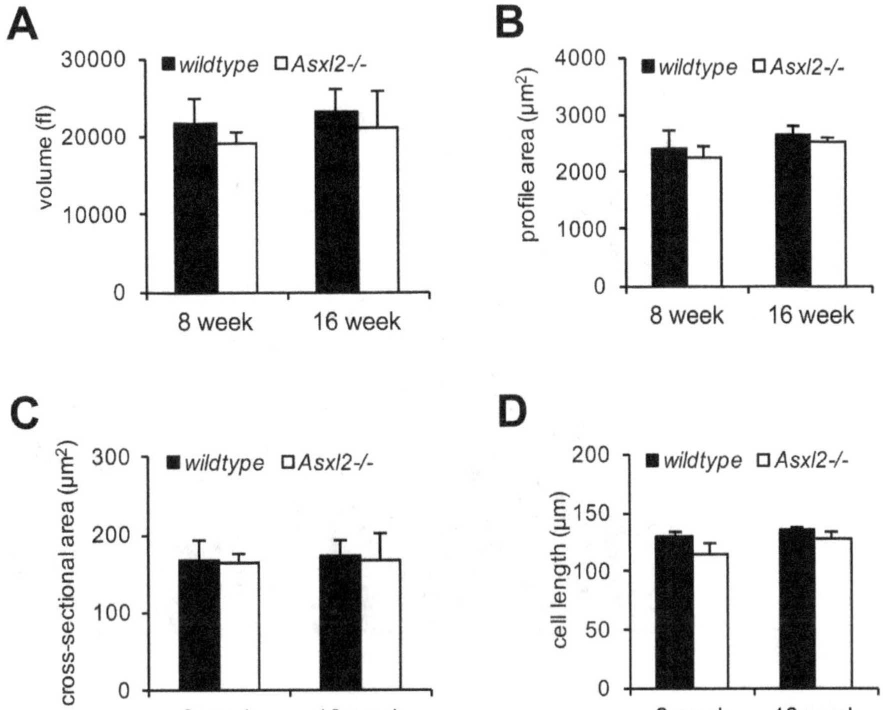

Figure 2. Lack of cardiomyocyte hypertrophy in *Asxl2*$^{-/-}$ heart. Isolated cardiomyocytes from 8- and 16-week old wildtype and *Asxl2*$^{-/-}$ hearts were assessed for: (**A**) volume; (**B**) profile area; (**C**) cross-sectional area; and (**D**) length. Sample size: 8-week wildtype, $n = 4$; 8-week *Asxl2*$^{-/-}$, $n = 3$; 16-week wildtype, $n = 5$; 16-week *Asxl2*$^{-/-}$, $n = 3$. Bars indicate standard deviation. No significant differences were observed (Student's *t*-test).

3.2. Asxl2$^{-/-}$ *Hearts Exhibit Elevated Proliferative Activity, but Not in Cardiomyocytes*

The mutation or transgenic expression of certain genes can induce elevated proliferation of cardiomyocytes in the adult heart, leading to heightened heart growth [11–13,15–19]. To determine whether this is the case for *Asxl2*$^{-/-}$ hearts, we labeled proliferating cells using thymidine analogs (BrdU or EdU) (Figure 3A–E) and Ki67 (Figure 3F–G). BrdU labeling revealed a ~3-fold higher proliferation index in the *Asxl2*$^{-/-}$ left ventricle at 12-weeks (*p*-value = 0.042, Student's *t*-test; Figure 3B). EdU labeling gave comparable results (Figure 3C). A similar trend was observed by Ki67 staining; ~5-fold more Ki67^{+} nuclei were found in the *Asxl2*$^{-/-}$ left ventricle compared to the wildtype (*p*-value = 0.028, Student's *t*-test; Figure 3F). Thus, both methods showed an elevated proliferative activity in the *Asxl2*$^{-/-}$ hearts during the time window of heart enlargement. However, very few of the BrdU-, EdU-, or Ki67-labeled cells appeared to be cardiomyocytes. In both *Asxl2*$^{-/-}$ and wildtype hearts, the vast majority of EdU^{+} cells were interstitial (98.6% ± 1.2% in *Asxl2*$^{-/-}$, compared to 98.7% ± 0.4% in wildtype) and cTnT^{-} (97.4% ± 2.4% in *Asxl2*$^{-/-}$, compared to 98.1% ± 1.3% in wildtype) (Figure 3A,E). Almost all EdU^{+} cells were vimentin^{+} (97.5% ± 3.2% in *Asxl2*$^{-/-}$, compared to 98.3% ± 1.8% in wildtype) (Figure 3D, Supplemental Figure S4A,B), and ~25% were CD31^{+} (Supplemental Figure S4C,D). Similar observations were made on Ki67-stained hearts (Figure 3G).

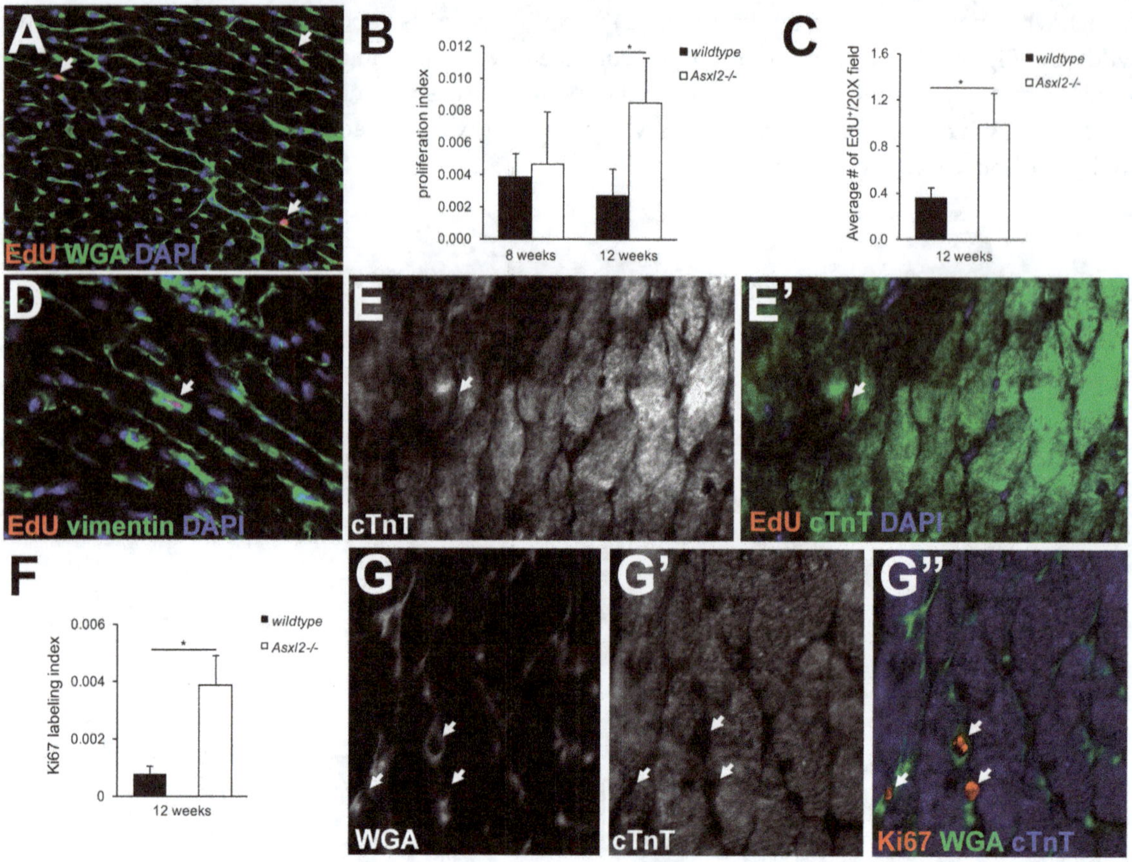

Figure 3. The *Asxl2*$^{-/-}$ hearts exhibit elevated proliferative activity in non-cardiomyocytes. (**A**) Representative image showing EdU, wheat germ agglutinin (WGA), and 4′,6-diamidino-2-phenylindole (DAPI); (**B**) Quantification of BrdU$^+$ nuclei in left ventricle (LV) at 8-weeks (wildtype: $n = 3$, 3 non-consecutive sections/heart, 116 BrdU$^+$, 30,045 total nuclei; *Asxl2*$^{-/-}$: $n = 3$, 3 non-consecutive sections/heart, 170 BrdU$^+$, 36,564 total nuclei) and 12-weeks (wildtype: $n = 3$, three non-consecutive sections/heart, 35 BrdU$^+$, 13,853 total nuclei; *Asxl2*$^{-/-}$: $n = 3$, three non-consecutive sections/heart, 127 BrdU$^+$, 15,250 total nuclei); (**C**) Quantification of average number of EdU$^+$ per 20× image field at 12-weeks (wildtype: $n = 3$, three non-consecutive sections/heart, 75 images; *Asxl2*$^{-/-}$: $n = 3$, three non-consecutive sections/heart, 75 images); (**D**) Representative image of EdU, vimentin and DAPI labeling; (**E**,**E′**) Representative image of EdU, cTnT and DAPI labeling; (**F**) Quantification of Ki67$^+$ nuclei in LV at 12-weeks (wildtype: $n = 3$, three non-consecutive sections/heart, 51 Ki67$^+$, 67,727 total nuclei; *Asxl2*$^{-/-}$: $n = 3$, three non-consecutive sections/heart, 236 Ki67$^+$, 62,124 total nuclei); (**G**–**G′′**) Representative image of Ki67, WGA and cTnT labeling. Arrows indicate proliferative cell nuclei. Bars indicate standard deviation. * p-Value (Student's t-test) < 0.05.

3.3. Expression of Cardiogenic Markers by Proliferative Cells in Asxl2$^{-/-}$ *Hearts*

Cardiomyocytes are by far the largest cells in the heart. The bulk of the size and weight of the heart comes from cardiomyocytes [52]. The moderate over-proliferation of non-cardiomyocytes in *Asxl2*$^{-/-}$ hearts is unlikely sufficient to cause an apparent increase in heart size, unless some proliferating cells are cardiogenic and give rise to cardiomyocytes. We pulse labeled proliferating cells in 12-week-old *Asxl2*$^{-/-}$ hearts by EdU or BrdU and asked whether labeled cells express NKX2-5, MEF2C, or GATA4, three transcription factors associated with cardiogenicity. Among all EdU$^+$cTnT$^-$ cells, the percentage of NKX2-5^{+} cells was ~7-fold higher in *Asx2*$^{-/-}$ hearts compared to wildtype (p-Value = 0.008, Student's t-test; Figure 4; distribution shown in Supplemental Figure S5). The percentage of MEF2C^{+} cells was ~2.9-fold higher (p = 0.018; Supplemental Figure S6A,B). Very few EdU^{+}GATA4^{+} cells and no EdU^{+}ISL1^{+} were observed in either genotype (Supplemental Figure S6C,D and data not shown).

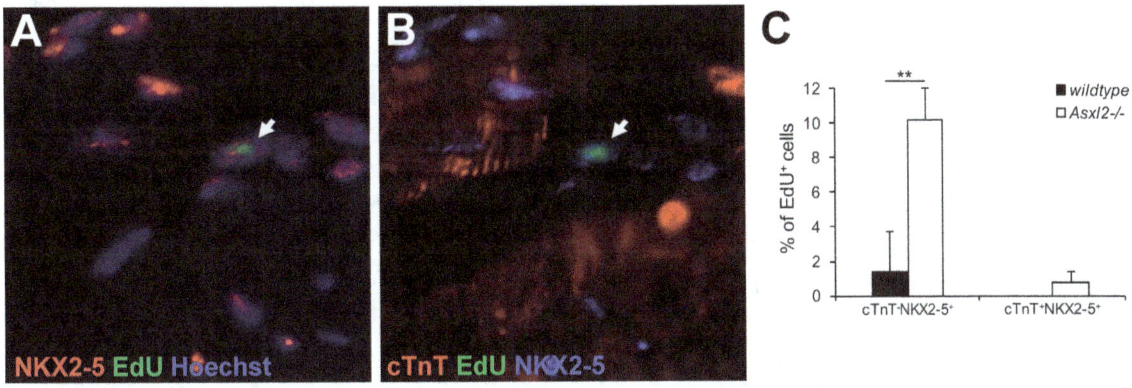

Figure 4. EdU^{+} cells in 12-week *Asxl2*$^{-/-}$ hearts show signs of being cardiogenic. (**A**,**B**) Paraffin heart sections from EdU-treated 12-week wildtype and *Asxl2*$^{-/-}$ animals were labeled for EdU, NKX2-5, and cTnT; (**C**) Quantification of the percentage of EdU-labeled cells that were EdU^{+}NKX2-5^{+}cTnT^{-} and EdU^{+}NKX2-5^{+}cTnT^{+}. Sample size: n = 3 animals per genotype; three non-consecutive sections/heart; analyzed fifteen 20× left ventricle images/section; 177 wildtype EdU^{+} cells, 252 *Asxl2*$^{-/-}$ EdU^{+} cells. Bars represent standard deviation. ** p-Value (Student's t-test) < 0.01.

3.4. EdU-Labeled Cells Give Rise to Cardiomyocytes in Asxl2$^{-/-}$ *Hearts after 4-Week Chase*

To definitively test whether the proliferative cells in *Asxl2*$^{-/-}$ hearts produce de novo cardiomyocytes, we pulse labeled *Asxl2*$^{-/-}$ and control hearts with EdU at 12-weeks and examined the fate of EdU-labeled cells after a 4-week chase (Figure 5A). While there is limited knowledge on the elimination half-life of EdU, for the closely related BrdU it is ~15–30 min [53,54]. Therefore, we expect that all cells that took up EdU did so during the pulse-labeling period and not during the chase. Cardiomyocytes were identified by markers (cTnT and Nkx2.5 staining) and by cell size (WGA staining). After the chase, there was a significant increase in the percentage of EdU^{+}cTnT^{+}NKX2-5^{+} cells (p-value = 0.018, Student's t-test) in *Asxl2*$^{-/-}$ and a concomitant trend toward a decrease in the percentage of EdU^{+}cTnT^{-}NKX2-5^{+} cells, but not in wildtype hearts (Figures 4C and 5B). EdU^{+} cardiomyocytes were observed readily in chased *Asxl2*$^{-/-}$ hearts (Figure 5C–E), but not in chased wildtype hearts. EdU^{+}NKX2-5^{+}cTnT^{+} cells represented 6.2% ± 1.6% of all EdU^{+} cells in the chased *Asxl2*$^{-/-}$ left ventricle (Figure 5B). The gap junction protein connexin 43 (Cx43) was detected on the membrane between EdU^{+} cardiomyocytes and neighboring EdU^{-} cardiomyocytes (Figure 5E), suggesting electrical coupling between the two. Concomitant to the appearance of EdU^{+} cardiomyocytes in *Asxl2*$^{-/-}$ hearts after the chase, there was a significant decrease of EdU^{+}vimentin^{+} cells (Supplemental Figure S4B). In contrast to EdU^{+}cTnT^{+}NKX2-5^{+} cells, the percentage of EdU^{+}CD31^{+} or EdU^{+}α-SMA^{+} cells did not increase after the chase period (Figure 5B and Supplemental Figure S4C–F), suggesting that there was no significant production of endothelial or smooth muscle cells. Finally, chased wildtype hearts showed neither appearance of EdU^{+} cardiomyocytes nor decrease of EdU^{+}vimentin^{+} cells (Figure 5B, Supplemental Figure S4B). Taken together, these data suggest that a significant number of proliferating interstitial cells differentiate into cardiomyocytes in the *Asxl2*$^{-/-}$ heart, but not in the wildtype heart.

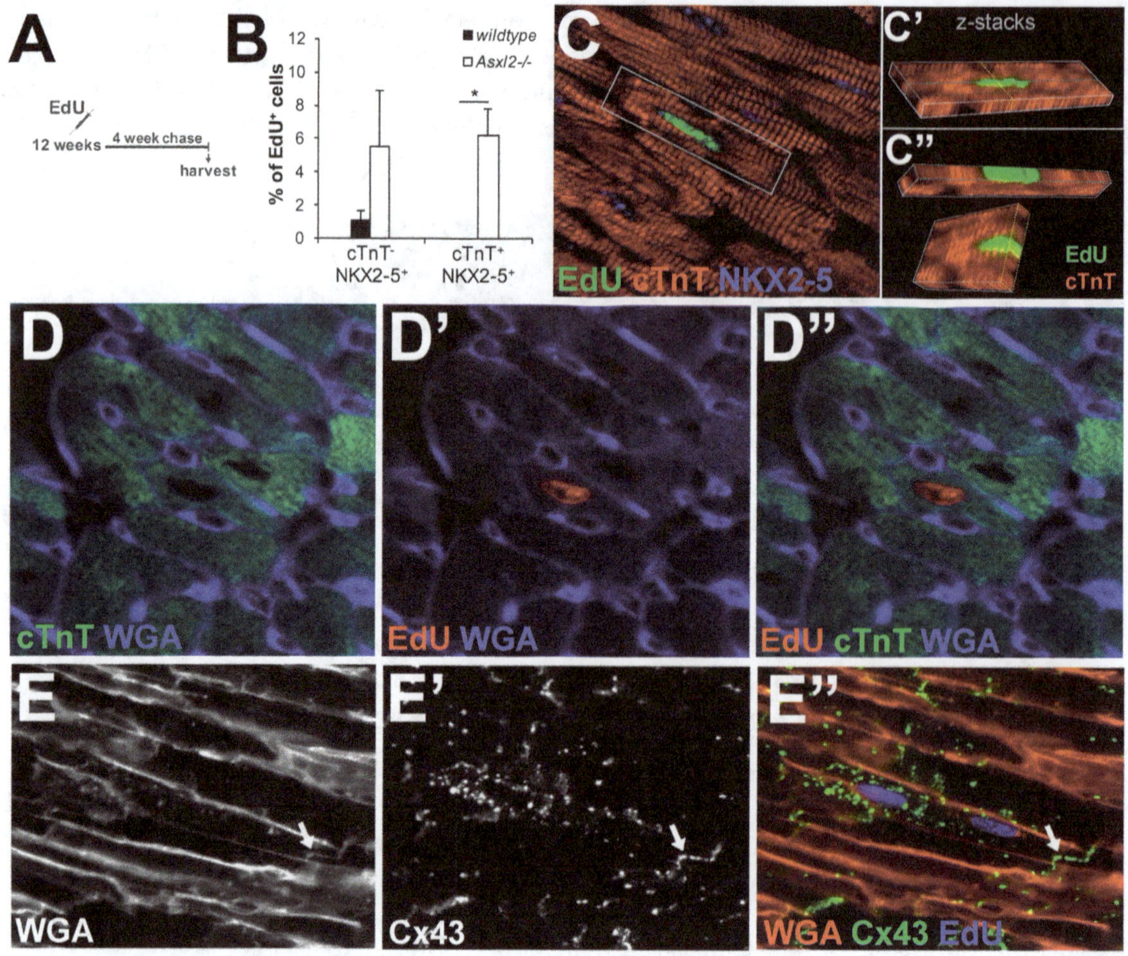

Figure 5. EdU-labeled cells give rise to cardiomyocytes in *Asxl2*$^{-/-}$ hearts after 4-week chase. (**A**) Schematic of the lineage tracing assay used to assess the differentiation potential of the proliferating non-cardiomyocytes observed at 12 weeks of age; (**B**) Quantification of the percentage of EdU$^+$ cells that are cTnT$^-$NKX2-5$^+$ and cTnT$^+$NKX2-5$^+$ in the left ventricle following the chase; (**C**) Representative image of a mononuclear, EdU$^+$cTnT$^+$NKX2-5$^+$ cell in a longitudinal orientation; (**C$'$**) The same cell in c, shown in a z-stack image; (**C$''$**) Cross-sections of the z-stack image in c$'$; (**D–D$''$**) Representative image of an EdU$^+$cTnT$^+$ cell, delineated with wheat germ agglutinin (WGA), in a transverse heart section; (**E–E$''$**) Representative image of EdU, WGA and labeling for the gap junction protein, connexin 43 (CX43). The arrow indicates the CX43$^+$ membrane between two cardiomyocytes. Sample size: $n = 3$ animals per genotype; three non-consecutive sections/heart; analyzed LV from whole section-stitched images; 617 and 755 EdU$^+$ cells were examined on wildtype and *Asxl2*$^{-/-}$ heart sections, respectively. Bars represent standard deviation. * *p*-Value (Student's *t*-test) < 0.05.

Because immunofluorescent identification of cardiomyocytes on heart sections could be ambiguous [55], we isolated cardiomyocytes from EdU pulse-chased hearts and repeated the analysis. EdU$^+$ cardiomyocytes were extremely rare among cardiomyocytes isolated from wildtype hearts, but were easily observed (at a frequency of 7.9–12.6 per 10,000 cardiomyocytes) in the *Asxl2*$^{-/-}$ isolates. Microscopic examination and measurement of 70 EdU$^+$ cardiomyocytes revealed a distribution of size, morphology, and nucleation status: 38 (54%) of EdU$^+$ cardiomyocytes were binuclear, rod-shaped, and with lengths comparable to EdU$^-$ cardiomyocytes (Figure 6A). In all the binuclear cells, both nuclei are labeled by EdU. Thirteen (19%) were mononuclear, rod-shaped with lengths comparable to EdU$^-$ cardiomyocytes (Figure 6B). The rest, 19 (27%), were mononuclear, spindle-shaped, and shorter than the shortest cardiomyocytes isolated from wildtype hearts (Figure 6C).

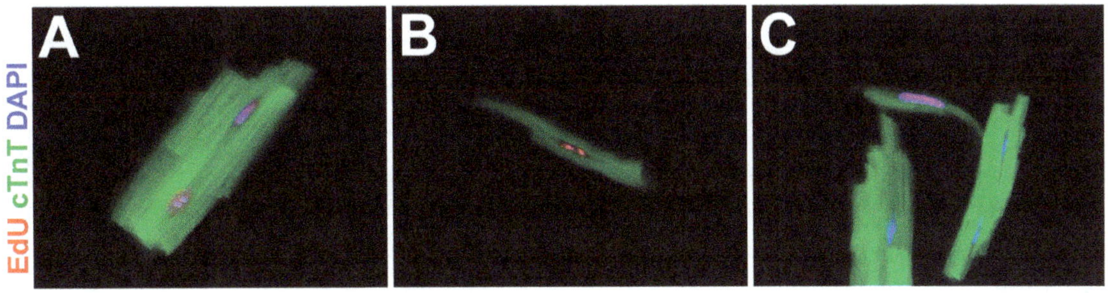

Figure 6. *Asxl2*$^{-/-}$ EdU^{+} cardiomyocytes display a range of size, morphology, and nucleation status. Cardiomyocytes were isolated and stained for EdU, cTnT, and DAPI. Representative images highlighting differences among *Asxl2*$^{-/-}$ EdU^{+} cardiomyocytes shown are (**A**) binuclear and rod-shaped with length comparable to EdU^{-} cardiomyocytes; (**B**) mononuclear and rod-shaped with normal length; and (**C**) mononuclear, spindle-shaped, and short.

3.5. The Proliferative Cardiogenic Cells in Asxl2$^{-/-}$ *Hearts are Distinct from c-kit*$^{+}$ *Cardiac Stem Cells and cCFU-Fs*

The above data suggest that a fraction of the proliferating interstitial cells detected in 12-week-old *Asx2*$^{-/-}$ hearts were cardiogenic, and that the abnormal growth of adult *Asx2*$^{-/-}$ hearts is due to production of de novo cardiomyocytes by these cells. The adult heart is known to contain resident cardiogenic cells, such as c-kit^{+} cardiac stem cells (CSCs) [26,29] and cardiac colony forming unit-fibroblasts (cCFU-Fs) [38]. We asked whether the cardiogenic phenomenon in the adult *Asxl2*$^{-/-}$ heart is due to activation of either of these cell types.

We pulse labeled proliferative cells in 12-week-old *Asxl2*$^{-/-}$ and control hearts with EdU, and co-stained heart sections for EdU and c-kit (Figure 7A). c-kit^{+} cells were readily detected (Figure 7B). However, we did not observe any EdU^{+}c-kit^{+} cells, nor was there a higher frequency of c-kit^{+} cells in the *Asxl2*$^{-/-}$ heart. Of 188 EdU^{+} cells identified on wildtype heart sections (n = 3, two sections/heart) and 265 EdU^{+} cells from *Asxl2*$^{-/-}$ sections (n = 3, two sections/heart), none was c-kit^{+}. cCFU-Fs are marked by a high level of PDGFRα expression and proximity to blood vessels [38]. We examined EdU^{+} cells in pulse-labeled 12-week-old hearts for PDGFRα expression. While many of the EdU^{+} cells express PDGFRα, EdU^{+}PDGFRα^{high} cells in *Asxl2*$^{-/-}$ hearts are neither more abundant nor more concentrated near blood vessels than those in wildtype hearts (Supplemental Figure S7). These results suggest that the proliferative cardiogenic cells in *Asx2*$^{-/-}$ hearts are distinct from c-kit^{+} CSCs and cCFU-Fs.

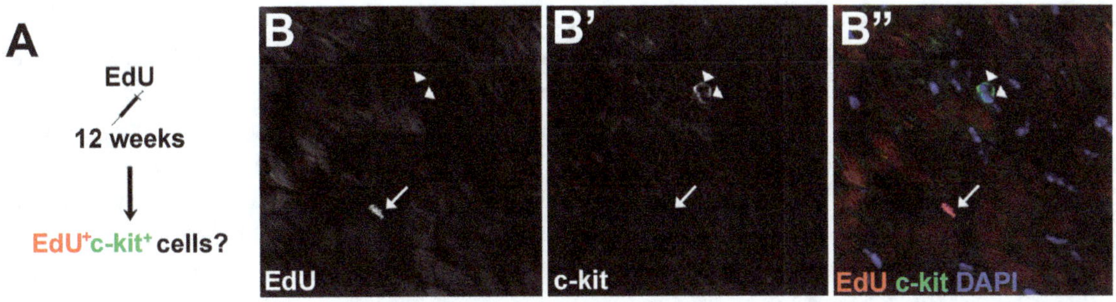

Figure 7. EdU^{+} cells are not c-kit^{+} at 12-weeks of age. (**A**) Wildtype and *Asxl2*$^{-/-}$ animals were treated with EdU at 12-weeks of age and frozen sections were assessed for EdU^{+}c-kit^{+} cells; (**B–B$''$**) Representative image of co-labeling for EdU and c-kit. EdU^{+}c-kit^{+} cells were not observed. Sample size: n = 3 animals per genotype; two non-consecutive sections/heart; ten 20× images/section; 188 wildtype EdU^{+} cells, 265 *Asxl2*$^{-/-}$ EdU^{+} cells. Arrows indicate the EdU^{+} nuclei and the arrowheads indicate c-kit^{+} cells.

4. Discussion

4.1. Adult *Asxl2*$^{-/-}$ Hearts Exhibit De Novo Cardiomyocyte Production

EdU$^+$ cardiomyocytes are readily observed in *Asxl2*$^{-/-}$ hearts that are pulse-labeled with EdU at 12 weeks and chased for 4 weeks, but not in similarly pulse-chased wildtype hearts. How did these EdU$^+$ cardiomyocytes arise? Proliferation of existing cardiomyocyte is the dominant mechanism for cardiomyocyte production in the adult mammalian heart [5,7]. The majority of studies estimated the rate of cardiomyocyte proliferation to be very low during adulthood, at or below ~1% per year [1,2,4–9]. Higher rates of cardiomyocyte proliferation have been reported when the activities of certain genes are manipulated [11–24]. However, this is not the case in *Asxl2*$^{-/-}$ mice. While the overall proliferation index is higher in 12-week *Asxl2*$^{-/-}$ hearts than in wildtype hearts (Figure 3), the vast majority (~98%) of EdU-labeled cells are non-cardiomyocytes: they are small, express vimentin, and do not express cTnT. Hence, the significantly higher frequency of EdU$^+$ cardiomyocytes in *Asxl2*$^{-/-}$ hearts after the chase (Figure 5, Supplemental Figure S5) is not due to proliferation, binucleation/polynucleation, or hypertrophic growth of existing cardiomyocytes. Neither did they arise via fusion between an EdU$^+$ cell and a pre-existing EdU$^-$ cardiomyocytes: EdU$^+$ cardiomyocytes isolated from chased *Asxl2*$^{-/-}$ hearts are either mononuclear or binuclear with both nuclei labeled by EdU (Figure 6). We did not observe any cardiomyocyte with one EdU$^+$ and one EdU$^-$ nucleus.

Our evidence strongly suggests that the EdU$^+$ cardiomyocytes in pulsed-chased *Asxl2*$^{-/-}$ hearts are progenies of non-cardiomyocytes that took up EdU during the time of the pulse label. We hypothesize that *Asxl2*$^{-/-}$ hearts harbor a population of proliferative and cardiogenic non-cardiomyocyte (PCN) cells, which can differentiate and produce de novo cardiomyocytes. Consistent with this scenario, EdU$^+$ cardiomyocytes isolated from pulse-chased *Asxl2*$^{-/-}$ hearts display morphological variations that may be correlated with which stage they are at during the differentiation process. While the largest of them are indistinguishable from mature cardiomyocytes (Figure 6A), the size and morphology of the smaller ones (Figure 6B,C) are suggestive of an intermediate or early stage of differentiation. The definitive testing of our hypothesis awaits the determination of the lineage origin of PCN cells and the genetic labeling and tracing of these cells.

4.2. What Are PCN Cells?

Many questions remain unanswered about PCN cells. First of all, the cellular identity of PCN cells remains unclear. Our evidence suggests that the PCN cells are not c-kit$^+$ CSCs, cCFU-Fs, endothelial cells, or ISL1$^+$ cardioblasts [26,38,56–58]. It has been shown that cardiac fibroblasts can be directly reprogrammed into cardiomyocytes with a trio of transcription factors, GATA4, MEF2C, and TBX5 [39,41]. More recently, increased reprogramming efficiency was reported by using relatively higher levels of MEF2C and lower levels of GATA4 and TBX5 [59]. The majority of freshly labeled EdU$^+$ cells are vimentin$^+$ and PDGFRα^+, which are known to be expressed by the fibroblast population. Interestingly, we observed a ~3-fold increase in BrdU$^+$MEF2C$^+$ cells in un-chased *Asxl2*$^{-/-}$ heart. These observations raise the possibility that PCN cells are spontaneously reprogramming fibroblasts. However, we did not detect an increase in EdU$^+$GATA4$^+$ cells. Moreover, neither vimentin nor PDGFRα exclusively mark fibroblasts [60,61]. Taken together, immunofluorescence-based marker analyses eliminated several scenarios but were not sufficient to pinpoint the identity of PCN cells. A more comprehensive approach is warranted to gain molecular insight into this highly intriguing cell population. While beyond the scope of this paper, a potential starting point could be to determine which transcripts are enriched in EdU$^+$ cells isolated from *Asxl2*$^{-/-}$ hearts compared to those from wildtype hearts. The correlation between select enriched transcripts and cardiogenic cells in *Asxl2*$^{-/-}$ hearts would need to be further examined by genetic lineage tracing and/or by cell isolation followed by in vitro differentiation assays.

Secondly, the exact time frame when PCN cells are active has not been determined. The heart size of *Asxl2*$^{-/-}$ mice is proportionally comparable to wildtype up to eight weeks, suggesting that if

PCN cells are active in young mice, the activity is too low to have a noticeable effect. Interestingly, ASXL2 is a member of Polycomb Group (PcG) proteins, which are best known for their roles in the longer-term epigenetic maintenance of lineage-specific expression pattern [62]. A number of studies of PcG mutants have reported a deterioration of gene expression pattern over time, after the correct pattern is established [63]. It is conceivable that the loss of ASXL2 resulted in a gradual disruption of gene expression in PCN cells, eventually resulting in their activation.

Finally, the cardiogenicity of PCN cells remains to be elucidated. In our EdU pulse-chase lineage tracing experiment, ~6% of EdU labeled cells in the *Asxl2*$^{-/-}$ left ventricle become cardiomyocytes after a 4-week chase (Figure 5B). In cardiomyocyte isolates from two chased *Asxl2*$^{-/-}$ hearts, the frequencies of EdU$^+$ cardiomyocytes are 0.079% and 0.126%, respectively. These data, while preliminary, suggest that the cardiogenicity of PCN cells is significantly higher than that reported for c-kit$^+$ cells: in a genetic lineage tracing experiment that continuously labels the c-kit lineage from embryonic stage to four weeks after birth, only 0.0027% of cardiomyocytes arise from c-kit$^+$ cells [29]. In the future, it will be exciting to fully assess PCN cells' cardiogenicity by genetic lineage tracing and to determine how PCN cells affect the ability of *Asxl2*$^{-/-}$ hearts to regenerate lost muscles in response to myocardial infarction.

4.3. Epigenetic Regulation of Cardiogenicity?

Epigenetic factors play crucial roles in the regulation of a cell's molecular signature and hence its cellular identity [64,65]. The three major epigenetic mechanisms—DNA methylation, histone modification, and chromatin remodeling—center around enzymes that modify DNA, histones, or chromatin organization, respectively. Several drugs that target epigenetic enzymes have been used in cancer treatment with some success [66–68].

In the heart, multiple epigenetic factors have been shown to regulate transcriptional activities, shape the process of cardiac morphogenesis, and modulate adult cardiac function [69–71]. ASXL2 is an essential regulator of histone H2A ubiquitination and H3K27 trimethylation [43,44,72,73]. Our discovery that the *Asxl2*$^{-/-}$ heart harbors a population of cardiogenic non-cardiomyocytes suggests that ASXL2 may be part of an epigenetic barrier that prevents a subset of non-cardiomyocytes from adopting a cardiomyocyte fate. Indeed, several recent reports suggest that epigenetic mechanisms are involved in regulating the adult heart's ability to produce cardiomyocytes both via fibroblast reprogramming and via progenitor differentiation [74–76]. These reports, along with our study, raise the exciting possibility of unlocking cardiogenicity in the adult heart by transiently modulating the activities of epigenetic enzymes. For example, if the wildtype heart harbors dormant PCN cells, modulating ASXL2 or its associated epigenetic activities may be an effective way to induce cardiogenic ability in situ, providing an alternative or a complementing approach to existing strategies of heart regeneration.

Acknowledgments: We thank the veterinary and animal care staff at the University of Illinois at Chicago; Anthony Martin Gerdes for assistance with morphometric analysis of isolated cardiomyocytes and helpful discussion; Pete Okkema, Elizaveta Benevolenskaya, and Simon Conway for valuable discussion; Teresa Orenic for thoughtful discussion and sharing reagents; Andrea McGinley for discussion and sharing her vast knowledge of immunofluorescence; Yanyang Li for help with mouse breeding; and Don Morrison for sharing equipment. These studies were supported by a research grant from the American Heart Association (13GRNT16970072).

Author Contributions: Q.T.W. and R.B. conceived and designed the experiments; R.B., H.-L.L. (Figure 1A and Supplemental Figure S1A,B), and Z.D. (isolated cardiomyocytes, Figure 6) performed the experiments; R.B., Q.T.W., and E.M. (contributed to Figure 4 and Supplemental Figure S5) analyzed the data; D.L.G. contributed reagents/materials/analysis tools; R.B. and Q.T.W. wrote the paper.

Conflicts of Interest: The authors declare no conflict of interest. The founding sponsors had no role in the design of the study; in the collection, analyses, or interpretation of data; in the writing of the manuscript, and in the decision to publish the results.

References

1. Quaini, F.; Cigola, E.; Lagrasta, C.; Saccani, G.; Quaini, E.; Rossi, C.; Olivetti, G.; Anversa, P. End-stage cardiac failure in humans is coupled with the induction of proliferating cell nuclear antigen and nuclear mitotic division in ventricular myocytes. *Circ. Res.* **1994**, *75*, 1050–1063. [CrossRef] [PubMed]
2. Soonpaa, M.H.; Field, L.J. Assessment of cardiomyocyte DNA synthesis in normal and injured adult mouse hearts. *Am. J. Physiol.* **1997**, *272*, H220–H226. [PubMed]
3. Kajstura, J.; Leri, A.; Finato, N.; di Loreto, C.; Beltrami, C.A.; Anversa, P. Myocyte proliferation in end-stage cardiac failure in humans. *Proc. Natl. Acad. Sci. USA* **1998**, *95*, 8801–8805. [CrossRef] [PubMed]
4. Bergmann, O.; Bhardwaj, R.D.; Bernard, S.; Zdunek, S.; Walsh, S.; Zupicich, J.; Alkass, K.; Buchholz, B.A.; Jovinge, S.; Frisén, J.; et al. Evidence for Cardiomyocyte Renewal in Humans. *Science* **2009**, *324*, 98–102. [CrossRef] [PubMed]
5. Malliaras, K.; Zhang, Y.; Seinfeld, J.; Galang, G.; Tseliou, E.; Cheng, K.; Sun, B.; Aminzadeh, M.; Marban, E. Cardiomyocyte proliferation and progenitor cell recruitment underlie therapeutic regeneration after myocardial infarction in the adult mouse heart. *EMBO Mol. Med.* **2013**, *5*, 191–209. [CrossRef] [PubMed]
6. Mollova, M.; Bersell, K.; Walsh, S.; Savla, J.; Das, L.T.; Park, S.-Y.; Silberstein, L.E.; dos Remedios, C.G.; Graham, D.; Colan, S.; et al. Cardiomyocyte proliferation contributes to heart growth in young humans. *Proc. Natl. Acad. Sci. USA* **2013**, *110*, 1446–51. [CrossRef] [PubMed]
7. Senyo, S.E.; Steinhauser, M.L.; Pizzimenti, C.L.; Yang, V.K.; Cai, L.; Wang, M.; Wu, T.-D.; Guerquin-Kern, J.-L.; Lechene, C.P.; Lee, R.T. Mammalian heart renewal by pre-existing cardiomyocytes. *Nature* **2013**, *493*, 433–436. [CrossRef] [PubMed]
8. Ali, S.R.; Hippenmeyer, S.; Saadat, L.V.; Luo, L.; Weissman, I.L.; Ardehali, R. Existing cardiomyocytes generate cardiomyocytes at a low rate after birth in mice. *Proc. Natl. Acad. Sci. USA* **2014**, *111*, 8850–8855. [CrossRef] [PubMed]
9. Bergmann, O.; Zdunek, S.; Felker, A.; Salehpour, M.; Alkass, K.; Bernard, S.; Sjostrom, S.L.; Szewczykowska, M.; Jackowska, T.; dos Remedios, C.; et al. Dynamics of Cell Generation and Turnover in the Human Heart. *Cell* **2015**, 1566–1575. [CrossRef] [PubMed]
10. Kimura, W.; Xiao, F.; Canseco, D.C.; Muralidhar, S.; Thet, S.; Zhang, H.M.; Abdulrahman, Y.; Chen, R.; Garcia, J.; Shelton, J.M.; et al. Hypoxia fate mapping identifies cycling cardiomyocytes in the adult heart. *Nature* **2015**, *523*, 226–230. [CrossRef] [PubMed]
11. Chaudhry, H.W.; Dashoush, N.H.; Tang, H.; Zhang, L.; Wang, X.; Wu, E.X.; Wolgemuth, D.J. Cyclin A2 mediates cardiomyocyte mitosis in the postmitotic myocardium. *J. Biol. Chem.* **2004**, *279*, 35858–35866. [CrossRef] [PubMed]
12. Sdek, P.; Zhao, P.; Wang, Y.; Huang, C.-J.; Ko, C.Y.; Butler, P.C.; Weiss, J.N.; Maclellan, W.R. Rb and p130 control cell cycle gene silencing to maintain the postmitotic phenotype in cardiac myocytes. *J. Cell Biol.* **2011**, *194*, 407–423. [CrossRef] [PubMed]
13. Eulalio, A.; Mano, M.; Ferro, M.D.; Zentilin, L.; Sinagra, G.; Zacchigna, S.; Giacca, M. Functional screening identifies miRNAs inducing cardiac regeneration. *Nature* **2012**, *492*, 376–381. [CrossRef] [PubMed]
14. Von Gise, A.; Lin, Z.; Schlegelmilch, K.; Honor, L.B.; Pan, G.M.; Buck, J.N. YAP1, the nuclear target of Hippo signaling, stimulates heart growth through cardiomyocyte proliferation but not hypertrophy. *Proc. Natl. Acad. Sci. USA* **2012**, *109*, 2–7. [CrossRef] [PubMed]
15. Chakraborty, S.; Sengupta, A.; Yutzey, K.E. Tbx20 promotes cardiomyocyte proliferation and persistence of fetal characteristics in adult mouse hearts. *J. Mol. Cell. Cardiol.* **2013**, *62*, 203–213. [CrossRef] [PubMed]
16. Chen, J.; Huang, Z.P.; Seok, H.Y.; Ding, J.; Kataoka, M.; Zhang, Z.; Hu, X.; Wang, G.; Lin, Z.; Wang, S.; et al. Mir-17-92 cluster is required for and sufficient to induce cardiomyocyte proliferation in postnatal and adult hearts. *Circ. Res.* **2013**, *112*, 1557–1566. [CrossRef] [PubMed]
17. Heallen, T.; Morikawa, Y.; Leach, J.; Tao, G.; Willerson, J.T.; Johnson, R.L.; Martin, J.F. Hippo signaling impedes adult heart regeneration. *Development* **2013**, *140*, 4683–4690. [CrossRef] [PubMed]
18. Mahmoud, A.I.; Kocabas, F.; Muralidhar, S.A.; Kimura, W.; Koura, A.S.; Thet, S.; Porrello, E.R.; Sadek, H.A. Meis1 regulates postnatal cardiomyocyte cell cycle arrest. *Nature* **2013**, *497*, 249–53. [CrossRef] [PubMed]
19. Xin, M.; Kim, Y.; Sutherland, L.B.; Murakami, M.; Qi, X.; McAnally, J.; Porrello, E.R.; Mahmoud, A.I.; Tan, W.; Shelton, J.M.; et al. Hippo pathway effector Yap promotes cardiac regeneration. *Proc. Natl. Acad. Sci. USA* **2013**, *110*, 13839–13844. [CrossRef] [PubMed]

20. Rochais, F.; Sturny, R.; Chao, C.M.; Mesbah, K.; Bennett, M.; Mohun, T.J.; Bellusci, S.; Kelly, R.G. FGF10 promotes regional foetal cardiomyocyte proliferation and adult cardiomyocyte cell-cycle re-entry. *Cardiovasc. Res.* **2014**, *104*, 432–442. [CrossRef] [PubMed]
21. Lin, Z.; von Gise, A.; Zhou, P.; Gu, F.; Ma, Q.; Jiang, J.; Yau, A.L.; Buck, J.N.; Gouin, K.A.; van Gorp, P.R.R.; et al. Cardiac-specific YAP activation improves cardiac function and survival in an experimental murine MI model. *Circ. Res.* **2014**, *115*, 354–363. [CrossRef] [PubMed]
22. Li, J.; Gao, E.; Vite, A.; Yi, R.; Gomez, L.; Goossens, S.; van Roy, F.; Radice, G.L. Alpha-Catenins Control Cardiomyocyte Proliferation by Regulating Yap Activity. *Circ. Res.* **2015**, *116*, 70–79. [CrossRef] [PubMed]
23. Tian, Y.; Liu, Y.; Wang, T.; Zhou, N.; Kong, J.; Chen, L.; Snitow, M.; Morley, M.; Li, D.; Petrenko, N.; et al. A microRNA-Hippo pathway that promotes cardiomyocyte proliferation and cardiac regeneration in mice. *Sci. Transl. Med.* **2015**, *7*, 279ra38. [CrossRef] [PubMed]
24. Xiang, F.; Guo, M.; Yutzey, K.E. Overexpression of Tbx20 in Adult Cardiomyocytes Promotes Proliferation and Improves Cardiac Function after Myocardial Infarction. *Circulation* **2016**, *133*, 1081–1092. [CrossRef] [PubMed]
25. Hsieh, P.C.H.; Segers, V.F.M.; Davis, M.E.; MacGillivray, C.; Gannon, J.; Molkentin, J.D.; Robbins, J.; Lee, R.T. Evidence from a genetic fate-mapping study that stem cells refresh adult mammalian cardiomyocytes after injury. *Nat. Med.* **2007**, *13*, 970–974. [CrossRef] [PubMed]
26. Beltrami, A.P.; Barlucchi, L.; Torella, D.; Baker, M.; Limana, F.; Chimenti, S.; Kasahara, H.; Rota, M.; Musso, E.; Urbanek, K.; et al. Adult Cardiac Stem Cells Are Multipotent and Support Myocardial Regeneration. *Cell* **2003**, *114*, 763–776. [CrossRef]
27. Ellison, G.M.; Vicinanza, C.; Smith, A.J.; Aquila, I.; Leone, A.; Waring, C.D.; Henning, B.J.; Stirparo, G.G.; Papait, R.; Scarfò, M.; et al. Adult c-kitpos cardiac stem cells are necessary and sufficient for functional cardiac regeneration and repair. *Cell* **2013**, *154*, 827–842. [CrossRef] [PubMed]
28. Hesse, M.; Fleischmann, B.K.; Kotlikoff, M.I. Concise review: The role of C-kit expressing cells in heart repair at the neonatal and adult stage. *Stem Cells* **2014**, *32*, 1701–1712. [CrossRef] [PubMed]
29. Van Berlo, J.H.; Kanisicak, O.; Maillet, M.; Vagnozzi, R.J.; Karch, J.; Lin, S.-C.J.; Middleton, R.C.; Marbán, E.; Molkentin, J.D. C-Kit$^+$ Cells Minimally Contribute Cardiomyocytes to the Heart. *Nature* **2014**, *509*, 337–341. [CrossRef] [PubMed]
30. Sultana, N.; Zhang, L.; Yan, J.; Chen, J.; Cai, W.; Razzaque, S.; Jeong, D.; Sheng, W.; Bu, L.; Xu, M.; et al. Resident c-kit(+) cells in the heart are not cardiac stem cells. *Nat. Commun.* **2015**, *6*, 8701. [CrossRef] [PubMed]
31. Smart, N.; Bollini, S.; Dubé, K.N.; Vieira, J.M.; Zhou, B.; Davidson, S.; Yellon, D.; Riegler, J.; Price, A.N.; Lythgoe, M.F.; et al. De novo cardiomyocytes from within the activated adult heart after injury. *Nature* **2011**, *474*, 640–644. [CrossRef] [PubMed]
32. Smart, N.; Bollini, S.; Dube, K.N.; Vieira, J.M.; Zhou, B.; Riegler, J.; Price, A.N.; Lythgoe, M.F.; Davidson, S.; Yellon, D.; et al. Myocardial regeneration: Expanding the repertoire of thymosin beta 4 in the ischaemic heart. *October* **2012**, *1269*, 92–101.
33. Chen, S.; Shimoda, M.; Chen, J.; Grayburn, P.A. Stimulation of adult resident cardiac progenitor cells by durable myocardial expression of thymosin beta 4 with ultrasound-targeted microbubble delivery. *Gene Ther.* **2013**, *20*, 225–233. [CrossRef] [PubMed]
34. Zhou, B.; Honor, L.B.; Ma, Q.; Oh, J.H.; Lin, R.Z.; Melero-Martin, J.M.; von Gise, A.; Zhou, P.; Hu, T.; He, L.; et al. Thymosin beta 4 treatment after myocardial infarction does not reprogram epicardial cells into cardiomyocytes. *J. Mol. Cell. Cardiol.* **2012**, *52*, 43–47. [CrossRef] [PubMed]
35. Oh, H.; Bradfute, S.B.; Gallardo, T.D.; Nakamura, T.; Gaussin, V.; Mishina, Y.; Pocius, J.; Michael, L.H.; Behringer, R.R.; Garry, D.J.; et al. Cardiac progenitor cells from adult myocardium: Homing, differentiation, and fusion after infarction. *Proc. Natl. Acad. Sci. USA* **2003**, *100*, 12313–12318. [CrossRef] [PubMed]
36. Pfister, O.; Mouquet, F.; Jain, M.; Summer, R.; Helmes, M.; Fine, A.; Colucci, W.S.; Liao, R. CD31CD31$^-$ but not CD31$^+$ cardiac side population cells exhibit functional cardiomyogenic differentiation. *Circ. Res.* **2005**, *97*, 52–61. [CrossRef] [PubMed]
37. Fuentes, T.I.; Appleby, N.; Tsay, E.; Martinez, J.J.; Bailey, L.; Hasaniya, N.; Kearns-Jonker, M. Human Neonatal Cardiovascular Progenitors: Unlocking the Secret to Regenerative Ability. *PLoS ONE* **2013**, *8*, e77464.

38. Chong, J.J.H.; Chandrakanthan, V.; Xaymardan, M.; Asli, N.S.; Li, J.; Ahmed, I.; Heffernan, C.; Menon, M.K.; Scarlett, C.J.; Rashidianfar, A.; et al. Adult cardiac-resident MSC-like stem cells with a proepicardial origin. *Cell Stem Cell* **2011**, *9*, 527–540. [PubMed]

39. Ieda, M.; Fu, J.D.; Delgado-Olguin, P.; Vedantham, V.; Hayashi, Y.; Bruneau, B.G.; Srivastava, D. Direct reprogramming of fibroblasts into functional cardiomyocytes by defined factors. *Cell* **2010**, *142*, 375–386. [CrossRef] [PubMed]

40. Song, K.; Nam, Y.-J.; Luo, X.; Qi, X.; Tan, W.; Huang, G.N.; Acharya, A.; Smith, C.L.; Tallquist, M.D.; Neilson, E.G.; et al. Heart repair by reprogramming non-myocytes with cardiac transcription factors. *Nature* **2012**, *485*, 599–604. [CrossRef] [PubMed]

41. Qian, L.; Huang, Y.; Spencer, C.I.; Foley, A.; Vedantham, V.; Liu, L.; Conway, S.J.; Fu, J.; Srivastava, D. In vivo reprogramming of murine cardiac fibroblasts into induced cardiomyocytes. *Nature* **2012**, *485*, 593–598. [CrossRef] [PubMed]

42. Furtado, M.B.; Costa, M.W.; Pranoto, E.A.; Salimova, E.; Pinto, A.R.; Lam, N.T.; Park, A.; Snider, P.; Chandran, A.; Harvey, R.P.; et al. Cardiogenic genes expressed in cardiac fibroblasts contribute to heart development and repair. *Circ. Res.* **2014**, *114*, 1422–1434. [CrossRef] [PubMed]

43. Baskind, H.A.; Na, L.; Ma, Q.; Patel, M.P.; Geenen, D.L.; Wang, Q.T. Functional conservation of Asxl2, a murine homolog for the Drosophila enhancer of trithorax and polycomb group gene Asx. *PLoS ONE* **2009**, *4*, e4750. [CrossRef] [PubMed]

44. Lai, H.L.; Wang, Q.T. Additional Sex Combs-Like 2 Is Required for Polycomb Repressive Complex 2 Binding at Select Targets. *PLoS ONE* **2013**, *8*, 1–13. [CrossRef] [PubMed]

45. Lai, H.L.; Grachoff, M.; McGinley, A.L.; Khan, F.F.; Warren, C.M.; Chowdhury, S.A.K.; Wolska, B.M.; Solaro, R.J.; Geenen, D.L.; Wang, Q.T. Maintenance of adult cardiac function requires the chromatin factor Asxl2. *J. Mol. Cell. Cardiol.* **2012**, *53*, 734–741. [CrossRef] [PubMed]

46. Schiller, N.B.; Shah, P.M.; Crawford, M.; DeMaria, A.; Devereux, R.; Feigenbaum, H.; Gutgesell, H.; Reicheck, N.; Sahn, D.; Schnittger, I.; et al. Recommendations for quantification of the left ventricle by two-dimensional echocardiography. American Society of Echocardiography Committee on Standards, Subcommittee on Quantification of Two-dimensional Echocardiograms. *J. Am. Soc. Echocardiogr.* **1989**, *2*, 358–367. [PubMed]

47. Louch, W.E.; Sheehan, K.A.; Wolska, B.M. Methods in cardiomyocyte isolation, culture, and gene transfer. *J. Mol. Cell. Cardiol.* **2011**, *51*, 288–298. [CrossRef] [PubMed]

48. Lamprecht, M.R.; Sabatini, D.M.; Carpenter, A.E. Short Technical Reports CellProfiler™: Free, versatile software for automated biological image analysis Short Technical Reports. *Biotechniques* **2007**, *42*, 71–75. [CrossRef] [PubMed]

49. Li, F.; Wang, X.; Capasso, J.M.; Gerdes, A.M. Rapid transition of cardiac myocytes from hyperplasia to hypertrophy during postnatal development. *J. Mol. Cell. Cardiol.* **1996**, *28*, 1737–1746. [CrossRef] [PubMed]

50. Bai, S.L.; Campbell, S.E.; Moore, J.A.; Morales, M.C.; Gerdes, A.M. Influence of age, growth, and sex on cardiac myocyte size and number in rats. *Anat. Rec.* **1990**, *226*, 207–212. [CrossRef] [PubMed]

51. Gerdes, A.M. A reliable, efficient, and comprehensive approach to assess myocyte remodeling in cardiac hypertrophy and failure. *J. Card. Fail.* **1997**, *3*, 63–68. [CrossRef]

52. Adler, C.P.; Friedburg, H.; Herget, G.W.; Neuburger, M.; Schwalb, H. Variability of cardiomyocyte DNA content, ploidy level and nuclear number in mammalian hearts. *Virchows Arch.* **1996**, *429*, 159–164. [CrossRef] [PubMed]

53. Matiasova, A.; Sevc, J.; Mikes, J.; Jendzelovsky, R.; Daxnerova, Z.; Fedorocko, P. Flow cytometric determination of 5-bromo-2′-deoxyuridine pharmokinetics in blood serum after intraperitoneal administration to rats and mice. *Histochem. Cell Biol.* **2014**, *142*, 703–12. [CrossRef] [PubMed]

54. Packard, D.; Menzies, R.; Skalko, R. Incorporation of thymidine and its analogue, bromodeoxyuridine, into embryos and maternal tissues of the mouse. *Differentiation* **1973**, *1*, 397–405. [CrossRef] [PubMed]

55. Soonpaa, M.H.; Rubart, M.; Field, L.J. Challenges measuring cardiomyocyte renewal. *Biochim. Biophys. Acta Mol. Cell Res.* **2013**, *1833*, 799–803. [CrossRef] [PubMed]

56. Fioret, B.A.; Heimfeld, J.D.; Paik, D.T.; Hatzopoulos, A.K. Endothelial cells contribute to generation of adult ventricular myocytes during cardiac homeostasis. *Cell Rep.* **2014**, *8*, 229–241. [CrossRef] [PubMed]

57. Cai, C.L.; Liang, X.; Shi, Y.; Chu, P.H.; Pfaff, S.L.; Chen, J.; Evans, S. Isl1 identifies a cardiac progenitor population that proliferates prior to differentiation and contributes a majority of cells to the heart. *Dev. Cell* **2003**, *5*, 877–889. [CrossRef]
58. Laugwitz, K.-L.; Moretti, A.; Lam, J.; Gruber, P.; Chen, Y.; Woodard, S.; Lin, L.-Z.; Cai, C.-L.; Lu, M.M.; Reth, M.; et al. Postnatal isl1$^+$ cardioblasts enter fully differentiated cardiomyocyte lineages. *Nature* **2005**, *433*, 647–53. [CrossRef] [PubMed]
59. Wang, L.; Liu, Z.; Yin, C.; Asfour, H.; Chen, O.; Li, Y.; Bursac, N.; Liu, J.; Qian, L. Stoichiometry of Gata4, Mef2c, and Tbx5 influences the efficiency and quality of induced cardiac myocyte reprogramming. *Circ. Res.* **2015**, *116*, 237–244. [CrossRef] [PubMed]
60. Dave, J.M.; Bayless, K.J. Vimentin as an integral regulator of cell adhesion and endothelial sprouting. *Microcirculation* **2014**, *21*, 333–344. [CrossRef] [PubMed]
61. Chong, J.J.H.; Reinecke, H.; Iwata, M.; Torok-Storb, B.; Stempien-Otero, A.; Murry, C.E. Progenitor Cells Identified by PDGFR-Alpha Expression in the Developing and Diseased Human Heart. *Stem Cells Dev.* **2013**, *22*, 1932–1943. [CrossRef] [PubMed]
62. Struhl, G.; Akam, M. Altered distributions of Ultrabithorax transcripts in extra sex combs mutant embryos of Drosophila. *EMBO J.* **1985**, *4*, 3259–3264. [PubMed]
63. Bantignies, F.; Cavalli, G. Cellular memory and dynamic regulation of polycomb group proteins. *Curr. Opin. Cell Biol.* **2006**, *18*, 275–283. [CrossRef] [PubMed]
64. Margueron, R.; Reinberg, D. Chromatin structure and the inheritance of epigenetic information. *Nat. Rev. Genet.* **2010**, *11*, 285–296. [CrossRef] [PubMed]
65. Barrero, M.J.; Boué, S.; Izpisúa Belmonte, J.C. Epigenetic Mechanisms that Regulate Cell Identity. *Cell Stem Cell* **2010**, *7*, 565–570. [CrossRef] [PubMed]
66. Griffiths, E.A.; Gore, S.D. *Epigenetic Therapies in MDS and AML*; Springer: New York, NY, USA, 2013; Volume 754.
67. Nie, J.; Liu, L.; Li, X.; Han, W. Decitabine, a new star in epigenetic therapy: The clinical application and biological mechanism in solid tumors. *Cancer Lett.* **2014**, *354*, 12–20. [CrossRef] [PubMed]
68. Nervi, C.; de Marinis, E.; Codacci-Pisanelli, G. Epigenetic treatment of solid tumours: A review of clinical trials. *Clin. Epigenet.* **2015**, *7*, 127. [CrossRef] [PubMed]
69. Van Weerd, J.H.; Koshiba-Takeuchi, K.; Kwon, C.; Takeuchi, J.K. Epigenetic factors and cardiac development. *Cardiovasc. Res.* **2011**, *91*, 203–211. [CrossRef] [PubMed]
70. Vallaster, M.; Vallaster, C.D.; Wu, S.M. Epigenetic mechanisms in cardiac development and disease Overview of Epigenetic Mechanisms. *Acta Biochim. Biophys. Hung.* **2012**, 92–102.
71. Wang, Q.T. Epigenetic regulation of cardiac development and function by polycomb group and trithorax group proteins. *Dev. Dyn.* **2012**, *241*, 1021–1033. [CrossRef] [PubMed]
72. Daou, S.; Hammond-Martel, I.; Mashtalir, N.; Barbour, H.; Gagnon, J.; Iannantuono, N.V.; Nkwe, N.S.; Motorina, A.; Pak, H.; Yu, H.; et al. The BAP1/ASXL2 histone H2A deubiquitinase complex regulates cell proliferation and is disrupted in cancer. *J. Biol. Chem.* **2015**, *290*, 28643–28663. [CrossRef] [PubMed]
73. Park, U.; Kang, M.; Kim, E.; Kwon, Y.; Hur, W.; Yoon, S.; Song, B.; Park, J.; Hwang, J.; Jeong, J.; et al. ASXL2 promotes proliferation of breast cancer cells by linking ERalpha to histone methylation. *Oncogene* **2016**, *14*, 3742–3752. [CrossRef] [PubMed]
74. Valiente-alandi, I.; Albo-castellanos, C.; Herrero, D.; Arza, E.; Garcia-gomez, M.; Segovia, J.C.; Capecchi, M.; Bernad, A. Cardiac Bmi1$^+$ cells contribute to myocardial renewal in the murine adult heart. *Stem Cell Res. Ther.* **2015**, *7*. [CrossRef] [PubMed]
75. Zhou, Y.; Wang, L.; Vaseghi, H.R.; Liu, Z.; Lu, R.; Alimohamadi, S.; Yin, C.; Fu, J.D.; Wang, G.G.; Liu, J.; et al. Bmi1 Is a Key Epigenetic Barrier to Direct Cardiac Reprogramming. *Cell Stem Cell* **2016**, *18*, 382–395. [CrossRef] [PubMed]
76. Valiente-alandi, I.; Albo-castellanos, C.; Herrero, D.; Sanchez, I.; Bernad, A. Bmi1$^+$ cardiac progenitor cells contribute to myocardial repair following acute injury. *Stem Cell Res. Ther.* **2016**. [CrossRef] [PubMed]

11

An Evolutionarily Conserved Network Mediates Development of the *zona limitans intrathalamica*, a Sonic Hedgehog-Secreting Caudal Forebrain Signaling Center

Elena Sena [1], Kerstin Feistel [2] and Béatrice C. Durand [1,*]

1 Institut Curie, Université Paris Sud, INSERM U1021, CNRS UMR3347, Centre Universitaire, Bâtiment 110, F-91405 Orsay Cedex, France; elena.sena@curie.fr

2 Institute of Zoology, University of Hohenheim, Garbenstr. 30, 70593 Stuttgart, Germany; k.feistel@uni-hohenheim.de

* Correspondence: Beatrice.Durand@curie.fr

Academic Editors: Henk Roelink and Simon J. Conway

Abstract: Recent studies revealed new insights into the development of a unique caudal forebrain-signaling center: the *zona limitans intrathalamica* (*zli*). The *zli* is the last brain signaling center to form and the first forebrain compartment to be established. It is the only part of the dorsal neural tube expressing the morphogen Sonic Hedgehog (Shh) whose activity participates in the survival, growth and patterning of neuronal progenitor subpopulations within the thalamic complex. Here, we review the gene regulatory network of transcription factors and *cis*-regulatory elements that underlies formation of a *shh*-expressing delimitated domain in the anterior brain. We discuss evidence that this network predates the origin of chordates. We highlight the contribution of Shh, Wnt and Notch signaling to *zli* development and discuss implications for the fact that the morphogen Shh relies on primary cilia for signal transduction. The network that underlies *zli* development also contributes to thalamus induction, and to its patterning once the *zli* has been set up. We present an overview of the brain malformations possibly associated with developmental defects in this gene regulatory network (GRN).

Keywords: developmental biology; embryogenesis; neural; segregation; compartment; sonic hedgehog; forebrain; thalamus; holoprosencephaly; patterning

1. Introduction

In the last decades, functional experiments in numerous model systems provided crucial insights into the early molecular, cellular, and morphological events underlying forebrain development. They revealed the importance of brain signaling centers: small groups of specialized cells acting as local sources of secreted factors. These secreted factors regulate survival, proliferation, and early patterning of neuroepithelial cells and facilitate compartmentalization of the neuroepithelium into functional histological units [1–7].

The forebrain (also called prosencephalon) is derived from the most anterior part of the neuroepithelium, the prosencephalic neural plate. Its rostral part, the telencephalon, develops into the cerebrum, which gets divided into the cerebral hemispheres. The caudal forebrain or diencephalon is situated between the cerebral cortex and the midbrain. It gives rise to the thalamic complex, a bilateral structure that contains the pre-thalamus (ventral thalamus) rostrally and the thalamus (dorsal thalamus) caudally.

Development of the thalamus is orchestrated by a brain signaling center called the *zona limitans intrathalamica* or *interthalamica* (*zli*), also referred to as the mid-diencephalic organizer (MDO) [3]. The *zli* emerges after the completion of neural tube closure and is characterized by a dorso-ventrally (DV) extending gene expression domain of the morphogen Sonic hedgehog (Shh). Shh secreted from the *zli* and the basal plate participates in the survival, growth, and patterning of neuronal progenitor subpopulations within the thalamic complex [8–15].

In-depth reviews detail the role of Shh in diencephalic patterning [3,13,16,17]. Here we focus on the principles underlying emergence of the *zli*. Specifically, we describe the network of transcription factors and *cis*-regulatory sequences that confer competence for *zli* establishment. We review work analyzing their evolutionary conservation. We also highlight the contribution of signaling pathways to *zli* formation, discuss implications for the fact that Shh relies on primary cilia for signal transduction and speculate about *zli* development for the etiology of brain malformations.

2. The Zli: A Unique Shh-Expressing Compartment in the Caudal Forebrain

2.1. Forebrain Morphology and Zli Positioning

On the basis of morphology [18–21] and gene expression [22,23], the diencephalic primordium is divided into three transverse segments called prosomeres (p) that generate three distinct histogenic fields: p3, or anterior parencephalon, corresponds to the presumptive prethalamus and the eminentia thalami; p2, or posterior parencephalon, gives rise to the epithalamus and the thalamus; and p1, or synencephalon, generates the presumptive pretectum (reviewed in [17,24–27]). Each prosomere is divided into a ventral (basal) and a dorsal (alar) part (Figure 1A).

The *zli* demarcates the boundary between pre-thalamus and thalamus, separating the posterior diencephalon (p1 and p2) from the anterior diencephalon (p3) (Figure 1; [28–30]). This position corresponds approximately to the junction between the prechordal neuraxis overlying the prechordal mesendoderm, and the epichordal neuraxis overlying the notochord [9]. The *zli* is defined by a dorsalward continuation of *shh* expression into the alar plate of the diencephalon (Figure 1A,Bb–d). The *zli* is the last brain signaling center to emerge: its dorsal progression within the alar plate starts at stage 24 in frog [12]), HH12 to HH26 in chicken [28,29,31], E9 in mouse, and between the 12- and 15-somite stage in zebrafish [32]. It is unique in brain regionalization because it represents the only neural area in which Shh, normally a DV patterning signal, regulates anterior-posterior (AP) regionalization [33,34].

2.2. Shh Expression Initiates and Demarcates Zli Development

During gastrulation and early neurulation, *shh* is expressed in the prechordal plate and notochord. Notochord-derived Shh activates the Hedgehog signaling pathway and its own neural midline expression in the overlying neuroepithelial cells ([35,36]; reviewed in [37]). In the posterior part of the neural tube, *shh* gene expression remains restricted to the medial floor plate, whereas in the mesencephalon and prosencephalon *shh* expression spreads from the floor plate to the basal plate. It is only at the p2 to p3 boundary that a triangle-shaped expression of *shh* extends dorsalward (i.e., more lateral) on either side of the diencephalic walls (Figure 1; reviewed in [37]). Studies in chicken revealed that a sequential induction process, initiated by Shh secreted by floor plate cells, underlies the appearance of the *shh*-expressing line of cells corresponding to the *zli* [38]. This homeotic induction of *shh* occuring from one cell to its neighbor is not understood. It could necessitate specific Shh maturation and secretion processes, together with modification at the level of the primary cilium that mediates Shh read-out. Because the *zli* is not formed through cell migration from the basal to the alar plate, the *zli* is considered an alar structure [38]. Observations in frog, chicken, and mouse demonstrate that a continuous source of Shh signals, provided *in vivo* by secretion of Shh initially from floor plate cells, and then from the basal plate, is strictly necessary both for induction of *shh* expression within the *zli* and for the correct segregation of *zli* cells from the thalamus [12,28,32,38,39].

In zebrafish, however, a small domain of *shh* expression at the position of the *zli* appears in *smoothened* (*smo*) mutants, in which signaling downstream of all Hedgehog family ligands is defective [40,41]. Similarly, a population of "*zli* cells" develops in a quarter of zebrafish *cyclops* mutants, which lack a floor plate in the diencephalon [32,38,39,42]. Taken together, these observations suggest that, contrary to observations in other model organisms, the continuous presence of a neural midline-derived Shh signal is not required for emergence of the *zli* in zebrafish.

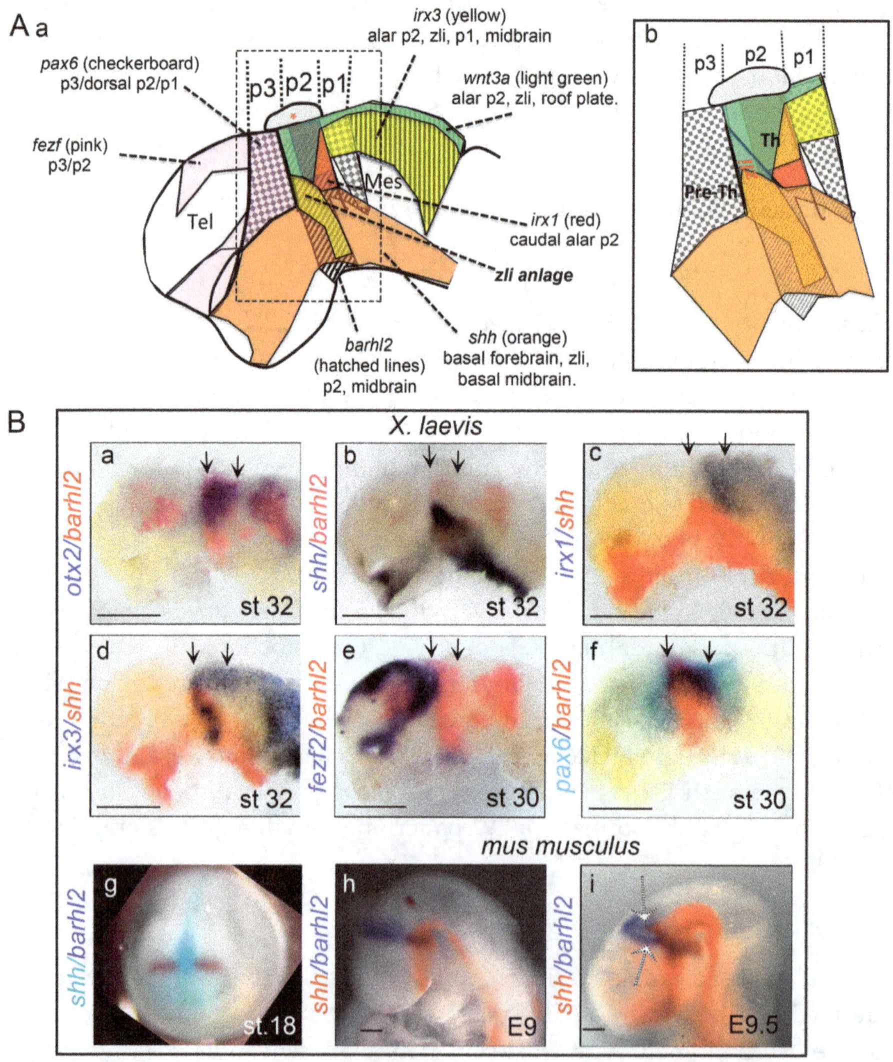

Figure 1. Schematic of *Xenopus* (*X.*) *laevis* forebrain markers and flat-mounted neural tubes from *X. laevis* and *Mus musculus* embryo after whole-mount double *in situ*-hybridization (WM-dISH). (**A**) Schematic of *X. laevis* forebrain markers at st. 30. (**a**) Areas of expression are shown and indicated for *fezf2* (pink), *pax6* (checkerboard pattern), *irx3* (yellow), *wnt3a* (light green), *shh* (orange), *barhl2* (hatched lines), and *irx1* (red). (**b**) Enlargement of the diencephalic part indicated by a rectangle in (**a**). The pineal gland located on top of p2 is marked by an asterisk. The future *zli* is indicated in (**b**). p: prosomere; Tel: telencephalon; Mes: mesencephalon; Pre-Th: pre-thalamus; *zli: zona limitans intrathalamica*; Th: thalamus. (**B**) WM-dISH is performed using *otx2*, *irx1*, *irx3*, *fezf2*, *pax6*, and *shh* or *barhl2* as probes on (**a**–**g**). *X. laevis* and mouse embryos at different developmental stages as indicated. Neural tubes (**a**–**f**) or heads (**h**,**i**) of representative embryos, dissected and flat-mounted, are shown from a side view, dorsal up, anterior left except for (**g**), shown in anterior view, dorsal side up. The markers and stages are indicated. Black arrows indicate the rostral and caudal boundaries of p2. White arrows indicate the bilateral *zli*. Scale bar: 0.5 mm.

2.3. *Physical Separation: The Zli as a Tissue Compartment*

To fulfill their localized patterning function, cells of a signaling center need to be arranged in a coherent compartment with clear boundaries and lineage restriction. This enables the signaling center to maintain its position relative to the surrounding tissue [24,28,29,43] (reviewed in [17,44]). An analysis of chick diencephalon development revealed that early on, the caudal forebrain territory is not overtly segmented [29]. Development of the *zli*, however, correlates with acquisition of cell lineage-specific properties. Cell movements become restricted at both the anterior and posterior limits of the *zli* and *zli* formation is associated with acquisition of properties observed at tissue boundaries: 1) interkinetic nuclear migration movements are disrupted and 2) markers of immiscible interfaces such as chondroitin sulfate proteoglycans, laminin, weakly polysialylated neural cell adhesion molecules, and vimentin are detected within the *zli* territory [28,29]. These studies established that the *zli* is a narrow compartment with cell lineage-restricted boundaries, introducing physical separation of the diencephalon into the prethalamus anteriorly and the thalamus posteriorly [24,28,29,31].

This physical separation brought about by lineage restriction within the *zli* may in addition be instructive for AP segmentation on a forebrain-wide scale. According to the prosomeric model the forebrain becomes subdivided into an anterior and a posterior part. Anteriorly, the so-called secondary prosencephalon contains the telencephalon, optic vesicles, and hypothalamus. More posteriorly, the alar plate of the diencephalon is divided into the anterior p3, p2 and p1 ([28]; reviewed in [17,25]). Note that whereas the prosomeric model differentiates p2 and p3 basal plates from the anterior basal hypothalamus, in *zebrafish* the hypothalamus is thought to develop within the basal diencephalon, a difference that may generate misinterpretations. Partitioning of the anterior forebrain from the remainder of the neural tube partly relies on the distinct processes inducing specification of the prechordal versus epichordal neural plate. It is still under debate whether the alar plate of p3, which generates the prethalamus, is under the influence of the chordal, or of the prechordal mesendoderm [25,29]. However, cell lineage analysis indicates that cells do not segregate between the prethalamus and the secondary prosencephalon [29]. Moreover, the competence of the prethalamus (p3) to respond to instructive factors such as fibroblast growth factor (Fgf) 8 is different from that of the other diencephalic prosomeres (p2 and p1) [45,46]. This raises the question whether the *zli* could represent a pivotal structure along the neuraxis segregating a "large prosencephalon" from the remainder of the neural tube. At minimum, the *zli* separates territories with differential competence to respond to morphogens [29,47].

In conclusion, the *zli* is a narrow compartment, located in the alar plate of the caudal forebrain that expresses *shh*. It is the last signaling center to form. It introduces segmentation within the diencephalon and has an important function in diencephalic patterning. The *zli* is characterized by the same positioning as similar gene expression patterns in all model organisms. However, the mode of its formation may vary between species. From studies in zebrafish, frog, chicken and mouse, it is now possible to propose a model for the GRN and the inductive cues allowing formation of this unique signaling center.

3. A Role for Wnt Ligands in Generating a *Zli*-Permissive Compartment

Observations in zebrafish, amphibian, and chicken have highlighted early and late roles for Wnt ligands in *zli* positioning, induction, and development [48,49] (Figure 2). Canonical Wnt signaling is involved in AP patterning of the forebrain [50] (reviewed in [51]). Receptors, ligands and modifiers of the Wnt pathway are expressed during caudal forebrain regionalization. Specifically, expression of the Wnt ligands Wnt3, Wnt3a, and Wnt8b mark the alar plate of p2 and the *zli* primordium [12,31,49].

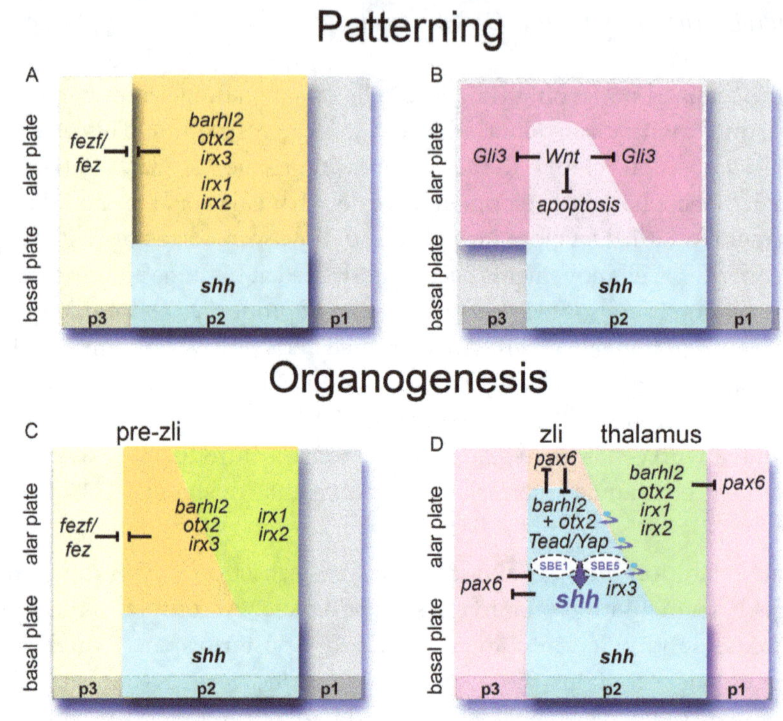

Figure 2. A model for *zli* formation. (**A**) During neural plate patterning the diencephalic primordium expresses *otx2, barhl2* and *irx1, irx2, irx3* (orange). Cross-repressive interactions between Fezf/Fez and Irx homeoproteins contribute to formation of the p2/p3 alar border. *shh* (light blue) is expressed in the p2 basal plate. (**B**) Concomitantly, a Wnt-mediated signal promotes survival of *zli* anlage cells, and *gli3* expression is repressed through a Wnt-mediated signal in a narrow band of cells that correspond to the *zli* primordium. (**C**) At the onset of *zli* formation a permissive environment is established in the future *zli*, and *zli* cells acquire the competence to express *shh*. The *zli* anlage is characterized by the expression of *barhl2, otx2, irx3* (orange), and the future thalamus by the expression of *barhl2, otx2, irx1/2/3* (green). At the onset of *zli* development, *irx3* is expressed in the future thalamus territory. (**D**) Using a sequential induction mechanism, Shh secreted by the basal plate initiates the spreading of *shh* expression into the pre-*zli* domain. Transcription factors (TFs) including Otx1, Otx2, Barhl2, Yap–Tead, and possibly unidentified TFs (TF-X) are recruited to the enhancers SBE1 and SBE5 to initiate *shh* transcription within the *zli*. Under the inductive influence of secreted N-Shh, some *irx3*-expressing cells located in caudal p2 sort out from thalamic cells. Shh signaling from the *zli* induces the repression of *pax6* within the mid-diencephalic furrow. Pax6 together with unidentified dorsal signals could in turn prevent *Shh* from being expressed beyond the *zli*. The thalamus (p2) and the prethalamus (p3) receive a combination of Shh signals coming from two orthogonally oriented signaling centers, the *zli* and the basal plate. Together, they establish a morphogenetic gradient of Shh protein that patterns the thalamus posterior to the *zli* and the prethalamus anterior to the *zli*. p: prosomere; *zli*: *zona limitans intrathalamica*.

In zebrafish, lack of canonical Wnt signaling brought about by a Wnt signaling antagonist induces loss of the pre-*zli* territory. Conversely, enhancement of Wnt signaling using the Glycogen Synthase Kinase 3β (GSK3β) inhibitor BIO leads to a broader expression domain of *shh* at the *zli* [48]. The depletion of both Wnt3 and Wnt3a leads to an increase of apoptosis and a loss of the diencephalic organizer primordium indicating that Wnt3 and Wnt3a are normally required for survival of *zli* anlage cells. Interestingly, the effect of canonical Wnt signaling on the survival of *zli* anlage cells is restricted to a time window of 4 hours during somitogenesis. In embryos depleted for Wnt3 and Wnt3a the size of the prethalamic and thalamic markers *fezf2, irx1b* and *otx2* expression domains are unaltered (Figures 1A and 2). In Wnt3/3a depleted embryos the concomitant depletion of *fezf2*, or of Irx1b, activity rescues *zli* formation, indicating that both Fezf2 and Irx1b normally restrict the *zli* territory [48].

Therefore, Wnt3/3a function is required for maintenance of the *zli* anlage, but not for the maintenance of the prethalamus and thalamus territories (Figure 2).

A study by Martinez-Ferre et al. reveals a requirement for a Wnt8b-mediated signal as a permissive step for the subsequent induction of *shh* expression and for emergence of the *zli* in the diencephalic primordium [49]. Gli3 is a Shh-regulated transcriptional repressor [52–54]. During early patterning stages of the neural plate *gli3* transcripts are detected within the alar neural plate and specifically in the alar diencephalon. In chick, this wide expression domain gets restricted to a narrow band of cells at the center of the *wnt8b* expression domain. The transverse alar stripe of *wnt8b* which is now devoid of *gli3* expression is the prospective *zli* anlage [11]. Over-expression of *gli3* in the *zli* inhibits *shh* induction, indicating that the local repression of *gli3* is necessary to allow *shh* homeotic induction during *zli* formation. Martinez-Ferre et al. demonstrated that the local downregulation of *gli3* at the future *zli* requires a Wnt-signal, which in chicken is mediated by the presence of Wnt8b. Wnt8b signal appears necessary to locally downregulate *gli3* at the center of the *wnt8b* expression domain. It is, however, not sufficient, as the entire domain that expresses *wnt8b* does not lose *gli3* expression. Finally, once Shh is expressed in the *zli*, inhibition of the Wnt pathway does not have any effect on its maintenance [49].

In conclusion, while Wnt-mediated signals are important for maintenance of the *zli* anlage, and to generate the permissive conditions for the activation of Shh in the *zli*, they must act in collaboration with a combination of factors at the prethalamic (Fezf/Fez) and thalamic (Barhl2, Irx3, and Otx2) forebrain borders.

4. Laying the Ground for the *zli*: Otx2 and Barhl2 Binding to *cis*-Regulatory Sequences Confers Competence for *zli* Formation

During neural induction, an underlying pre-pattern, partly encoded by TFs, emerges in the neural plate. These early patterning cues contribute to the specification of forebrain territories and influence the way in which neighboring cell populations respond differentially to similar morphogens [46,55,56]. Analysis of TF expression dynamics in the anterior neural plate provides crucial information about the cues involved in emergence of the *zli* and reveals that *zli* induction starts during neurulation.

Two TFs are especially involved in conferring competence to the future *zli* tissue to express *shh*. Orthodenticle homeobox (Otx) 1 and 2 are homeodomain-containing proteins involved in specification and regionalization of the forebrain (reviewed in [57,58]). The *otx2* expression territory marks the anterior neural plate from gastrulation onwards [59]. In both zebrafish and frog *otx* expression decreases in the telencephalic territory during neurulation. At the onset of *zli* development *otx* expression is restricted to the p2 and midbrain territories ([12,32]; Figure 1Ba). The Bar-class homeodomain-containing (BarH) Barh1 and Barh2 are also homeodomain-containing TFs (reviewed in [60–62]). Transcripts encoding BarH-like (*barhl) 2* are detected in the diencephalic primordium of amphibian [63,64], zebrafish [65], and mouse [66,67]. Similar to *otx* genes, at the onset of *zli* development *barhl1* and *barhl2* expression are restricted to prosomere p2. While *barhl2* is expressed in the entire p2, *barhl1* expression is restricted to basal p2 ([68,69]; Figure 1A,Bb).

In zebrafish, the lack of Otx1l/2 (the zebrafish ortholog of Otx2) function leads to absence of the *zli* and subsequently of *zli*-dependent target genes. With a lack of Otx function, the thalamus is mis-specified prior to, and independently from, *zli* formation. The adjacent territories of the prethalamus and pretectum expand into the mis-specified territory and form a new interface. Mouse embryos with reduced *otx1/2* transcripts similarly show a lack of *shh zli* expression [67]. Therefore, the presence of Otx1/2 is required to establish a competence area, allowing induction of *shh* in the *zli* [32,67,70]. In Otx-depleted zebrafish embryos, ectopic expression of Otx rescues formation of the *zli* solely within the diencephalic territory, anterior to the presumptive thalamus [32]. In *X. laevis* embryos depleted for Barhl2, development of the *zli* is abolished [12]. Similarly, in *barhl2*$^{-/-}$ mice, *shh* expression within the *zli* is significantly reduced [67,71]. Barhl2-depleted *X. laevis* embryos resemble zebrafish Otx1l/2-depleted embryos: although most forebrain markers are unaltered, the embryos exhibit defects in *shh* expression in the *zli* and in the formation of the mid-diencephalic furrow [12,71].

Noteworthy, in zebrafish loss of Otx2 induces a loss of *barhl2* expression while *otx2* expression is maintained in Barhl2-depleted *X. laevis*. Therefore, Otx1l/2 proteins are necessary for maintenance of *barhl2* in the future *zli* territory, and the loss of *barhl2* contributes to the *zli* defects observed in Otx-deficient zebrafish [12,32].

Synergistic activities of Barhl2 and Otx2 in *zli* formation are confirmed by analysis of the *cis*-regulatory-motifs controlling *shh* expression within the mouse *zli*. Yao et al. identified two enhancers, SBE1 and SBE5, driving *shh* expression within the *zli*. Both enhancers function in a partially redundant manner and their activity relies on six position-independent motifs directly regulated by a combination of Otx2, Barhl2 and the TEA-domain family member 2 (Tead2), a key mediator of Hippo signaling, and its co-transcriptional activation partner Yap. Using large-scale genomic approaches associated with bioinformatic analytical tools, Yao et al. characterized a 116-bp homology block, referred to as SBE1-like-enhancer, present in enhancers scattered throughout the mouse and human genomes. The SBE1-like-enhancer sequence is conserved from human to zebrafish. Using luciferase reporter assays, chromatin immunoprecipitation (ChIP), and transgenic mouse reporter assays, Yao et al. showed that the six motifs are necessary and sufficient for full enhancer activity and paired three motifs with cognate transcription factors. Motifs 1 and 6 correspond to recognition sequences for Otx1/Otx2 and Barhl2, respectively. Indeed, the combined action of Barhl2 and Otx2 resulted in a synergistic induction of reporters whose activity is under the control of SBE1-like enhancers containing motifs 1 and 6. Similarly, Tead2 and Yap recognize and contribute to *shh* expression through binding on motif 2. Embryos with conditional loss of Yap1 showed a selective reduction in *zli shh* expression. Tead2 and Yap activities on *zli* formation have not been assessed in species others than mouse. The second enhancer SBE5 contains a cluster of permuted motifs similar to those identified in the SBE1 enhancer in the absence of any other overt sequence homology. This second enhancer, located in the vicinity of the *shh* locus, performs equivalently to SBE1 in cell-based reporter and ChIP assays. Deletion of both enhancers in mouse entirely abolishes expression of *shh* within the *zli* [67].

Taken together, these studies reveal the presence of a "*zli* developmental cassette" that uses two main TFs, Otx2 and Barhl2, in combination, at least in mice, with the Tead2-Yap1 activation complex and conserved *cis*-regulatory motifs to induce *shh* expression in a narrow band of cells within the anterior brain.

5. Hedgehog Sequential Induction Process in *zli* Formation: Shh Is a Secreted Signal Read out through a Primary Cilium

During *zli* formation the N-terminal part of Shh secreted by basal plate cells is released into the ventricular lumen and N-Shh activates its own expression only in its neighboring cell [38]. Pre-*zli* cells respond differently to Shh compared to neighboring prethalamus and thalamus cells. The cues controlling sequential induction and the differential response to Shh are not fully deciphered. Modifications in Shh maturation and secretion, and/or in the primary cilium that mediates Shh read-out, could participate in this unique process.

5.1. Shh Maturation and Secretion

The secretion of Hedgehog (HH) proteins into the ventricular lumen necessitates that HH proteins undergo an autocatalytic internal cleavage associated with the addition of lipid molecules, specifically cholesterol and palmitic acid moieties (reviewed in [72]). Cleavage confers a hydrophobic character to HH that is required for its association with the cell membrane. HH proteins truncated at the site of internal cleavage diffused more largely [73]. The internal cleavage occurs in the endoplasmic reticulum and produces a 20 kD amino-terminal domain (N-HH) and a 25 kD carboxy-terminal domain (C-HH) [74–76]. The C-HH part recruits a cholesterol molecule that binds to the C-terminus of N-HH [76,77]. The binding of cholesterol to N-HH is essential in limiting the range of HH signaling. In mouse, N-HH lacking cholesterol has an extended signaling range in the limb bud [78].

Besides cholesterol, the N-HH protein is also modified by the attachment of a palmitic acid group on its N-terminal part that has proved important for HH activity [79]. The transfer is mediated by an acyltransferase named Skinny Hedgehog (SKI) in *Drosophila* and mouse and Hedgehog acyltransferase (HHAT) in humans [80–82]. Mouse embryos lacking N-HH palmitoylation exhibit decreased Shh signaling in limb buds [83] as well as in ventral forebrain formation [84].

Different mechanisms have been described for the release of processed N-HH (pN-HH) into the extracellular space. pN-HH is strongly hydrophobic. It can be secreted as a monomer but needs to bind to other secreted proteins. In *Drosophila*, the transmembrane protein Dispatched-1 (Disp1) [85] binds to pN-HH along with the secreted protein Scube2. Disp1 and Scube2 bind to different parts of the cholesterol attached to N-HH and promote the release of pN-HH from the cell surface [86,87]. There is also evidence that pN-HH secretion is mediated through a lipoprotein complex [88,89]. Lipoproteins consist of a phospholipid monolayer that embraces the lipids present in the pN-HH leaving the hydrophilic areas outside, decreasing the strong hydrophobicity of the pN-HH molecule. In *Caenorhabditis elegans* (*C. elegans*) [90] and *Drosophila* [91], an exosome-mediated release has been described. In vertebrates pN-HH can also be released in a vesicular form [92].

5.2. The Shh Signal Is Read through a Primary Cilium

Shh-related patterning defects in mice carrying mutations in genes essential for cilia function revealed that Shh signaling in vertebrates requires the presence of primary cilia [93]. Cilia are small, membrane-sheathed cell protrusions that occur on almost all cells during development and adulthood. Motile cilia can be bent by dynein motors, which results in rotational or beating movement, whereas primary cilia are immotile. Owing to the fact that their membranes are studded with a range of different mechano- and chemo-receptors, primary cilia are considered highly specialized sensory organelles [94]. Signal transduction from the sensory primary cilium requires motor protein-driven transport along axonemal microtubules, termed intraflagellar transport (IFT). IFT is required for Hedgehog signal transduction since Patched (Ptc), a twelve-transmembrane domain receptor, Smo, a G-protein coupled receptor, and the zinc finger containing Gli family of TFs are moving into, within, and out of the cilium, depending on pathway activation status ([95]; reviewed in [72,96–98]).

During embryogenesis, neural tube cells in mouse [93,99], chicken [100], *X. laevis* [101], and zebrafish [102] carry primary cilia on their apical surface. These are crucial for Shh signal transduction governing DV patterning of the neural tube [93]. Cilia have also been identified on ventricle-contacting cells specifically in the forebrain region in mouse [103] and on diencephalic cells corresponding to the region of the *zli* in *X. laevis* (Figure 3; [104]). Interestingly, cilia on cells within *shh*-expressing brain signaling centers including the floor plate, rhombomere boundaries, and *zli* are specifically elongated compared to surrounding primary cilia on non-*shh*-expressing cells ([100,101,104] and Figure 3). This could indicate a way of differentially adapting signaling pathway activation, as longer cilia seem to correlate with reduced pathway activation [100].

In conclusion, the mechanisms controlling Shh secretion and diffusion rate in the context of *zli* development are still being investigated. The strong hydrophobic character of N-Shh suggests that it diffuses very slowly and tends to stay close to the membrane surface. This could at least partly explain the sequential induction process observed during *zli* development. Moreover, a primary cilium is present in diencephalic progenitor cells. The primary cilium is structurally different on pre-*zli* cells versus prethalamic and thalamic cells and little is known on the biological consequences of such differences.

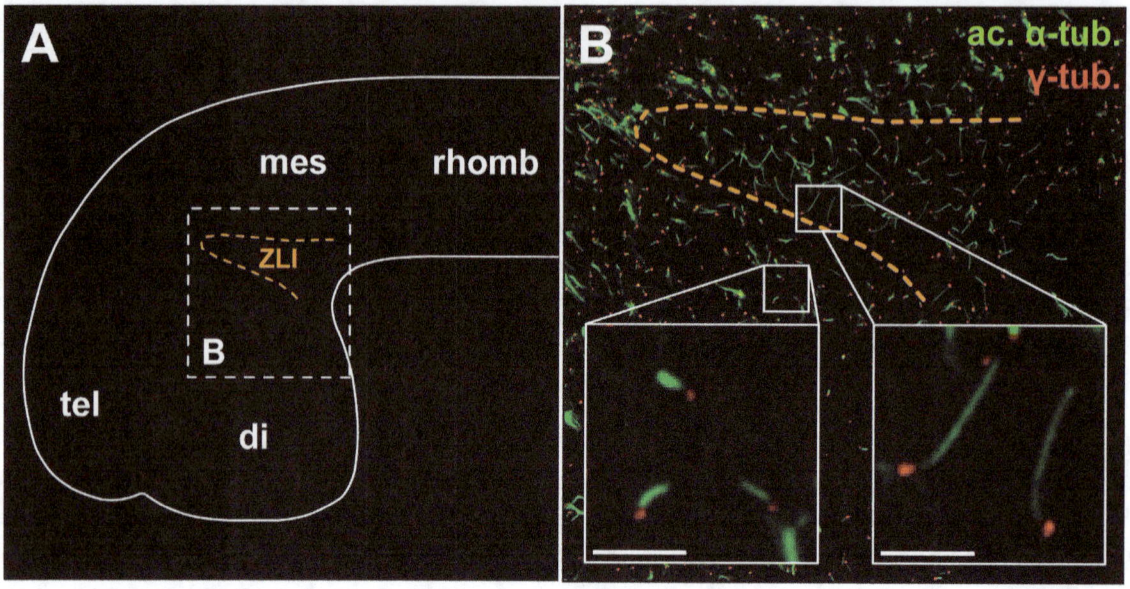

Figure 3. Elongated cilia on the *zli*. (**A**) Schematic overview of the *Xenopus* tadpole brain comprised of (from anterior to posterior) telencephalon (tel), diencephalon (di), mesencephalon (mes) and rhombencephalon (rhomb). The position of the *zli* is indicated by orange dashed line. (**B**) Magnification of area indicated by white dashed line in (**A**). Whole mount immunofluorescence staining on the ventricular surface of a tadpole brain. Antibodies detecting acetylated alpha-tubulin (ac. α-tub.) and gamma-tubulin (γ-tub.) mark cilia axonemes and basal bodies, respectively. Note that *zli* cilia are markedly elongated compared to cilia on neighboring cells (scale bars: 5 μm).

6. Building Compartments: The Iroquois Genes Refine the *Zli* Borders

The Iroquois (*irx*) genes encode for homeodomain-containing TFs, highly conserved from *Drosophila* to mammals (reviewed in [105,106]). Most vertebrates contain six *irx* genes grouped in two paralog clusters of three genes each [107,108]. The *irxA* cluster contains *irx1*, *irx2* and *irx4*, while the *irxB* cluster corresponds to *irx3*, *irx5* and *irx6*, and *irx7* in zebrafish. The Irx proteins participate in defining territories and in specifying cell identity. *irx* gene activities in boundary formation have been previously described (reviewed in [105]).

Observations in zebrafish and amphibian promote the idea that Irx factors participate in acquisition of *zli* compartment identity [12,32]. In all species studied, *irx1*, *irx2*, and *irx3* are co-expressed in the anterior neural plate during the early stages of neural patterning and mark the future p2 (Figures 1 and 3; [63,109,110]). In zebrafish, *irx3b* is strongly expressed in the developing *zli* together with *irx5a*, *irx5b* and *irx7*. The *irx3b/5/7* expression pattern subdivides the *zli* into distinct DV domains. In contrast, *irx1a*, *irx1b*, *and irx2a* expression domains abut the *zli* posteriorly [110]. In zebrafish, *irx1b* function is dispensable for *zli* development; however, depletion of the *irx1* orthologs *irx1b and irx7* induces a posterior shift of the *zli* caudal border and an expansion of the *shh* expression domain at the expense of the thalamic field [32,70].

Expression analysis of *irx* genes during *X. laevis zli* development reveals a temporally dynamic process [12]. At neural plate stages, *irx1*, *2* and *3* are co-expressed in the future p2 domain expressing *barhl2* and *otx2*. At the onset of *zli* formation the rostral p2, i.e., the future *zli*, expresses solely *irx3*, whereas the caudal p2, i.e., future thalamus, expresses *irx1*, *2* and 3. After the *zli* has fully developed, the rostral p2 expresses *irx3* and *shh*, and the caudal p2 expresses *irx1* and *irx2* (Figure 1A,Bc,d and Figure 2 and [12]). Furthermore, the mis-expression of the Irx factors within the *barhl2*/*otx2*-expressing field in frog indicates that the ratio of Irx3 to Irx1/2 is essential for *zli* specification and for establishment of the *zli* posterior boundary. Both overexpression of *irx3* as well as *irx1*/*2* depletion promotes the acquisition of a *zli* fate at the expense of a thalamic fate. It is unknown whether Irx1, Irx2, and Irx3 proteins regulate one another's expression.

Observations in chick embryos are divergent from those described in other vertebrates. Chicken *irx3* is described as being expressed posterior to the *zli* and appears to have a repressive function on *zli* formation [10]. In chicken, Irx3 does not appear to participate in *zli* development but in establishment of the differential cellular competence to respond to Shh signaling from the *zli* [46]. Misexpression experiments in the caudal forebrain through in ovo electroporation approaches indicate that Irx3 together with Pax6 alters the competence of caudal forebrain cells to respond to both Fgf8 and Shh. Irx3 ectopic expression in the prethalamic anlage induces expression of thalamic markers and represses prethalamic markers [46].

In conclusion, numerous observations indicate that Irx activities are important to establish the anterior and posterior borders of the *zli*, and to acquire cell segregation properties. Irx activities may vary depending on species. Whether the Irx proteins are strictly required for the expression of *shh* within the *zli* territory, and what is their exact function in *zli* formation, remain open questions.

7. Sticking Together: GRNs and Signaling Pathways Modulate *Zli* Cell Differential Adhesive Properties and Proliferation Rate

Compartments exhibit specific features: they are characterized by a reduced rate of cell proliferation and cells of neighboring compartments separate along boundaries. Cells segregate from one another based on differences in the adhesive properties (affinity) of their cell surfaces. At the boundary, cells deposit an extracellular matrix that acts as a mechanical barrier between different cell populations. Establishment of adhesive differences constitutes the first step of lineage restriction, whereas fence-type mechanisms might stabilize compartments at later stages (reviewed in [44,111]).

7.1. Canonical Wnt Signaling Regulates Thalamic Cell Adhesiveness and Segregation

Various canonical Wnt ligands are markers of the alar plate of p2. Analysis of caudal forebrain proliferation kinetics in chicken and mouse reveal that *zli* cells divide slowly relative to cells in their flanking territories [43,112]. β-catenin, the main effector of the canonical Wnt pathway, controls neuroepithelial cell proliferation (reviewed in [113–115]). β-catenin mediates the interactions between the intracellular cytoskeleton and the cadherins, a group of cell-cell adhesion proteins important in the formation of neural boundaries. Members of the cadherin superfamily mediate differential adhesive properties and are expressed differentially in the forebrain subdivisions. Therefore, it is thought that cadherins participate in formation of compartment boundaries (reviewed in [116–118]). In the zebrafish thalamus, β-catenin regulates the expression of protocadherin 10b (Pcdh10b, formerly known as OL-protocadherin; [118]). Alteration of *pcdh10b* expression in the thalamus territory leads to an intermingling of thalamic cells with the neighboring brain areas, predominantly with the pretectum, indicating an important role for *pcdh10b* in thalamus separation. Stabilization and nuclear translocation of β-catenin leads to a broadening of the expression domain of *pcdh10b*, whereas inhibition of Wnt signaling decreases *pcdh10b* expression. Therefore, the Wnt canonical pathway participates in acquisition of the adhesive differences supporting differential cell segregation behaviors within p2.

7.2. The Notch/Delta Pathway Contributes to Correct Separation of the Zli and Thalamic Fields

Notch signaling mediates lateral inhibition in embryonic tissues and during neural development (reviewed in [119]). *delta* as well as *radical fringe (rfng)* and *lunatic fringe* (*lfng*), two glycosyltransferases that regulate Notch signaling, are expressed and have been involved in the establishment of boundaries. In zebrafish, *in vivo* activation of the Notch pathway directs cells to the rhombomere boundaries, whereas inhibition of Notch activity excludes cells from boundaries [120]. In avian embryos, a forebrain wedge-shaped domain referred to as the "pre-*zli*" is characterized by a gap in the expression of *lfng*. Ectopic expression of *lfng* in this "pre-*zli*" compartment results in sorting of the electroporated cells into the *lfng*-expressing border regions, indicating that *lfng* contributes to specification of the *zli* borders and acquisition of its compartment properties [31,121]. Of note, cell lineage analysis suggests that this "pre-*zli*" domain has compartment properties and is thought to collapse during development to form the final *zli* [31]. However, the mechanism allowing this allosteric growth are not yet explained, and are not attributable to either cellular movements in the epithelium [28] or to cell death [31].

7.3. The TFs Barhl2 and Irx Facilitate Acquisition of Zli Compartment Properties

As described above, Barhl2 participates in *shh* induction within the *zli* territory. Moreover, Barhl2 acts as a brake on p2 neuroepithelial cell proliferation and plays a key role in the maintenance of diencephalic neuroepithelial architecture by limiting activation of the Wnt canonical pathway [64]. In fly imaginal discs, the reduced activity of *irx* genes promotes cell proliferation by accelerating the G1 to S transition whereas their increased expression causes cell-cycle arrest contributing to the size determination of the imaginal disc [122]. Irx proteins act on establishment of distinct cell affinities in dorsal versus ventral cells in the *Drosophila* eye [123]. In zebrafish, *iro7* is required for the proper positioning of the prospective r4/r5 hindbrain rhombomeric boundary [110,124]. Taken together, these observations indicate that Barhl2 along with the Irx factors could participate in limiting *zli* cell proliferation as well as enabling adhesion and facilitating acquisition of *zli* compartment properties (reviewed in [62]).

In conclusion, the Wnt, Shh, and Notch signaling pathways act in a coordinate way to drive segregation of *zli* cells. The identification of genetic and epigenetic modifications initiated by the co-expression of *otx2, barhl2,* and *irx* genes in the presence or absence of Shh should identify the molecular cues driving segregation of *zli* and thalamic cells.

8. The Programming of A "*Zli*-Like" Structure in *X. laevis* Animal Cap Explants Confirms Otx2, Barhl2 and Iroquois Cell Autonomous and Non-Autonomous Activities in *Zli* Formation

In amphibian, cells from the roof of the blastocoel are pluripotent. These cells can be isolated and programmed to generate tissues through manipulation of gene expression—injection of synthetic mRNA or morpholino (MO)—or induction by secreted factors. The investigation of the GRN underlying *zli* formation was performed using such explants [12,125].

In the presence of Shh, explants co-expressing *barhl2, otx2* and *irx3* acquire a *zli*-like identity. The shape of the *shh*-expressing domain observed in the programmed explants—wide where in contact with the source of Shh signal and pointed at its other extremity—is consistent with an induction process occurring sequentially from one cell expressing *shh* to the next. Concomitant in time with acquisition of a Shh-expression program, *zli*-like cells acquire cell segregation properties that depend on the presence or absence of a Shh signal: explant cells mimic their *in vivo* segregation behaviors. Moreover, when grafted into a developing neural plate and continuously exposed to a Shh signal, neuroepithelial cells co-expressing *barhl2, otx2,* and *irx3* form an ectopic *zli in vivo* [12,125].

These approaches confirmed that the GRN which governs competence and refinement of the *zli* can be reconstructed in an *ex vivo* system, and a cell-autonomous role for Otx2 and Barhl2 [12,125]. Moreover, these data indicate that there could be an ongoing recruitment of *irx3* expressing cells from the thalamus into the *zli* during *zli* formation. It is not known whether Shh induces the segregation of *zli* and thalamus cells *in vivo,* however, such recruitment would participate in giving the *zli* its wedged shape, wide at the bottom and pointed at its extremity, a morphology observed in most vertebrate species (Figures 1Bb and 2D). Finally, these data also reveal that the efficiency of *shh* induction—i.e., the average size of the *shh*-expressing area—is increased in the presence of a thalamus-like explant [12,125]. Further experiments are necessary to determine whether the thalamus territory through a cell non-autonomous mechanism facilitates the induction of *shh* and contributes to the dorsal progression of *shh* expression.

9. Inductive Cues and *Zli* Positioning: A Role for the p2/p3 Border and for Fezf/Fez TFs

Using grafting experiments in chicken embryos, Vieira et al. showed that an ectopic border between neural tissue from a prechordal (defined as being *six3*-positive) and epichordal (defined as being *irx3*-positive) origin is sufficient to induce an ectopic *zli* [9]. Indeed, comparison of *barhl2* and *shh* expression patterns during neurulation support the idea that the *zli* forms in immediate proximity to the interface between the prechordal neuraxis, induced by prechordal plate mesoderm, and the epichordal neuraxis, induced by the chordamesoderm ([29,30]; Figure 1Bg). In any case, the *zli* anterior border develops at the interface between the expression domains of the FEZ family zinc finger 2 (*fezf2*), which marks the alar plate of p3 rostral to the *zli*, and *irx3* that marks p2 Figure 1Be; [30,38,39,49]).

fez-like (fezl) genes are highly conserved during evolution from flies to men. They are expressed in the presumptive pre-thalamus during early segmentation stages but are not detected within the *zli* territory (Figures 1A,Be and 2). Functional analysis of Fez and Fezl in zebrafish and mouse indicate an evolutionarily conserved role for these TFs in *zli* formation. Analysis of E12.5 mouse embryos in which both Fez and Fezl genes have been disrupted reveals defects in both prethalamus and diencephalic development, including loss of *shh* expression in the *zli*, ectopic expression of *pax6* in the mid-diencephalic furrow, and extension of *wnt3a* expression within the p2 alar plate. In contrast, in zebrafish the knock down of *fezl* results in anterior expansion of the *zli*, associated with concomitant expansion of *irx3a*. This difference in loss of function phenotypes remains unexplained. It could be species-dependent or it could be due to a difference in levels of expression. The zebrafish *fezl* morphant represents a weaker loss of function of *fezl* with *fez* being intact, whereas *fez*/*fezl* double-mutant mice represent true null conditions. Conversely, in both zebrafish and mouse the over-expression of *fezl*, or *fez* abolishes expression of *shh* in the *zli*. *fezl* over-expression in late gastrula zebrafish embryos expands the prethalamus and hypothalamus territories at the expense of the *zli* and posterior forebrain and/or mid-brain regions. Therefore *fezl*/*fez* gene activity, and specifically *fezf2* activity, plays a key role in formation of the p2/p3 boundary and co-localizes with the *zli* anterior boundary [70,126].

By analogy with induction of the midbrain–hindbrain boundary, which develops at the interface of *otx2* and *gbx2* expression, it has been suggested, but not tested, that the interface between the *fezf2* and *irx3* expression domains is the inductive cue at the origin of *zli* formation (reviewed in [3,13,17,44,127]). Indeed, the p2/p3 boundary localizes at the interface between the expression domains of *fezf2* and *irx3*, but also *fezf2* and *otx2*, and *fezf2* and *barhl2* (Figures 1Be and 3). In amphibian, neither over-expression of *irx3*, nor down-regulation of *barhl2*, affects formation of the p2/p3 border [12,32]. Similarly, in zebrafish, the loss of *otx* does not affect *zli* anterior border location [32]. It is possible, however not tested, that Otx2, Barhl2 and Irx TFs have redundant activities in establishment of the interface with the Fezf/Fez TFs that set up the p2/p3 border. Results in frog explants suggest that whereas the p2/p3 limit is important in setting up the *zli* anterior border, it is the co-expression of *barhl2*, *otx2* and *irx3* that initiates *zli* development. This observation is important, as it indicates that the *zli* developmental mode may be different from that of the midbrain–hindbrain boundary.

Finally, grafts of dorsal diencephalic tissue in chicken inhibit *zli* propagation, arguing that the progression of *shh* expression is limited dorsally by inhibitory factors [38]. The identity of such signals is unknown. However, since *shh* signaling from the *zli* represses *pax6* within the mid-diencephalic furrow [128], Pax6 together with unidentified dorsal signals could in turn prevent *shh* from being expressed beyond the *zli* (Figure 1A,Bf).

In conclusion, the molecular cues initiating *zli* positioning remain partially unknown. In this regard, the determination of signals controlling *otx*, *barhl2* and *irx* neural plate expression will provide important information.

10. The GRN Involved in *Zli* Formation is Evolutionarily Conserved

Yao et al. analyzed whether the *cis*- and *trans*-regulatory landscape underlying *zli* development was conserved in the chordate phylum. They investigated conservation of the "*zli* developmental cassette" in the hemichordate *Saccoglossus kowalevskii (S. kowalevskii)* that is closest to the central basic reference animal at the root of the chordate phylogenetic tree [129,130]. In *S. kowalevskii*, the narrow band of cells at the proboscis-collar boundary is considered *zli*-like as it expresses *hh*. A study of expression patterns for *barhl2* (*barH*), *otx*, and *irx* orthologs was performed in this hemichordate. Indeed, the proboscis-collar boundary expresses the orthologs of *barhl2*, *otx* and *irx* at the right time and place to perform their patterning function [129,130]. Yao et al. identified in *S. kowalevskii* a 1.1-kb region containing a *cis*-regulatory element containing the six motifs of the mouse SBE1: skSBE1. Functional experiments in mouse and in *S. kowalevskii* demonstrate that skSBE1 is a functional ortholog of the mmSBE1 enhancer [67]. Similar SBE1 *cis*-regulatory elements, intact or with a shuffled motif arrangement, were discovered in lamprey and all jawed vertebrates that display "*zli*-like" structures. In contrast, similar SBE1-like motifs were not found in amphioxus (cephalochordate) or in ascidian (tunicate) that both lack an *hh*-expressing domain in the anterior brain.

In conclusion, these studies support the hypothesis that early chordates inherited an *hh cis*-regulatory-motif from a deuterostome ancestor that was subsequently lost in the invertebrate chordate lineages. A conserved *hh cis*-regulatory-motif (SBE1-like) was maintained in the vertebrate *shh* gene and used to activate its transcription in the *zli*, paving the way for the establishment of this brain-signaling center more than 500 millions years ago [67].

11. Forebrain Malformations Associated with Defects in the GRN and Signaling Pathways Supporting *Zli* Formation

The GRN controlling *zli* development is also involved in thalamic primordium induction and in thalamic growth, patterning, and organogenesis. Indeed, depletion of *otx* genes impairs forebrain development (reviewed in [57]). In both frog and mouse, targeted-depletion of *barhl2* generates thalamic developmental defects [12,71]. In *barhl2*$^{-/-}$ mice, p2 thalamic progenitors acquire a pretectal fate and there is an absence of thalamo-cortical axon projections [71]. Finally, *irx* gene depletion in frog, chicken, and zebrafish disrupts thalamus development [3,12,16,17,46,109,131]. Some genetic disorders such as holoprosencephaly (HPE) or the ciliopathies, which are characterized or accompanied by brain malformations in humans and model organisms, are associated with thalamic and thalamo-cortical developmental defects (reviewed in [132,133]). Neuropsychiatric disorders such as obsessive-compulsive disorder and attention deficit hyperactivity disorder have been associated with abnormal neuronal activity within the thalamus (reviewed in [131,134–137]).

The most prominent pathological condition in which development of the thalamus is affected is HPE. HPE is a rare developmental disorder with an occurrence of 1 case in about 16,000 live births. However, approximately 1/250 conceptuses are thought to be affected by HPE which makes it the most common forebrain defect in humans. HPE presents itself as variable degrees of fusion between the left and right halves of the cerebral hemispheres, basal ganglia and, interestingly, also the thalamus [132,138]. However, rather than representing a fusion event, HPE arises from a failure to separate the two halves of the forebrain along the midline. Most studies focus on correct separation between the cerebral hemispheres (reviewed in [139]). However, little is known on the thalamic contribution to the etiology of HPE.

Mutations in genes linked to HPE have been identified in animal models of HPE and human patients. All of them play important roles in major brain developmental signaling pathways [140–145]. Whereas defects in the Shh pathway are the most frequent cause of HPE, genetic screening of HPE patients and studies in animal models also involved the Fgf, Nodal and Notch pathways as major contributors to HPE etiology [140–145]. Most mutations found in patients with HPE are in genes that participate in midline formation and early neural plate patterning, strengthening the hypothesis that developmental defects at the origin of these malformations take place early during neural development.

As described above, the Shh pathway relies on primary cilia for signal transduction. It is thus not surprising that mutations in certain cilia-related genes affect diencephalon morphology and induce HPE or HPE-like phenotypes. In mouse, mutants carrying deletions in the anterograde intraflagellar transport (IFT) 172 gene, Shh signaling in the forebrain is down-regulated. Consequently, the mutant diencephalon is severely reduced and embryos exhibit lobar or semilobar HPE [146]. The hypomorphic *cobblestone* allele of another anterograde IFT gene, IFT88, also affects diencephalon development. In this mouse mutant, the border between tel- and diencephalon is dissolved and cells mix freely between the two parts [99], a feature reminiscent of a failure to establish *zli*-mediated compartmentalization and therefore correct establishment of segregation properties. Loss of function of *forkhead box J1 (foxj1)*, a TF regulating the biogenesis of motile cilia, caused striking forebrain defects in *X. laevis* [104]. While cilia on *zli* cells in *Xenopus* are longer than other primary cilia in the wildtype ([104]; Figure 2), *zli* cilia are shortened in *foxj1* morphants and diencephalon size is massively reduced. Taken together, cilia dysfunction in the forebrain leads to variable molecular and morphological defects, pinpointing the dependence of diencephalon development on cilia-based Shh signaling.

In conclusion, even though thalamic fusion in human patients has not yet been connected to diencephalic primordium and *zli* development, it appears possible that malformation of p2 and concurrent mis-patterning may prevent separation of the thalamic complex and plays an as yet unrecognized role in the etiology of HPE. Specifically, failure to establish the anterior medial source of Shh, a lack of competence in the presumptive *zli* region, defective spreading and/or signaling of Shh can cause problems with setting up the p2/p3 border, the *zli*, and correct Shh, Notch and Fgf signaling in the thalamic field. All these events can lead to brain malformations.

12. Conclusions

The morphogenetic events supporting brain development are controlled and coordinated by a handful of extracellular signaling networks. Shh is one of the most prominent and collaborates with the Wnt, Bone Morphogenetic Protein, Fgf, Nodal, and retinoid acid (RA) pathways in neurogenesis. The patterning and growth of the diencephalon is, in that sense, particularly meaningful as Wnt, Shh, Fgf, and RA signaling act together in a strictly regulated chronology and topology to orchestrate the development and neurogenesis of the diencephalon and specifically of the thalamus (reviewed in [3,13,17,127]). Whereas recent efforts have allowed partial identification of the *zli* developmental *cis*- and *trans*-determinants, important questions remain unresolved. Specifically, it remains to identify: (i) the molecular cues initiating *zli* emergence and positioning; (ii) the molecular cues driving segregation of *zli* and thalamic cells; and (iii) the contribution of *otx*, *barhl*, and *irx* genes to thalamic development. It will also be interesting to investigate: (i) how the Shh signal is distributed and read in the developing *zli* and the thalamus; and (ii) which signals generate the splitting of the midline in the forebrain region, knowing that this process is defective in HPE. In the years to come, an important focus should be to better understand the gene-environment interactions involved in thalamic organogenesis and to develop animal models (chicken, zebrafish, mouse, *Xenopus*) mimicking human thalamic developmental disorders, a strategy that shall provide critical help into possible clinical interventions.

Acknowledgments: We thank Paul Johnson for his editing work on the manuscript and Elisabeth Schuster and Nathalie Rocques for technical assistance. B. Durand is funded by the Centre National de la Recherche Scientifique (Institut Curie-CNRS UMR3387, INSERM U1021), and the Fondation Pierre Gilles de Gennes (FPGG0039). E. Sena is supported by a fellowship from the "Ministère de la Recherche".

Conflicts of Interest: The authors declare no conflict of interest. The funding sponsors had neither a role in the design of the study, nor in the collection, analyses, or interpretation of data, nor in the writing of the manuscript, nor in the decision to publish the results.

References

1. Jessell, T.M.; Sanes, J.R. Development. The decade of the developing brain. *Curr. Opin. Neurobiol.* **2000**, *10*, 599–611. [CrossRef]
2. Hebert, J.M.; Fishell, G. The genetics of early telencephalon patterning: Some assembly required. *Nat. Rev. Neurosci.* **2008**, *9*, 678–685. [CrossRef] [PubMed]

3. Scholpp, S.; Lumsden, A. Review: Building a bridal chamber: Development of the thalamus. *Trends Neurosci.* **2010**, *33*, 373–380. [CrossRef] [PubMed]
4. Cavodeassi, F.; Houart, C. Brain regionalization: Of signaling centers and boundaries. *Dev. Neurobiol.* **2012**, *72*, 218–233. [CrossRef] [PubMed]
5. Kiecker, C.; Lumsden, A. The role of organizers in patterning the nervous system. *Annu. Rev. Neurosci.* **2012**, *35*, 347–367. [CrossRef] [PubMed]
6. Echevarría, D.; Vieira, C.; Gimeno, L.; Martinez, S. Neuroepithelial secondary organizers and cell fate specification in the developing brain. *Brain Res. Brain Res. Rev.* **2003**, *43*, 179–191. [CrossRef] [PubMed]
7. Anderson, C.; Stern, C.D. Organizers in development. *Curr. Top. Dev. Biol.* **2016**, *117*, 435–454. [PubMed]
8. Scholpp, S.; Wolf, O.; Brand, M.; Lumsden, A. Hedgehog signalling from the zona limitans intrathalamica orchestrates patterning of the zebrafish diencephalon. *Development* **2006**, *133*, 855–864. [CrossRef] [PubMed]
9. Vieira, C.; Garda, A.L.; Shimamura, K.; Martinez, S. Thalamic development induced by shh in the chick embryo. *Dev. Biol.* **2005**, *284*, 351–363. [CrossRef] [PubMed]
10. Kiecker, C.; Lumsden, A. Hedgehog signaling from the *zli* regulates diencephalic regional identity. *Nat. Neurosci.* **2004**, *7*, 1242–1249. [CrossRef] [PubMed]
11. Hashimoto-Torii, K.; Motoyama, J.; Hui, C.C.; Kuroiwa, A.; Nakafuku, M.; Shimamura, K. Differential activities of sonic hedgehog mediated by gli transcription factors define distinct neuronal subtypes in the dorsal thalamus. *Mech. Dev.* **2003**, *120*, 1097–1111. [CrossRef] [PubMed]
12. Juraver-Geslin, H.A.; Gomez-Skarmeta, J.L.; Durand, B.C. The conserved barh-like homeobox-2 gene barhl2 acts downstream of orthodentricle-2 and together with iroquois-3 in establishment of the caudal forebrain signaling center induced by sonic hedgehog. *Dev. Biol.* **2014**, *396*, 107–120. [CrossRef] [PubMed]
13. Epstein, D.J. Regulation of thalamic development by sonic hedgehog. *Front. Neurosci.* **2012**, *6*, 57. [CrossRef] [PubMed]
14. Jeong, Y.; Dolson, D.K.; Waclaw, R.R.; Matise, M.P.; Sussel, L.; Campbell, K.; Kaestner, K.H.; Epstein, D.J. Spatial and temporal requirements for sonic hedgehog in the regulation of thalamic interneuron identity. *Development* **2011**, *138*, 531–541. [CrossRef] [PubMed]
15. Szabo, N.E.; Zhao, T.; Zhou, X.; Alvarez-Bolado, G. The role of sonic hedgehog of neural origin in thalamic differentiation in the mouse. *J. Neurosci.* **2009**, *29*, 2453–2466. [CrossRef] [PubMed]
16. Chatterjee, M.; Li, J.Y. Patterning and compartment formation in the diencephalon. *Front. Neurosci.* **2012**, *6*, 66. [CrossRef] [PubMed]
17. Martinez-Ferre, A.; Martinez, S. Molecular regionalization of the diencephalon. *Front. Neurosci.* **2012**, *6*, 73. [CrossRef] [PubMed]
18. Bergquist, H.; Kallen, B. Notes on the early histogenesis and morphogenesis of the central nervous system in vertebrates. *J. Comp. Neurol.* **1954**, *100*, 627–659. [CrossRef] [PubMed]
19. Coggeshall, R.E. A study of diencephalic development in the albino rat. *J. Comp. Neurol.* **1964**, *122*, 241–269. [CrossRef] [PubMed]
20. Keyser, A. The development of the diencephalon of the chinese hamster. An investigation of the validity of the criteria of subdivision of the brain. *Acta Anat. Suppl.* **1972**, *59*, 1–178. [CrossRef]
21. Vaage, S. The segmentation of the primitive neural tube in chick embryos (gallus domesticus): A morphological, histochemical and autoradiographical investigation. *Ergeb. Anat. Entwicklungsgesch.* **1969**, *41*, 3–87. [PubMed]
22. Bulfone, A.; Puelles, L.; Porteus, M.H.; Frohman, M.A.; Martin, G.R.; Rubenstein, J.L. Spatially restricted expression of dlx-1, dlx-2 (tes-1), gbx-2, and wnt-3 in the embryonic day 12.5 mouse forebrain defines potential transverse and longitudinal segmental boundaries. *J. Neurosci.* **1993**, *13*, 3155–3172. [PubMed]
23. Rubenstein, J.L.; Martinez, S.; Shimamura, K.; Puelles, L. The embryonic vertebrate forebrain: The prosomeric model. *Science* **1994**, *266*, 578–580. [CrossRef] [PubMed]
24. Figdor, M.C.; Stern, C.D. Segmental organization of embryonic diencephalon. *Nature* **1993**, *363*, 630–634. [CrossRef] [PubMed]
25. Puelles, L.; Rubenstein, J.L.R. Forebrain gene expression domains and the evolving prosomeric model. *Trends Neurosci.* **2003**, *26*, 469–476. [CrossRef]
26. Thompson, C.L.; Ng, L.; Menon, V.; Martinez, S.; Lee, C.K.; Glattfelder, K.; Sunkin, S.M.; Henry, A.; Lau, C.; Dang, C.; et al. A high-resolution spatiotemporal atlas of gene expression of the developing mouse brain. *Neuron* **2014**, *83*, 309–323. [CrossRef] [PubMed]

27. Ferran, J.L.; Puelles, L.; Rubenstein, J.L. Molecular codes defining rostrocaudal domains in the embryonic mouse hypothalamus. *Front. Neuroanat.* **2015**, *9*, 46. [CrossRef] [PubMed]
28. Garcia-Lopez, R.; Vieira, C.; Echevarria, D.; Martinez, S. Fate map of the diencephalon and the zona limitans at the 10-somites stage in chick embryos. *Dev. Biol.* **2004**, *268*, 514–530. [CrossRef] [PubMed]
29. Larsen, C.W.; Zeltser, L.M.; Lumsden, A. Boundary formation and compartition in the avian diencephalon. *J. Neurosci.* **2001**, *21*, 4699–4711. [PubMed]
30. Shimamura, K.; Hartigan, D.J.; Martinez, S.; Puelles, L.; Rubenstein, J.L. Longitudinal organization of the anterior neural plate and neural tube. *Development* **1995**, *121*, 3923–3933. [PubMed]
31. Zeltser, L.M.; Larsen, C.W.; Lumsden, A. A new developmental compartment in the forebrain regulated by lunatic fringe. *Nat. Neurosci.* **2001**, *4*, 683–684. [CrossRef] [PubMed]
32. Scholpp, S.; Foucher, I.; Staudt, N.; Peukert, D.; Lumsden, A.; Houart, C. Otx1l, otx2 and irx1b establish and position the zli in the diencephalon. *Development* **2007**, *134*, 3167–3176. [CrossRef] [PubMed]
33. Ericson, J.; Morton, S.; Kawakami, A.; Roelink, H.; Jessell, T.M. Two critical periods of sonic hedgehog signaling required for the specification of motor neuron identity. *Cell* **1996**, *87*, 661–673. [CrossRef]
34. Watanabe, Y.; Nakamura, H. Control of chick tectum territory along dorsoventral axis by sonic hedgehog. *Development* **2000**, *127*, 1131–1140. [PubMed]
35. Placzek, M.; Tessier-Lavigne, M.; Yamada, T.; Jessell, T.; Dodd, J. Mesodermal control of neural cell identity: Floor plate induction by the notochord. *Science* **1990**, *250*, 985–988. [CrossRef] [PubMed]
36. Chamberlain, C.E.; Jeong, J.; Guo, C.; Allen, B.L.; McMahon, A.P. Notochord-derived shh concentrates in close association with the apically positioned basal body in neural target cells and forms a dynamic gradient during neural patterning. *Development* **2008**, *135*, 1097–1106. [CrossRef] [PubMed]
37. Placzek, M.; Briscoe, J. The floor plate: Multiple cells, multiple signals. *Nat. Rev. Neurosci.* **2005**, *6*, 230–240. [CrossRef] [PubMed]
38. Zeltser, L.M. Shh-dependent formation of the zli is opposed by signals from the dorsal diencephalon. *Development* **2005**, *132*, 2023–2033. [CrossRef] [PubMed]
39. Vieira, C.; Martinez, S. Sonic hedgehog from the basal plate and the zona limitans intrathalamica exhibits differential activity on diencephalic molecular regionalization and nuclear structure. *Neuroscience* **2006**, *143*, 129–140. [CrossRef] [PubMed]
40. Mathieu, J.; Barth, A.; Rosa, F.M.; Wilson, S.W.; Peyrieras, N. Distinct and cooperative roles for nodal and hedgehog signals during hypothalamic development. *Development* **2002**, *129*, 3055–3065. [PubMed]
41. Holzschuh, J.; Hauptmann, G.; Driever, W. Genetic analysis of the roles of hh, fgf8, and nodal signaling during catecholaminergic system development in the zebrafish brain. *J. Neurosci.* **2003**, *23*, 5507–5519. [PubMed]
42. Sampath, K.; Rubinstein, A.L.; Cheng, A.M.; Liang, J.O.; Fekany, K.; Solnica-Krezel, L.; Korzh, V.; Halpern, M.E.; Wright, C.V. Induction of the zebrafish ventral brain and floorplate requires cyclops/nodal signalling. *Nature* **1998**, *395*, 185–189. [PubMed]
43. Martinez, S.; Puelles, L. Neurogenetic compartments of the mouse diencephalon and some characteristic gene expression patterns. *Results Probl. Cell Differ.* **2000**, *30*, 91–106. [PubMed]
44. Kiecker, C.; Lumsden, A. Compartments and their boundaries in vertebrate brain development. *Nat. Rev. Neurosci.* **2005**, *6*, 553–564. [CrossRef] [PubMed]
45. Crossley, P.H.; Martinez, S.; Martin, G.R. Midbrain development induced by fgf8 in the chick embryo. *Nature* **1996**, *380*, 66–68. [CrossRef] [PubMed]
46. Robertshaw, E.; Matsumoto, K.; Lumsden, A.; Kiecker, C. Irx3 and pax6 establish differential competence for shh-mediated induction of gabaergic and glutamatergic neurons of the thalamus. *Proc. Natl. Acad. Sci. USA* **2013**, *110*, E3919–E3926. [CrossRef] [PubMed]
47. Rubenstein, J.L.; Shimamura, K.; Martinez, S.; Puelles, L. Regionalization of the prosencephalic neural plate. *Annu. Rev. Neurosci.* **1998**, *21*, 445–477. [CrossRef] [PubMed]
48. Mattes, B.; Weber, S.; Peres, J.; Chen, Q.; Davidson, G.; Houart, C.; Scholpp, S. Wnt3 and wnt3a are required for induction of the mid-diencephalic organizer in the caudal forebrain. *Neural Dev.* **2012**, *7*, 12. [CrossRef] [PubMed]
49. Martinez-Ferre, A.; Navarro-Garberi, M.; Bueno, C.; Martinez, S. Wnt signal specifies the intrathalamic limit and its organizer properties by regulating shh induction in the alar plate. *J. Neurosci.* **2013**, *33*, 3967–3980. [CrossRef] [PubMed]

50. Houart, C.; Westerfield, M.; Wilson, S.W. A small population of anterior cells patterns the forebrain during zebrafish gastrulation. *Nature* **1998**, *391*, 788–792. [CrossRef] [PubMed]
51. Wilson, S.W.; Houart, C. Review: Early steps in the development of the forebrain. *Dev. Cell* **2004**, *6*, 167–181. [CrossRef]
52. Ruiz i Altaba, A.; Nguyen, V.; Palma, V. The emergent design of the neural tube: Prepattern, shh morphogen and gli code. *Curr. Opin. Genet. Dev.* **2003**, *13*, 513–521. [CrossRef] [PubMed]
53. Jacob, J.; Briscoe, J.; Britto, J.; Tannahill, D.; Keynes, R. Gli proteins and the control of spinal-cord patterning. *EMBO Rep.* **2003**, *4*, 761–765. [CrossRef] [PubMed]
54. Jessell, T.M. Neuronal specification in the spinal cord: Inductive signals and transcriptional codes. *Nat. Rev. Genet.* **2000**, *1*, 20–29. [CrossRef] [PubMed]
55. Kobayashi, D.; Kobayashi, M.; Matsumoto, K.; Ogura, T.; Nakafuku, M.; Shimamura, K. Early subdivisions in the neural plate define distinct competence for inductive signals. *Development* **2002**, *129*, 83–93. [PubMed]
56. Gomez-Skarmeta, J.L.; Campuzano, S.; Modolell, J. Half a century of neural prepatterning: The story of a few bristles and many genes. *Nat. Rev. Neurosci.* **2003**, *4*, 587–598. [CrossRef] [PubMed]
57. Acampora, D.; Postiglione, M.P.; Avantaggiato, V.; di Bonito, M.; Simeone, A. The role of otx and otp genes in brain development. *Int. J. Dev. Biol.* **2000**, *44*, 669–677. [PubMed]
58. Beby, F.; Lamonerie, T. The homeobox gene otx2 in development and disease. *Exp. Eye Res.* **2013**, *111*, 9–16. [CrossRef] [PubMed]
59. Pannese, M.; Polo, C.; Andreazzoli, M.; Vignali, R.; Kablar, B.; Barsacchi, G.; Boncinelli, E. The xenopus homologue of otx2 is a maternal homeobox gene that demarcates and specifies anterior body regions. *Development* **1995**, *121*, 707–720. [PubMed]
60. Schuhmacher, L.N.; Albadri, S.; Ramialison, M.; Poggi, L. Evolutionary relationships and diversification of barhl genes within retinal cell lineages. *BMC Evol. Biol.* **2011**, *11*, 340. [CrossRef] [PubMed]
61. Reig, G.; Cabrejos, M.E.; Concha, M.L. Functions of barh transcription factors during embryonic development. *Dev. Biol.* **2007**, *302*, 367–375. [CrossRef] [PubMed]
62. Juraver-Geslin, H.A.; Durand, B.C. Early development of the neural plate: New roles for apoptosis and for one of its main effectors caspase-3. *Genesis* **2015**, *53*, 203–224. [CrossRef] [PubMed]
63. Offner, N.; Duval, N.; Jamrich, M.; Durand, B. The pro-apoptotic activity of a vertebrate bar-like homeobox gene plays a key role in patterning the xenopus neural plate by limiting the number of chordin- and shh-expressing cells. *Development* **2005**, *132*, 1807–1818. [CrossRef] [PubMed]
64. Juraver-Geslin, H.A.; Ausseil, J.J.; Wassef, M.; Durand, B.C. Barhl2 limits growth of the diencephalic primordium through caspase3 inhibition of beta-catenin activation. *Proc. Natl. Acad. Sci. USA* **2011**, *108*, 2288–2293. [CrossRef] [PubMed]
65. Staudt, N.; Houart, C. The prethalamus is established during gastrulation and influences diencephalic regionalization. *PLoS Biol.* **2007**, *5*, e69. [CrossRef] [PubMed]
66. Mo, Z.; Li, S.; Yang, X.; Xiang, M. Role of the barhl2 homeobox gene in the specification of glycinergic amacrine cells. *Development* **2004**, *131*, 1607–1618. [CrossRef] [PubMed]
67. Yao, Y.; Minor, P.J.; Zhao, Y.T.; Jeong, Y.; Pani, A.M.; King, A.N.; Symmons, O.; Gan, L.; Cardoso, W.V.; Spitz, F.; et al. *cis*-Regulatory architecture of a brain signaling center predates the origin of chordates. *Nat. Genet.* **2016**, *48*, 575–580. [CrossRef] [PubMed]
68. Patterson, K.D.; Cleaver, O.; Gerber, W.V.; White, F.G.; Krieg, P.A. Distinct expression patterns for two xenopus bar homeobox genes. *Dev. Genes Evol.* **2000**, *210*, 140–144. [CrossRef] [PubMed]
69. Colombo, A.; Reig, G.; Mione, M.; Concha, M.L. Zebrafish barh-like genes define discrete neural domains in the early embryo. *Gene Expr. Patterns* **2006**, *6*, 347–352. [CrossRef] [PubMed]
70. Hirata, T.; Nakazawa, M.; Muraoka, O.; Nakayama, R.; Suda, Y.; Hibi, M. Zinc-finger genes fez and fez-like function in the establishment of diencephalon subdivisions. *Development* **2006**, *133*, 3993–4004. [CrossRef] [PubMed]
71. Ding, Q.; Balasubramanian, R.; Zheng, D.; Liang, G.; Gan, L. Barhl2 determines the early patterning of the diencephalon by regulating shh. *Mol. Neurobiol.* **2016**. [CrossRef] [PubMed]
72. Ramsbottom, S.A.; Pownall, M.E. Regulation of hedgehog signalling inside and outside the cell. *J. Dev. Biol.* **2016**, *4*, 23. [CrossRef] [PubMed]

73. Porter, J.A.; Ekker, S.C.; Park, W.J.; von Kessler, D.P.; Young, K.E.; Chen, C.H.; Ma, Y.; Woods, A.S.; Cotter, R.J.; Koonin, E.V.; et al. Hedgehog patterning activity: Role of a lipophilic modification mediated by the carboxy-terminal autoprocessing domain. *Cell* **1996**, *86*, 21–34. [CrossRef]
74. Porter, J.A.; von Kessler, D.P.; Ekker, S.C.; Young, K.E.; Lee, J.J.; Moses, K.; Beachy, P.A. The product of hedgehog autoproteolytic cleavage active in local and long-range signalling. *Nature* **1995**, *374*, 363–366. [CrossRef] [PubMed]
75. Bumcrot, D.A.; Takada, R.; McMahon, A.P. Proteolytic processing yields two secreted forms of sonic hedgehog. *Mol. Cell. Biol.* **1995**, *15*, 2294–2303. [CrossRef] [PubMed]
76. Chen, X.; Tukachinsky, H.; Huang, C.H.; Jao, C.; Chu, Y.R.; Tang, H.Y.; Mueller, B.; Schulman, S.; Rapoport, T.A.; Salic, A. Processing and turnover of the hedgehog protein in the endoplasmic reticulum. *J. Cell Biol.* **2011**, *192*, 825–838. [CrossRef] [PubMed]
77. Porter, J.A.; Young, K.E.; Beachy, P.A. Cholesterol modification of hedgehog signaling proteins in animal development. *Science* **1996**, *274*, 255–259. [CrossRef] [PubMed]
78. Li, Y.; Zhang, H.; Litingtung, Y.; Chiang, C. Cholesterol modification restricts the spread of shh gradient in the limb bud. *Proc. Natl. Acad. Sci. USA* **2006**, *103*, 6548–6553. [CrossRef] [PubMed]
79. Pepinsky, R.B.; Zeng, C.; Wen, D.; Rayhorn, P.; Baker, D.P.; Williams, K.P.; Bixler, S.A.; Ambrose, C.M.; Garber, E.A.; Miatkowski, K.; et al. Identification of a palmitic acid-modified form of human sonic hedgehog. *J. Biol. Chem.* **1998**, *273*, 14037–14045. [CrossRef] [PubMed]
80. Buglino, J.A.; Resh, M.D. Hhat is a palmitoylacyltransferase with specificity for N-palmitoylation of sonic hedgehog. *J. Biol. Chem.* **2008**, *283*, 22076–22088. [CrossRef] [PubMed]
81. Chamoun, Z.; Mann, R.K.; Nellen, D.; von Kessler, D.P.; Bellotto, M.; Beachy, P.A.; Basler, K. Skinny hedgehog, an acyltransferase required for palmitoylation and activity of the hedgehog signal. *Science* **2001**, *293*, 2080–2084. [CrossRef] [PubMed]
82. Hardy, R.Y.; Resh, M.D. Identification of n-terminal residues of sonic hedgehog important for palmitoylation by hedgehog acyltransferase. *J. Biol. Chem.* **2012**, *287*, 42881–42889. [CrossRef] [PubMed]
83. Chen, M.H.; Li, Y.J.; Kawakami, T.; Xu, S.M.; Chuang, P.T. Palmitoylation is required for the production of a soluble multimeric hedgehog protein complex and long-range signaling in vertebrates. *Genes Dev.* **2004**, *18*, 641–659. [CrossRef] [PubMed]
84. Kohtz, J.D.; Lee, H.Y.; Gaiano, N.; Segal, J.; Ng, E.; Larson, T.; Baker, D.P.; Garber, E.A.; Williams, K.P.; Fishell, G. N-Terminal fatty-acylation of sonic hedgehog enhances the induction of rodent ventral forebrain neurons. *Development* **2001**, *128*, 2351–2363. [PubMed]
85. Burke, R.; Nellen, D.; Bellotto, M.; Hafen, E.; Senti, K.A.; Dickson, B.J.; Basler, K. Dispatched, a novel sterol-sensing domain protein dedicated to the release of cholesterol-modified hedgehog from signaling cells. *Cell* **1999**, *99*, 803–815. [CrossRef]
86. Creanga, A.; Glenn, T.D.; Mann, R.K.; Saunders, A.M.; Talbot, W.S.; Beachy, P.A. Scube/you activity mediates release of dually lipid-modified hedgehog signal in soluble form. *Genes Dev.* **2012**, *26*, 1312–1325. [CrossRef] [PubMed]
87. Tukachinsky, H.; Kuzmickas, R.P.; Jao, C.Y.; Liu, J.; Salic, A. Dispatched and scube mediate the efficient secretion of the cholesterol-modified hedgehog ligand. *Cell Rep.* **2012**, *2*, 308–320. [CrossRef] [PubMed]
88. Eugster, C.; Panakova, D.; Mahmoud, A.; Eaton, S. Lipoprotein-heparan sulfate interactions in the hh pathway. *Dev. Cell* **2007**, *13*, 57–71. [CrossRef] [PubMed]
89. Panakova, D.; Sprong, H.; Marois, E.; Thiele, C.; Eaton, S. Lipoprotein particles are required for hedgehog and wingless signalling. *Nature* **2005**, *435*, 58–65. [CrossRef] [PubMed]
90. Liegeois, S.; Benedetto, A.; Garnier, J.M.; Schwab, Y.; Labouesse, M. The v0-atpase mediates apical secretion of exosomes containing hedgehog-related proteins in caenorhabditis elegans. *J. Cell Biol.* **2006**, *173*, 949–961. [CrossRef] [PubMed]
91. Callejo, A.; Bilioni, A.; Mollica, E.; Gorfinkiel, N.; Andres, G.; Ibanez, C.; Torroja, C.; Doglio, L.; Sierra, J.; Guerrero, I. Dispatched mediates hedgehog basolateral release to form the long-range morphogenetic gradient in the drosophila wing disk epithelium. *Proc. Natl. Acad. Sci. USA* **2011**, *108*, 12591–12598. [CrossRef] [PubMed]
92. Vyas, N.; Walvekar, A.; Tate, D.; Lakshmanan, V.; Bansal, D.; Lo Cicero, A.; Raposo, G.; Palakodeti, D.; Dhawan, J. Vertebrate hedgehog is secreted on two types of extracellular vesicles with different signaling properties. *Sci. Rep.* **2014**, *4*, 7357. [CrossRef] [PubMed]

93. Huangfu, D.; Liu, A.; Rakeman, A.S.; Murcia, N.S.; Niswander, L.; Anderson, K.V. Hedgehog signalling in the mouse requires intraflagellar transport proteins. *Nature* **2003**, *426*, 83–87. [CrossRef] [PubMed]
94. Hilgendorf, K.I.; Johnson, C.T.; Jackson, P.K. The primary cilium as a cellular receiver: Organizing ciliary gpcr signaling. *Curr. Opin. Cell Biol.* **2016**, *39*, 84–92. [CrossRef] [PubMed]
95. Bai, C.B.; Stephen, D.; Joyner, A.L. All mouse ventral spinal cord patterning by hedgehog is gli dependent and involves an activator function of gli3. *Dev. Cell* **2004**, *6*, 103–115. [CrossRef]
96. Briscoe, J. Making a grade: Sonic hedgehog signalling and the control of neural cell fate. *EMBO J.* **2009**, *28*, 457–465. [CrossRef] [PubMed]
97. Fuccillo, M.; Joyner, A.L.; Fishell, G. Morphogen to mitogen: The multiple roles of hedgehog signalling in vertebrate neural development. *Nat. Rev. Neurosci.* **2006**, *7*, 772–783. [CrossRef] [PubMed]
98. Mukhopadhyay, S.; Rohatgi, R. G-Protein-coupled receptors, hedgehog signaling and primary cilia. *Semin. Cell. Dev. Biol.* **2014**, *33*, 63–72. [CrossRef] [PubMed]
99. Willaredt, M.A.; Hasenpusch-Theil, K.; Gardner, H.A.; Kitanovic, I.; Hirschfeld-Warneken, V.C.; Gojak, C.P.; Gorgas, K.; Bradford, C.L.; Spatz, J.; Wolfl, S.; et al. A crucial role for primary cilia in cortical morphogenesis. *J. Neurosci.* **2008**, *28*, 12887–12900. [CrossRef] [PubMed]
100. Cruz, C.; Ribes, V.; Kutejova, E.; Cayuso, J.; Lawson, V.; Norris, D.; Stevens, J.; Davey, M.; Blight, K.; Bangs, F.; et al. Foxj1 regulates floor plate cilia architecture and modifies the response of cells to sonic hedgehog signalling. *Development* **2010**, *137*, 4271–4282. [CrossRef] [PubMed]
101. Chung, M.I.; Peyrot, S.M.; LeBoeuf, S.; Park, T.J.; McGary, K.L.; Marcotte, E.M.; Wallingford, J.B. Rfx2 is broadly required for ciliogenesis during vertebrate development. *Dev. Biol.* **2012**, *363*, 155–165. [CrossRef] [PubMed]
102. Kramer-Zucker, A.G.; Olale, F.; Haycraft, C.J.; Yoder, B.K.; Schier, A.F.; Drummond, I.A. Cilia-driven fluid flow in the zebrafish pronephros, brain and kupffer's vesicle is required for normal organogenesis. *Development* **2005**, *132*, 1907–1921. [CrossRef] [PubMed]
103. Christ, A.; Christa, A.; Kur, E.; Lioubinski, O.; Bachmann, S.; Willnow, T.E.; Hammes, A. Lrp2 is an auxiliary shh receptor required to condition the forebrain ventral midline for inductive signals. *Dev. Cell.* **2012**, *22*, 268–278. [CrossRef] [PubMed]
104. Hagenlocher, C.; Walentek, P.; Müller, C.; Thumberger, T.; Feistel, K. Ciliogenesis and cerebrospinal fluid flow in the developing xenopus brain are regulated by foxj1. *Cilia* **2013**, *2*, 12. [CrossRef] [PubMed]
105. Gomez-Skarmeta, J.L.; Modolell, J. Iroquois genes: Genomic organization and function in vertebrate neural development. *Curr. Opin. Genet. Dev.* **2002**, *12*, 403–408. [CrossRef]
106. Cavodeassi, F.; Modolell, J.; Gomez-Skarmeta, J.L. The iroquois family of genes: From body building to neural patterning. *Development* **2001**, *128*, 2847–2855. [PubMed]
107. Peters, T.; Dildrop, R.; Ausmeier, K.; Ruther, U. Organization of mouse iroquois homeobox genes in two clusters suggests a conserved regulation and function in vertebrate development. *Genome Res.* **2000**, *10*, 1453–1462. [CrossRef] [PubMed]
108. De la Calle-Mustienes, E.; Modolell, J.; Gomez-Skarmeta, J.L. The xiro-repressed gene corest is expressed in xenopus neural territories. *Mech. Dev.* **2002**, *110*, 209–211. [CrossRef]
109. Rodriguez-Seguel, E.; Alarcon, P.; Gomez-Skarmeta, J.L. The xenopus irx genes are essential for neural patterning and define the border between prethalamus and thalamus through mutual antagonism with the anterior repressors fezf and arx. *Dev. Biol.* **2009**, *329*, 258–268. [CrossRef] [PubMed]
110. Lecaudey, V.; Anselme, I.; Rosa, F.; Schneider-Maunoury, S. The zebrafish iroquois gene iro7 positions the r4/r5 boundary and controls neurogenesis in the rostral hindbrain. *Development* **2004**, *131*, 3121–3131. [CrossRef] [PubMed]
111. Dahmann, C.; Oates, A.C.; Brand, M. Boundary formation and maintenance in tissue development. *Nat. Rev. Genet.* **2011**, *12*, 43–55. [CrossRef] [PubMed]
112. Baek, J.H.; Hatakeyama, J.; Sakamoto, S.; Ohtsuka, T.; Kageyama, R. Persistent and high levels of hes1 expression regulate boundary formation in the developing central nervous system. *Development* **2006**, *133*, 2467–2476. [CrossRef] [PubMed]
113. Grigoryan, T.; Wend, P.; Klaus, A.; Birchmeier, W. Deciphering the function of canonical wnt signals in development and disease: Conditional loss- and gain-of-function mutations of beta-catenin in mice. *Genes Dev.* **2008**, *22*, 2308–2341. [CrossRef] [PubMed]

114. Van Amerongen, R.; Nusse, R. Towards an integrated view of wnt signaling in development. *Development* **2009**, *136*, 3205–3214. [CrossRef] [PubMed]
115. Noelanders, R.; Vleminckx, K. How wnt signaling builds the brain: Bridging development and disease. *Neuroscientist.* **2016**. [CrossRef] [PubMed]
116. Redies, C. Cadherins in the central nervous system. *Prog. Neurobiol.* **2000**, *61*, 611–648. [CrossRef]
117. Puelles, E. Genetic control of basal midbrain development. *J. Neurosci. Res.* **2007**, *85*, 3530–3534. [CrossRef] [PubMed]
118. Peukert, D.; Weber, S.; Lumsden, A.; Scholpp, S. Lhx2 and lhx9 determine neuronal differentiation and compartition in the caudal forebrain by regulating wnt signaling. *PLoS Biol.* **2011**, *9*, e1001218. [CrossRef] [PubMed]
119. Pinto-Teixeira, F.; Desplan, C. Notch activity in neural progenitors coordinates cytokinesis and asymmetric differentiation. *Sci. Signal.* **2014**, *7*. [CrossRef] [PubMed]
120. Cheng, Y.C.; Amoyel, M.; Qiu, X.; Jiang, Y.J.; Xu, Q.; Wilkinson, D.G. Notch activation regulates the segregation and differentiation of rhombomere boundary cells in the zebrafish hindbrain. *Dev. Cell* **2004**, *6*, 539–550. [CrossRef]
121. Tossell, K.; Kiecker, C.; Wizenmann, A.; Lang, E.; Irving, C. Notch signalling stabilises boundary formation at the midbrain-hindbrain organiser. *Development* **2011**, *138*, 3745–3757. [CrossRef] [PubMed]
122. Barrios, N.; Gonzalez-Perez, E.; Hernandez, R.; Campuzano, S. The homeodomain iroquois proteins control cell cycle progression and regulate the size of developmental fields. *PLoS Genet.* **2015**, *11*, e1005463. [CrossRef] [PubMed]
123. Villa-Cuesta, E.; Gonzalez-Perez, E.; Modolell, J. Apposition of iroquois expressing and non-expressing cells leads to cell sorting and fold formation in the drosophila imaginal wing disc. *BMC Dev. Biol.* **2007**, *7*, 106. [CrossRef] [PubMed]
124. Jimenez-Guri, E.; Pujades, C. An ancient mechanism of hindbrain patterning has been conserved in vertebrate evolution. *Evol. Dev.* **2011**, *13*, 38–46. [CrossRef] [PubMed]
125. Durand, B.C. Stem cell-like xenopus embryonic explants to study early neural developmental features in vitro and in vivo. *J. Vis. Exp.* **2016**, e53474.
126. Jeong, J.Y.; Einhorn, Z.; Mathur, P.; Chen, L.; Lee, S.; Kawakami, K.; Guo, S. Patterning the zebrafish diencephalon by the conserved zinc-finger protein fezl. *Development* **2007**, *134*, 127–136. [CrossRef] [PubMed]
127. Hagemann, A.I.; Scholpp, S. The tale of the three brothers—Shh, wnt, and fgf during development of the thalamus. *Front. Neurosci.* **2012**, *6*, 76. [CrossRef] [PubMed]
128. Caballero, I.M.; Manuel, M.N.; Molinek, M.; Quintana-Urzainqui, I.; Mi, D.; Shimogori, T.; Price, D.J. Cell-autonomous repression of shh by transcription factor pax6 regulates diencephalic patterning by controlling the central diencephalic organizer. *Cell Rep.* **2014**, *8*, 1405–1418. [CrossRef] [PubMed]
129. Pani, A.M.; Mullarkey, E.E.; Aronowicz, J.; Assimacopoulos, S.; Grove, E.A.; Lowe, C.J. Ancient deuterostome origins of vertebrate brain signalling centres. *Nature* **2012**, *483*, 289–294. [CrossRef] [PubMed]
130. Lowe, C.J.; Wu, M.; Salic, A.; Evans, L.; Lander, E.; Stange-Thomann, N.; Gruber, C.E.; Gerhart, J.; Kirschner, M. Anteroposterior patterning in hemichordates and the origins of the chordate nervous system. *Cell* **2003**, *113*, 853–865. [CrossRef]
131. Llinas, R.R. Thalamo-cortical dysrhythmia syndrome: Neuropsychiatric features. *Anales de la Real Academia Nacional de Medicina* **2003**, *120*, 267–290; discussion 290–265. [PubMed]
132. Marcorelles, P.; Laquerriere, A. Neuropathology of holoprosencephaly. *Am. J. Med. Genet. Part. C Semin. Med. Genet.* **2010**, *154*, 109–119. [CrossRef] [PubMed]
133. Willaredt, M.A.; Tasouri, E.; Tucker, K.L. Primary cilia and forebrain development. *Mech. Dev.* **2013**, *130*, 373–380. [CrossRef] [PubMed]
134. Xavier, G.M.; Seppala, M.; Barrell, W.; Birjandi, A.A.; Geoghegan, F.; Cobourne, M.T. Hedgehog receptor function during craniofacial development. *Dev. Biol.* **2016**, *415*, 198–215. [CrossRef] [PubMed]
135. Guerrini, R.; Dobyns, W.B. Malformations of cortical development: Clinical features and genetic causes. *Lancet Neurol.* **2014**, *13*, 710–726. [CrossRef]
136. Hu, W.F.; Chahrour, M.H.; Walsh, C.A. The diverse genetic landscape of neurodevelopmental disorders. *Ann. Rev. Genom. Hum. Genet.* **2014**, *15*, 195–213. [CrossRef] [PubMed]
137. Shepherd, G.M. Corticostriatal connectivity and its role in disease. *Nat. Rev. Neurosci.* **2013**, *14*, 278–291. [CrossRef] [PubMed]

138. Hahn, J.S.; Barnes, P.D. Neuroimaging advances in holoprosencephaly: Refining the spectrum of the midline malformation. *Am. J. Med. Genet. Part C Semin. Med. Genet.* **2010**, *154C*, 120–132. [CrossRef] [PubMed]
139. Munke, M. Clinical, cytogenetic, and molecular approaches to the genetic heterogeneity of holoprosencephaly. *Am. J. Med. Genet.* **1989**, *34*, 237–245. [CrossRef] [PubMed]
140. Nanni, L.; Ming, J.E.; Bocian, M.; Steinhaus, K.; Bianchi, D.W.; Die-Smulders, C.; Giannotti, A.; Imaizumi, K.; Jones, K.L.; Campo, M.D.; et al. The mutational spectrum of the sonic hedgehog gene in holoprosencephaly: Shh mutations cause a significant proportion of autosomal dominant holoprosencephaly. *Hum. Mol. Genet.* **1999**, *8*, 2479–2488. [CrossRef] [PubMed]
141. Gongal, P.A.; French, C.R.; Waskiewicz, A.J. Aberrant forebrain signaling during early development underlies the generation of holoprosencephaly and coloboma. *Biochim. Biophys. Acta* **2011**, *1812*, 390–401. [CrossRef] [PubMed]
142. Petryk, A.; Graf, D.; Marcucio, R. Holoprosencephaly: Signaling interactions between the brain and the face, the environment and the genes, and the phenotypic variability in animal models and humans. *Wiley Interdiscip. Rev. Dev. Biol.* **2015**, *4*, 17–32. [CrossRef] [PubMed]
143. Dubourg, C.; Carre, W.; Hamdi-Roze, H.; Mouden, C.; Roume, J.; Abdelmajid, B.; Amram, D.; Baumann, C.; Chassaing, N.; Coubes, C.; et al. Mutational spectrum in holoprosencephaly shows that fgf is a new major signaling pathway. *Hum. Mutat.* **2016**. [CrossRef] [PubMed]
144. Dupe, V.; Rochard, L.; Mercier, S.; Le Petillon, Y.; Gicquel, I.; Bendavid, C.; Bourrouillou, G.; Kini, U.; Thauvin-Robinet, C.; Bohan, T.P.; et al. Notch, a new signaling pathway implicated in holoprosencephaly. *Hum. Mol. Genet.* **2011**, *20*, 1122–1131. [CrossRef] [PubMed]
145. Mercier, S.; David, V.; Ratie, L.; Gicquel, I.; Odent, S.; Dupe, V. Nodal and shh dose-dependent double inhibition promotes an hpe-like phenotype in chick embryos. *Dis. Models Mech.* **2013**, *6*, 537–543. [CrossRef] [PubMed]
146. Gorivodsky, M.; Mukhopadhyay, M.; Wilsch-Braeuninger, M.; Phillips, M.; Teufel, A.; Kim, C.; Malik, N.; Huttner, W.; Westphal, H. Intraflagellar transport protein 172 is essential for primary cilia formation and plays a vital role in patterning the mammalian brain. *Dev. Biol.* **2009**, *325*, 24–32. [CrossRef] [PubMed]

12

An Overview of *Hox* Genes in Lophotrochozoa: Evolution and Functionality

Marco Barucca *, Adriana Canapa and Maria Assunta Biscotti

Dipartimento di Scienze della Vita e dell'Ambiente, Università Politecnica delle Marche, via Brecce Bianche, 60131 Ancona, Italy; a.canapa@univpm.it (A.C.); m.a.biscotti@univpm.it (M.A.B.)

* Correspondence: m.barucca@univpm.it

Academic Editor: Vincenzo Zappavigna

Abstract: *Hox* genes are regulators of animal embryonic development. Changes in the number and sequence of *Hox* genes as well as in their expression patterns have been related to the evolution of the body plan. Lophotrochozoa is a clade of Protostomia characterized by several phyla which show a wide morphological diversity. Despite that the works summarized in this review emphasize the fragmentary nature of the data available regarding the presence and expression of *Hox* genes, they also offer interesting insight into the evolution of the *Hox* cluster and the role played by *Hox* genes in several phyla. However, the number of genes involved in the cluster of the lophotrochozoan ancestor is still a question of debate. The data presented here suggest that at least nine genes were present while two other genes, *Lox4* and *Post-2*, may either have been present in the ancestor or may have arisen as a result of duplication in the Brachiopoda-Mollusca-Annelida lineage. Spatial and temporal collinearity is a feature of *Hox* gene expression which was probably present in the ancestor of deuterostomes and protostomes. However, in Lophotrochozoa, it has been detected in only a few species belonging to Annelida and Mollusca.

Keywords: *Hox*; Lophotrochozoa; Mollusca; Annelida; Nemertea; Platyhelminthes

---

1. Introduction

Hox genes, a subfamily of homeobox genes, encode transcription factors containing a highly conserved 60 amino acid homeodomain characterized by a helix-turn-helix motif [1]. The members of this gene subfamily are regulators of animal embryonic development and play a role in the patterning of the anterior-posterior body axis of Bilateria [2].

Hox genes were first discovered in *Drosophila melanogaster* where they are organized in the split *Antennapedia-Bithorax* complex located on chromosome 3 [3,4]. In some cases, these genes are arranged in clusters, and, hence, they are physically linked on the same chromosome. The number of clusters varies in agreement with the genome duplications that the organisms experienced during evolution, ranging from one in protostomes and invertebrate deuterostomes to four in sarcopterygians which experienced two rounds of whole genome duplication (WGD), with as many as seven in teleosts which experienced a third lineage specific event of WGD (Teleost Specific Genome Duplication, TSGD) [5].

Generally, one feature of the *Hox* cluster is spatial collinearity: the gene order on chromosomes reflects the order of gene expression and function. The genes at the 3′ end of the cluster are expressed in the anterior part of the body while those at the 5′ end in the posterior part. Collinearity may also be temporal, meaning that the genes at the 3′ end are expressed before those at the 5′ end [6]. This feature is more evident in bilaterian organisms displaying an unbroken cluster than in others which show dispersed or broken clusters [7–10].

On the basis of gene position and gene function, the *Hox* cluster can be subdivided into four classes [11]: anterior class, Paralog Group 3, central class, and posterior class. The composition of these classes vary across taxa due to duplication, inversion, or gene loss events that occurred during evolution [10,12–14]. Moreover, fragmented clusters may also be related to the presence of transposable elements that could promote chromosomal rearrangements [15].

Although *Hox* genes show a high sequence similarity, they play a remarkable role in the wide morphological diversity of animals [10]. One of the major groups of bilaterian organisms is Protostomia, which is subdivided into two clades: Ecdysozoa and Lophotrochozoa. The peculiarity of the former is the ability to undergo ecdysis under the hormonal control of ecdysteroids. The latter clade is characterized by the trochophore, the free-swimming ciliated larvae, and/or the lophophore, the feeding structure made up of tentacles surrounding the mouth of adults.

Regarding *Hox* cluster composition, besides the *Hox* genes belonging to the *Paralog Group-1 (PG-1)*, *PG-2*, *PG-3*, *PG-4*, and *PG-5*, Ecdysozoa also exhibit *ftz*, *Antp*, *Ubx*, *abd-A*, and *Abd-B*, while Lophotrochozoa also include *Lox5*, *Lox2*, *Lox4*, *Post-1* and *Post-2* [16,17].

This paper focuses on Lophotrochozoa which are characterized by a high diversity of body architecture, and are, therefore, ideal for studying the evolution of development. This review presents an overview of the presence and expression patterns of *Hox* genes in 12 lophotrochozoan phyla. The data obtained allowed different hypotheses to be delineated regarding the evolution of the *Hox* gene subfamily within Lophotrochozoa and its implications on development.

2. Hox Presence in Lophotrochozoa

Body plan evolution and diversification in metazoans have not only been related to changes in *Hox* cluster composition such as cluster and gene duplications, and gene loss, but also to gene expression and regulatory interactions [14]. An understanding of *Hox* gene cluster composition can provide insight into the evolutionary history that these genes have undergone within Lophotrochozoa.

Although internal relationships within the Lophotrochozoa clade are still controversial [18–22], we focused on works concerning the presence of *Hox* genes in 12 phyla, and the evolution of the genes composing the *Hox* cluster was discussed in relation to the evolutionary relationships between lophotrochozoan phyla. The works summarized here clearly indicate that data on *Hox* genes are rather scarce for Lophotrochozoa, and a limited number of works on complete genomes have provided insight into *Hox* gene cluster composition [20,22–33]. About half of the lophotrochozoan phyla have never been investigated while only one species has been analyzed in Brachiopoda and Bryozoa, two species in Rotifera, and three species in Nemertea (Table 1) [16,22,23,34–37]. More information is available for the three major lophotrochozoan phyla: Mollusca (about 30 species, Figure 1, Table S1) [16,20,24,26,38–58], Annelida (about 20 species, Figure 2, Table S2) [8,16,20,46,59–73] and Platyhelminthes (about 30 species, Figure 3, Table S3) [27–33,74–90].

Table 1. *Hox* genes identified to date in Brachiopoda, Bryozoa, Rotifera, and Nemertea.

Phylum	Species	*Hox* Genes											References
		PG-1	*PG-2*	*PG-3*	*PG-4*	*PG-5*	*Lox5*	*Antp*	*Lox2*	*Lox4*	*Post-1*	*Post-2*	
Brachiopoda	*Lingula anatina*	√	√	√	√	√	√	√	√	√	√	√	[16,22]
Bryozoa	*Bugula turrita*		√	√	√*		√					√	[34]
Rotifera	*Adineta vaga*	√*	√		√*	√*	√*	√*		√*			[23]
	Philodina roseola					√*	√*						[91]
Nemertea	*Lineus sanguineus*	√		√	√		√	√				√	[35]
	Micrura alaskensis	√	√	√	√	√	√	√		√		√	[36]
	Pantinonemertes californiensis	√	√	√	√		√					√	[37]

* indicates duplicated genes.

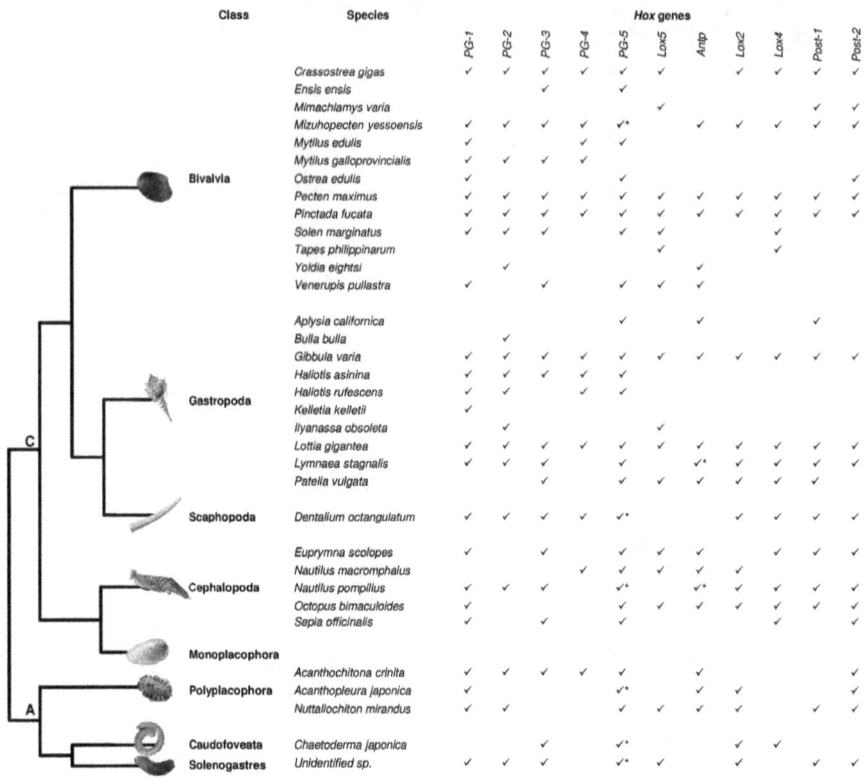

Figure 1. *Hox* genes in Mollusca. *Hox* genes identified in the Mollusca phylum are reported. Tree topology following Smith *et al.* [92,93]. C: Conchifera; A: Aculifera. * indicates duplicated genes probably due to erroneous attribution [94]. The graphics of the figure were modified from Biscotti *et al.* [94]. For references see Table S1.

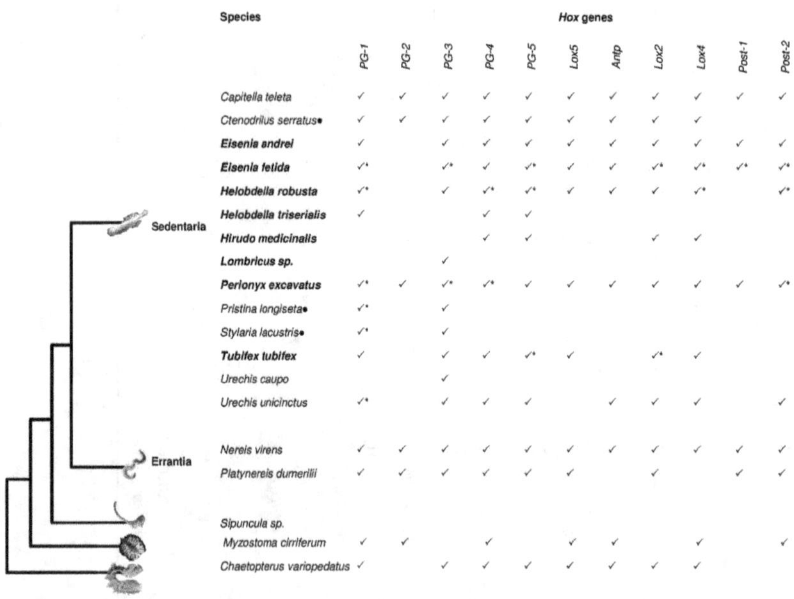

Figure 2. *Hox* genes in Annelida. *Hox* genes identified in the Annelida phylum are reported. Tree topology following Struck *et al.* [95]. * indicates duplicated genes. Black dots indicate species not included in the analysis carried out by Struck *et al.* [95]. Species in bold belong to Clitellata. For references see Table S2.

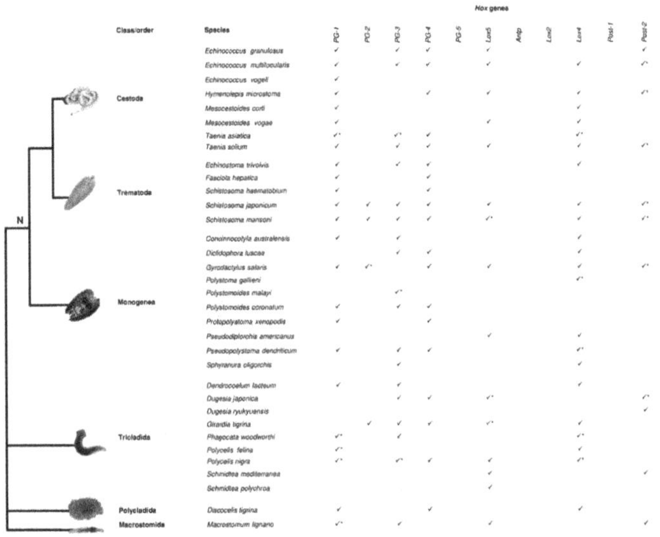

Figure 3. *Hox* genes in Platyhelminthes. *Hox* genes identified in the Platyhelminthes phylum are reported. The tree is modified from Hahn *et al.* [32]. N: Neodermata. * indicates duplicated genes. The duplication reported for *PG-2* gene in *Gyrodactylus salaris* might be due to an erroneous attribution of *PG-3* gene. For references see Table S3.

The Mollusca phylum is divided into eight classes: Solenogastres, Caudofoveata, and Polyplacophora belonging to Aculifera and Bivalvia, Cephalopoda, Gastropoda, Monoplacophora, and Scaphopoda belonging to the sister taxa Conchifera [92,93]. Mollusks include organisms living in different ecological niches from marine to freshwater and terrestrial environments. Their morphology is extremely variable, ranging from Aplacophora with a wormlike appearance and no shell to Cephalopoda which have a well-developed cephalic region and have co-opted the mantle for locomotion. The muscular foot also presents different morphologies as it is adapted for a variety of functions. With the exception of Monoplacophora, for which no data are available, *Hox* genes have been identified in a single species of Scaphopoda [40], two species of Aplacophora [40], three species of Polyplacophora [40,57,58], five species of Cephalopoda [26,40,46,55,56], 10 species of Gastropoda [16,20,40,46–54], and 13 species of Bivalvia [24,38–45] (Figure 1, Table S1). Our previous review [94] provided evidence that all 11 genes of the *Hox* lophotrochozoan cluster are found within Bivalvia, Cephalopoda, and Gastropoda, which are the three most studied molluskan classes. Moreover, genome sequencing has revealed, for the first time in lophotrochozoans, that the gastropod *Lottia gigantea* presents an intact cluster [20]. In the genome assembly of the Pacific oyster, *Crassostrea gigas*, the *Antp* gene is clearly missing and *Hox* genes are located on four scaffolds [24,38]. In the pearl oyster *Pinctada fucata,* all 11 *Hox* genes are present and clustered on three scaffolds. Moreover, non-*Hox* genes are present in the *Hox* clusters of both the above-mentioned oysters, thereby suggesting that this is a feature which occurred in their common ancestor [25].

In the genome of the cephalopod, *Octopus bimaculoides* eight of the 11 *Hox* genes (with the exception of *Hox2*, *Hox3* and *PG-4*) have been identified and located on eight separate scaffolds [26].

Annelida is another large and morphologically diverse taxon of Lophotrochozoa. Traditionally, this phylum includes segmented worms but currently Myzostomida [95], Echiura, and Sipuncula are also considered as Annelida. The two latter taxa show segmentation only at the larval stage [96,97].

A recent phylogenomic analysis divides Annelida into two main groups reflecting different lifestyles, Errantia and Sedentaria, with Sipuncula and Myzostomida which occupy a basal position (Figure 2) [95].

Within the Sedentaria group, the species belonging to Clitellata exhibit a highly dynamic *Hox* gene cluster which is broken into several genomic regions and characterized by gene duplication and loss events. In fact, extensive studies on the leech *Helobdella robusta* [20] and two earthworms,

Eisenia fetida [61] and *Perionyx excavatus* [67], suggest that individual duplications, large segmental duplications and/or whole genome duplications could have played a pivotal role in determining the current *Hox* gene number and arrangement in these lineages. Moreover, the ortholog *Post-1* gene has not been retrieved in leeches and the *PG-2* gene seems be absent in Clitellata with the exception of *P. excavates* [67]. These organisms exhibit a different number of segments, internal anatomy, and shapes. Therefore, it has been suggested that differences in presence, absence, and the arrangement of *Hox* genes may be responsible for the evolution of the annelid body plan diversity [20].

The genomic survey performed in *Capitella teleta* indicates that *Hox* genes are located on three scaffolds although duplication events have not been detected [8,20].

In general, the common ancestor of Sedentaria and Errantia had all 11 genes of the *Hox* cluster. The species *Myzostoma cirriferum* and *Chaetopterus variopedatus*, which are basal in the annelid phylogeny, do not show a complete cluster although whole genome sequencing has not yet been performed for these species.

Within Lophotrochozoa, the phylogenetic relationships of Annelida and Mollusca with respect to Brachiopoda, Phoronida, and Nemertea are still a question of debate. Recently, three hypotheses have been proposed regarding the phylogenetic relationships between these taxa (Figure 4): according to the first hypothesis, Brachiozoa (Brachiopoda + Phoronida) is a sister group to Mollusca while Nemertea occupies a basal position in all the taxa considered [22,98]; the second hypothesis suggests that Brachiozoa is a sister group to Mollusca + Annelida [99]; the third hypothesis considers Annelida as a sister group to the clade including Brachiozoa and Nemertea [21].

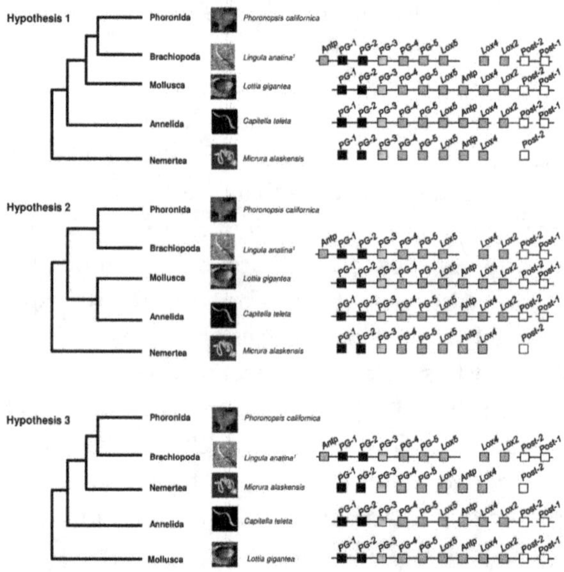

Figure 4. Hypotheses on phylogenetic relationships between the major lophotrochozoan phyla. In the trees displaying the three hypotheses [22], the species in which the largest numbers of *Hox* genes have been identified are reported for Nemertea and Brachiopoda, for Mollusca *Lottia gigantea*, the only species showing the intact cluster in Lophotrochozoa, for Annelida *Capitella teleta*, the species to date showing the least fragmented cluster. Lines underneath boxes indicate syntenic arrangement. Split lines indicate that the cluster is located on different chromosomes or scaffolds. The length of the depicted cluster is not proportional to the effective length in genomes. [1]: in *Lingula anatina*, *Lox4* and *Lox2* genes have been identified in de Rosa *et al.* [16]. Dark boxes indicate genes belonging to the Anterior class; green boxes indicate genes belonging to the *Paralog Group 3*; grey boxes indicate genes belonging to the Central class; white boxes indicate genes belonging to the Posterior class. Gene abbreviations: *Antp*: *Antennapedia*; *lab*: *labial*; *Lox2*: *Lophotrochozoa Hox2*; *Lox4*: *Lophotrochozoa Hox4*; *Lox5*: *Lophotrochozoa Hox5*; *pb*: *Proboscipedia*; *PG-1*: *paralog group 1*; *PG-2*: *paralog group 2*; *PG-3*: *paralog group 3*; *PG-4*: *paralog group 4*; and *PG-5*: *paralog group 5*; *Post-1*: *Posterior-1*; *Post-2*: *Posterior-2*. The graphics of the figure were modified from Biscotti *et al.* [94]. For references see Table 1, Tables S1 and S2.

Lox2 and *Post-1* genes have not yet been recovered in three nemertean species, *Lineus sanguineus* [35], *Micrura alaskensis* [36], *Pantinonemertes californiensis* [37]. According to the evolutionary hypotheses 1 [22,98] and 2 [99], the duplication events leading to these two genes occurred after the separation of Nemertea and other lophotrochozoan phyla, thereby suggesting that the common ancestor of Nemertea and Annelida-Mollusca-Brachiopoda had nine genes [36,37,94]. Alternatively, considering evolutionary hypothesis 3, the absence of *Lox2* and *Post-1* might be due to a secondary loss.

The phylum Platyhelminthes is the fourth largest animal phylum after arthropods, mollusks and chordates. Members of the phylum Platyhelminthes have a simple bilateral body plan, characterized by the absence of traits found in most bilaterians (e.g., coelom, anus, circulatory and respiratory systems) and for this reason they were previously considered as basal bilaterians [100]. However, the presence in many flatworms of spiral embryonic cleavage, which is a mode of development present in other invertebrate phyla, such as annelids or mollusks, has made their classification very controversial. Platyhelminthes are currently considered as part of Lophotrochozoa, with the acoel and nemertodermatid flatworms separate from the Platyhelminthes and considered as the earliest branching extant bilaterians. However, some phylogenetic studies regard this taxon as separate from Lophotrochozoa [21,98]. The Platyhelminthes are divided into three classes, the free-living class "Turbellaria" and two parasitic classes, Cestoda and Trematoda. Subsequently, the parasitic species are subdivided into three classes, Trematoda, Cestoda, and Monogenea (together called Neodermata), while the class "Turbellaria", a paraphyletic group, is divided into about 11 orders (depending on the authors) [100].

In Platyhelminthes, the *Hox* gene set is reduced (Figure 3, Table S3), with *PG-5*, *Antp*, *Lox2*, and *Post-1* being absent in all the considered taxa [27–29,31,32]. The lack of these genes even in free-living flatworms suggests that their absence may not be related to the parasitic lifestyle of many species of this phylum. In Cestoda, the *PG-2* gene seems to be absent and, among the other species examined, orthologs have only been reported in four species: *Girardia tigrina*, *Schistosoma mansoni*, *S. japonicum*, *Gyrodactylus salaris* (Figure 3). *PG-1*, *PG-3*, *Lox5*, *Lox4*, and *Post-2* genes show duplications in several species belonging to all groups. A duplication has been detected for *PG-2* in *G. salaris* although one of the two genes might belong to *PG-3*; indeed, in Platyhelminthes, *PG-2* and *PG-3* genes share several conserved amino acidic residues that might make correct gene attribution difficult [28,75].

Moreover, as regards *Hox* gene arrangement, genome surveys have shown that the cluster is disrupted. The developmental strategies of some of these organisms with a parasitic lifestyle [75] and/or the presence of transposable elements [79,101] have been proposed in order to explain this dynamic arrangement.

In Rotifera, the genome survey of *Adineta vaga* demonstrated that *PG-3*, *Lox2* and posterior genes are absent while the other genes, with the exception of *PG-2*, present multiple copies. Moreover, not all the genes identified are arranged in a cluster [23]. The study in *Philodina roseola*, even if fragmentary, to date has identified only two central genes which present multiple copies (Table 1).

3. Expression Patterns

Insight into the presence of *Hox* genes and their expression patterns might be useful for better understanding the morphological innovations and the wide variety of lophotrochozoan species.

Most of the expression data on mollusks are limited to three species of gastropods [47–49,51,52], two species of cephalopods [46,56,102], one species of bivalves [24,38], and one species of chitons [57]. Data on the conchiferan species suggest the role of *Hox* genes in the central nervous system, their involvement in shell formation, tentacles, and the funnel; their expression has also been identified in sensory organs such as the apical organ and statocyst and in the light organ in the cephalopod *Euprymna scolopes*. Expression analysis in the bivalve *C. gigas* has shown that *Hox* genes are not activated according to temporal collinearity but that the peak in *PG-4*, *PG-1*, and *Lox4* expression occurs before gastrulation, while *Lox5* and *Post-2* are expressed during the trochophore stage, and the other genes in late development [24,38]. Recent findings in the aculiferan *Acanthochitona crinita* reveal a collinearity of *Hox* gene expression [57]. This is the first evidence in mollusks of the ancestral role of *Hox* genes in patterning the structures along the anterior-posterior body axis in the same way as most other bilaterian animals. Therefore, after the aculiferan-conchiferan split, *Hox* genes were co-opted into the formation of novelties in gastropods and cephalopods and perhaps in all the other conchiferan classes [57].

A comparison of expression patterns during trochophore larvae development in mollusks and annelids reveals that chitons share comparable expression patterns with annelids along the anterior-posterior body axis [57].

Expression surveys in the annelid Sedentaria confirm the ancestral role of *Hox* genes in determining the structures along the anterior-posterior body axis [8,62,67,70,73,103]. In particular, the study performed in *Capitella teleta* has shown that *Hox* genes exhibit spatial and temporal collinearity in line with the ancestral role that these genes had in the deuterostome-protostome ancestor [8]. However, the absence of collinearity observed in *H. robusta*, that shows a disrupted *Hox* cluster, indicates that collinearity is not conserved within Sedentaria [62]. The duplicated genes present in *P. excavatus* also seem to be expressed from the anterior to the posterior body region [67].

In the Errantia *N. virens* and *Platynereis dumerilii*, the expression patterns are very similar: all the *Hox* genes are involved in vectorial regionalization with the exception of *Post-1*; *PG-1*, *PG-4*, *PG-5*, *Lox5*, and *Post-2* are expressed in spatial collinearity while temporal collinearity is not respected since the Nereid body plan lacks unique segmental identities [70]. On the contrary, temporal collinearity has been detected in *Chaetopterus variopedatus* which occupies the basal position in Annelida [73].

The expression analysis of the nine *Hox* genes in the nemertean *M. alaskensis* clearly indicated that *Hox* genes are expressed not in the pilidial larva but in the juvenile stage of pilidiophorans [36]. The gene expression studies also performed on the hoplonemerteans *P. californiensis* revealed the homology between the imaginal discs of the pilidium and the paired larval invaginations in hoplonemerteans and showed that pilidial development evolved before the split between the two nemertean groups [37]. This sheds light on how *Hox* genes might be useful for understanding the evolution of embryonic development.

A high expression of *Hox* genes belonging to the paralog groups *PG-2*, *PG-4*, *Lox5*, *Lox4* and, to a lesser extent, *PG-1*, has been identified inside the eggs and at the miracidium stage in *Schistosoma* and in monogenean parasites [31,79,104], thus suggesting that these genes could be involved in embryo development. *Lox5* and *PG-4* genes have also exhibited a higher expression in the schistosomulum stage of *S. japonicum*, thereby highlighting the possible involvement of these two genes in multiple ontogenetic development stages in schistosomes [31,79]. Moreover, high expression of *PG-4*, *Lox5*, and *Lox4* has also been observed in sporocysts, another stage at which the determination of cell fate along the anterior-posterior axis may be important [79]. *Lox4* could be involved in the process of development in the monogenean parasite, given its high expression in the early developmental stages of the branchial phenotype [104]. During embryogenesis and embryo patterning in the planarian *Schmidtea polychroa*, the *Spol-hoxD* (*Lox5*) gene transcripts started to be detected at early stage 6 (8–10 days), in a strip of cells on the side of the embryo containing the definitive pharynx; the pattern spreads from the definitive pharynx to the posterior end, as reported by Iglesias *et al.* [89] in the adult. These data suggest that the adult anterior-posterior axis is established after yolk ingestion and the proliferation of the blastomeres in the germ band (stage 5) [90].

Studies on *Hox* expression have also been performed in adults. The patterns in adult specimens of *Girardia tigrina*, *Discocelis tigrina*, and *Dugesia japonica* have shown two types of *Hox* genes, that either conserve or lose their typical differential spatial expression whereas duplicated genes may show both patterns of expression [84,105]. In the planarian *D. japonica*, Nogi and Watanabe [83] reported a similar expression pattern for the two *Post-2* genes: *DjAbd-Ba* is expressed from the posterior pharyngeal region to the entire tail region suggesting that this gene is involved in the specification of the tail region, while *DjAbd-Bb* is expressed in several types of cells throughout the body. However, counter to the rule of spatial collinearity, the anterior boundary of the expression domain of the posterior gene *DjAbd-Ba* is anterior to the domains of the central genes *PG-4* and *Lox5*. Different functions have also been reported for *Hox* genes, apart from their involvement in embryonic development; for example, it has been hypothesized that *Lox4*, being highly expressed in adult males and down-regulated in adult females, could be involved in tissue differentiation of the male reproductive tract [104]. Another feature is the permanent *Hox* expression in adult organisms; the high morphological plasticity of these organisms, related to the presence in adult organisms of neoblasts (undifferentiated and totipotent cell types, [106]), could explain their activity, suggesting that pattern formation in planarians may occur continuously [82,105].

As well as embryonic development, the anterior-posterior positional values of *Hox* genes are also involved in regeneration. After wounding, regenerative tissue is quickly formed [106], accompanied by cell proliferation and *Hox* expression, even if the two processes are not necessarily related [86]. It has been pointed out that during regeneration, only *Lox5* and *Post-2* genes have a differential axial nested expression, while the other genes are ubiquitously expressed. These genes are activated during the first day of tail regeneration and down-regulated during head regeneration [82,83,105,107,108]. In particular, the presence of a system that maintains anterior-posterior axial polarity and regulates the expression of *DjAbd-Ba* rapidly after amputation could be confirmed by the rapid expression of *DjAbd-Ba* (*Post-2a*) in the head piece after amputation, with the anterior boundary of the *DjAbd-Ba* expression domain shifting rapidly and dynamically toward the posterior in the tail pieces [83]. Moreover, *Hox* gene expression has also been detected during lateral regeneration underlining the importance of these genes in regenerative processes in order to specify positional information on any axis [86].

The extra copies of *Hox* genes identified in some Platyhelminthes may have lost their typical anterior-posterior axial patterning role as a result of independent duplication and may, on the contrary, have acquired a function in cell differentiation [79,82,86].

In general, flatworm *Hox* genes have been studied for many years, and, in some cases, distinct spatial domains of expression have been highlighted, although their specific functions have not yet been identified. Further research studies, especially on embryo development, are required in order to draw some conclusions on the role and the possible expression collinearity of *Hox* genes in Platyhelminthes.

4. Conclusions

This review provides clear evidence that insight into *Hox* cluster composition and expression patterns is limited to a few phyla and that information is completely lacking for about half the taxa belonging to Lophotrochozoa.

The overview of the literature considered in this work indicates that *Lox2* and *Post-1* genes have only been recovered in some taxa of Lophotrochozoa. In addition to the hypothesis of a common ancestor with 11 genes, a further suggestion is that these two genes originated as a result of secondary duplications and that the common ancestor had nine genes [36,37,94].

If the latter hypothesis is valid the duplications of the central *Lox4* and *Lox2* genes and the posterior *Post-1* and *Post-2* genes must have occurred in the ancestor of Mollusca-Annelida-Brachiopoda. As regards Nemerteans, if their position is within this clade, the lack of *Lox2* and *Post-1* genes (since they have not been identified to date) may be due to a secondary loss. Moreover, in Platyhelminthes, *PG-5* and *Antp* have also been lost, thereby suggesting that additional gene losses have occurred compared with the common ancestor.

In Mollusca and Annelida, a complete spatial collinearity involving all the genes that make up the *Hox* cluster has only been detected in the polyplacophoran *Acanthochitona* and in the annelid *Capitella*, while in other species of annelids, such as *Helobdella triserialis*, *Nereis virens*, and *Platynereis dumerilii*, only some *Hox* genes maintain spatial collinearity [62,70].

This suggests that, during evolution, the collinearity of *Hox* gene expression has been maintained only in some taxa of these phyla. In flatworms, in which the breakage of the cluster and the loss of some genes occurred, spatial-temporal collinearity seems to have been lost. Since *Hox* genes are expressed in spatial collinearity in lophotrochozoans, ecdysozoans, and deuterostames, the most parsimonious conclusion is that the spatial collinearity of *Hox* genes was already a feature of the last common bilaterian ancestor.

Duplicated genes have also been identified in Clitellata and in Platyhelminthes, and the extra copies of *Hox* genes show different expression patterns suggesting neofunctionalizations.

The comparison of *Hox* cluster composition and expression patterns in various animal groups is a pivotal step toward understanding the mechanisms by which body plan modifications occurred determining animal radiation. Therefore, future studies should be focused on the identification and expression of *Hox* genes in phyla and classes which either have not yet been analyzed or which have received scarce attention (Micrognathozoa, Acanthocephala, Rotifera, Entoprocta, Cycliophora, and Phoronida). Furthermore, the information about some taxa is restricted to a single species and, therefore, more species need to be investigated so as to have a better overview.

Acknowledgments: We apologize to colleagues whose work could not be cited because of space limitations. We thank Jean-François Flot and Andreas Hejnol for data on the rotifer *Adineta vaga*. We also wish to thank the anonymous reviewers whose comments and suggestions helped to improve the manuscript.

Author Contributions: All the authors have made contributions to the conception and design of the manuscript and have given final approval of the version to be published.

Conflicts of Interest: The authors declare no conflicts of interest.

References

1. Gehring, W.J.; Affolter, M.; Burglin, T. Homeodomain proteins. *Annu. Rev. Biochem.* **1994**, *63*, 487–526. [CrossRef] [PubMed]
2. McGinnis, W.; Krumlauf, R. Homeobox genes and axial patterning. *Cell* **1992**, *68*, 283–302. [CrossRef]
3. Lewis, E.B. A gene complex controlling segmentation in *Drosophila*. *Nature* **1978**, *276*, 565–570. [CrossRef] [PubMed]
4. Dessain, S.; Gross, C.T.; Kuziora, M.A.; McGinnis, W. Antp-type homeodomains have distinct DNA binding specificities that correlate with their different regulatory functions in embryos. *EMBO J.* **1992**, *11*, 991–1002. [PubMed]
5. Meyer, A.; van de Peer, Y. From 2R to 3R: Evidence for a fish-specific genome duplication (FSGD). *BioEssays* **2005**, *27*, 937–945. [CrossRef] [PubMed]
6. Duboule, D. Temporal colinearity and phylotypic progression: A basis for the stability of a vertebrate Bauplan and the evolution of morphologies through heterochrony. *Dev. Suppl.* **1994**, 135–142.
7. Ikuta, T.; Yoshida, N.; Satoh, N.; Saiga, H. *Ciona intestinalis Hox* gene cluster: Its dispersed structure and residual collinear expression in development. *Proc. Natl. Acad. Sci. USA* **2004**, *101*, 15118–15123. [CrossRef] [PubMed]
8. Fröbius, A.C.; Matus, D.Q.; Seaver, E.C. Genomic organization and expression demonstrate spatial and temporal *Hox* gene colinearity in the lophotrochozoan *Capitella sp I*. *PLoS ONE* **2008**, *3*, e4004. [CrossRef] [PubMed]
9. Moreno, E.; Permanyer, J.; Martinez, P. The origin of patterning systems in bilateria—Insights from the *Hox* and *ParaHox* genes in Acoelomorpha. *Genom. Proteom. Bioinf.* **2011**, *9*, 65–76. [CrossRef]
10. Garcia-Fernàndez, J. The genesis and evolution of Homeobox gene clusters. *Nat. Rev. Genet.* **2005**, *6*, 881–892. [CrossRef] [PubMed]
11. Kourakis, M.J.; Martindale, M.K. *Hox* gene duplication and deployment in the annelid leech *Helobdella*. *Evol. Dev.* **2001**, *3*, 145–153. [PubMed]
12. Arthur, W. *The Origin of Animal Body Plans: A Study in evolutionary Developmental Biology*, 1st ed.; Cambridge University Press: Cambridge, UK, 1997.

13. Garcia-Fernàndez, J. *Hox, ParaHox, ProtoHox*: Facts and guesses. *Heredity* **2005**, *94*, 145–152. [CrossRef] [PubMed]
14. Holland, P.W. Evolution of homeobox genes. *Wiley Interdiscip. Rev. Dev. Biol.* **2013**, *2*, 31–45. [CrossRef] [PubMed]
15. Moreno, E.; Nadal, M.; Baguñà, J.; Martínez, P. Tracking the origins of the bilaterian *Hox* patterning system: Insights from the acoel flatworm *Symsagittifera roscoffensis*. *Evol. Dev.* **2009**, *11*, 574–581. [CrossRef] [PubMed]
16. De Rosa, R.; Grenier, J.K.; Andreeva, T.; Cook, C.E.; Adoutte, A.; Akam, M.; Carroll, S.B.; Balavoine, G. *Hox* genes in brachiopods and priapulids and protostome evolution. *Nature* **1999**, *399*, 772–776. [PubMed]
17. Balavoine, G.; de Rosa, R.; Adoutte, A. *Hox* clusters and bilaterian phylogeny. *Mol. Phylogenet. Evol.* **2002**, *24*, 366–373. [CrossRef]
18. Dunn, C.W.; Hejnol, A.; Matus, D.Q.; Pang, K.; Browne, W.E.; Smith, S.A.; Seaver, E.; Rouse, G.W.; Obst, M.; Edgecombe, G.D.; *et al.* Broad phylogenomic sampling improves resolution of the animal tree of life. *Nature* **2008**, *452*, 745–749. [CrossRef] [PubMed]
19. Paps, J.; Baguñà, J.; Riutort, M. Lophotrochozoa internal phylogeny: New insights from an up-to-date analysis of nuclear ribosomal genes. *Proc. Biol. Sci.* **2009**, *276*, 1245–1254. [CrossRef] [PubMed]
20. Simakov, O.; Marletaz, F.; Cho, S.-J.; Edsinger-Gonzales, E.; Havlak, P.; Hellsten, U.; Kuo, D.H.; Larsson, T.; Lv, J.; Arendt, D.; *et al.* Insights into bilaterian evolution from three spiralian genomes. *Nature* **2013**, *493*, 526–531. [CrossRef] [PubMed]
21. Laumer, C.E.; Bekkouche, N.; Kerbl, A.; Goetz, F.; Neves, R.C.; Sørensen, M.V.; Kristensen, R.M.; Hejnol, A.; Dunn, C.W.; Giribet, G.; *et al.* Spiralian phylogeny informs the evolution of microscopic lineages. *Curr. Biol.* **2015**, *25*, 2000–2006. [CrossRef] [PubMed]
22. Luo, Y.J.; Takeuchi, T.; Koyanagi, R.; Yamada, L.; Kanda, M.; Khalturina, M.; Fujie, M.; Yamasaki, S.; Endo, K.; Satoh, N. The *Lingula* genome provides insights into brachiopod evolution and the origin of phosphate biomineralization. *Nat. Commun.* **2015**, *6*. [CrossRef] [PubMed]
23. Flot, J.F.; Hespeels, B.; Li, X.; Noel, B.; Arkhipova, I.; Danchin, E.G.; Hejnol, A.; Henrissat, B.; Koszul, R.; Aury, J.M.; *et al.* Genomic evidence for ameiotic evolution in the bdelloid rotifer Adineta vaga. *Nature* **2013**, *500*, 453–457. [CrossRef] [PubMed]
24. Zhang, G.; Fang, X.; Guo, X.; Li, L.; Luo, R.; Xu, F.; Yang, P.; Zhang, L.; Wang, X.; Qi, H.; *et al.* The oyster genome reveals stress adaptation and complexity of shell formation. *Nature* **2012**, *490*, 49–54. [CrossRef] [PubMed]
25. Takeuchi, T.; Koyanagi, R.; Gyoja, F.; Kanda, M.; Hisata, K.; Fujie, M.; Goto, H.; Yamasaki, S.; Nagai, K.; Morino, Y.; *et al.* Bivalve-specific gene expansion in the pearl oyster genome: Implications of adaptation to a sessile lifestyle. *Zool. Lett.* **2016**, *2*. [CrossRef] [PubMed]
26. Albertin, C.B.; Simakov, O.; Mitros, T.; Wang, Z.Y.; Pungor, J.R.; Edsinger-Gonzales, E.; Brenner, S.; Ragsdale, C.W.; Rokhsar, D.S. The octopus genome and the evolution of cephalopod neural and morphological novelties. *Nature* **2015**, *524*, 220–224. [CrossRef] [PubMed]
27. Zheng, H.; Zhang, W.; Zhang, L.; Zhang, Z.; Li, J.; Lu, G.; Zhu, Y.; Wang, Y.; Huang, Y.; Liu, J.; *et al.* The genome of the hydatid tapeworm *Echinococcus granulosus*. *Nat. Genet.* **2013**, *45*, 1168–1175. [CrossRef] [PubMed]
28. Koziol, U.; Lalanne, A.I.; Castillo, E. *Hox* genes in the parasitic Platyhelminthes *Mesocestoides corti*, *Echinococcus multilocularis*, and *Schistosoma mansoni*: Evidence for a reduced *Hox* complement. *Biochem. Genet.* **2009**, *47*, 100–116. [CrossRef] [PubMed]
29. Tsai, I.J.; Zarowiecki, M.; Holroyd, N.; Garciarrubio, A.; Sánchez-Flores, A.; Brooks, K.L.; Tracey, A.; Bobes, R.J.; Fragoso, G.; Sciutto, E.; *et al.* The genomes of four tapeworm species reveal adaptations to parasitism. *Nature* **2013**, *496*, 57–63. [CrossRef] [PubMed]
30. Young, N.D.; Jex, A.R.; Li, B.; Liu, S.; Yang, L.; Xiong, Z.; Li, Y.; Cantacessi, C.; Hall, R.S.; Xu, X.; *et al.* Whole-genome sequence of *Schistosoma haematobium*. *Nat. Genet.* **2012**, *44*, 221–225. [CrossRef] [PubMed]
31. Gu, J.L.; Chen, S.X.; Dou, T.H.; Xu, M.J.; Xu, J.X.; Zhang, L.; Hu, W.; Wang, S.J.; Zhou, Y. *Hox* genes from the parasitic flatworm *Schistosoma japonicum*. *Genomics* **2012**, *99*, 59–65. [CrossRef] [PubMed]
32. Hahn, C.; Fromm, B.; Bachmann, L. Comparative genomics of flatworms (Platyhelminthes) reveals shared genomic features of ecto- and endoparastic Neodermata. *Genome Biol. Evol.* **2015**, *6*, 1105–1117. [CrossRef] [PubMed]

33. Wasik, K.; Gurtowski, J.; Zhou, X.; Ramos, O.M.; Delás, M.J.; Battistoni, G.; El Demerdash, O.; Falciatori, I.; Vizoso, D.B.; Smith, A.D.; *et al.* Genome and transcriptome of the regeneration-competent flatworm, *Macrostomum lignano*. *Proc. Natl. Acad. Sci. USA* **2015**, *112*, 12462–12467. [CrossRef] [PubMed]

34. Passamaneck, Y.J.; Halanych, K.M. Evidence from *Hox* genes that bryozoans are lophotrochozoans. *Evol. Dev.* **2004**, *6*, 275–281. [CrossRef] [PubMed]

35. Kmita-Cunisse, M.; Loosli, F.; Bièrne, J.; Gehring, W.J. Homeobox genes in the ribbonworm *Lineus sanguineus*: Evolutionary implications. *Proc. Natl. Acad. Sci. USA* **1998**, *95*, 3030–3035. [CrossRef] [PubMed]

36. Hiebert, L.S.; Maslakova, S.A. *Hox* genes pattern the anterior-posterior axis of the juvenile but not the larva in a maximally indirect developing invertebrate, *Micrura alaskensis* (Nemertea). *BMC Biol.* **2015**, *13*. [CrossRef] [PubMed]

37. Hiebert, L.S.; Maslakova, S.A. Expression of Hox, Cdx, and Six3/6 genes in the hoplonemertean *Pantinonemertes californiensis* offers insight into the evolution of maximally indirect development in the phylum Nemertea. *Evodevo* **2015**, *6*. [CrossRef] [PubMed]

38. Paps, J.; Xu, F.; Zhang, G.; Holland, P.W. Reinforcing the egg-timer: Recruitment of novel lophotrochozoa homeobox genes to early and late development in the pacific oyster. *Genome Biol. Evol.* **2015**, *7*, 677–688. [CrossRef] [PubMed]

39. Barucca, M.; Olmo, E.; Canapa, A. *Hox* and *ParaHox* genes in bivalve molluscs. *Gene* **2003**, *317*, 97–102. [CrossRef]

40. Iijima, M.; Akiba, N.; Sarashina, I.; Kuratani, S.; Endo, K. Evolution of *Hox* genes in molluscs: A comparison among seven morphologically diverse classes. *J. Molluscan Stud.* **2006**, *72*, 259–266. [CrossRef]

41. Mesías-Gansbiller, C.; Sánchez, J.L.; Pazos, A.J.; Lozano, V.; Martínez-Escauriaza, R.; Luz Pérez-Parallé, M. Conservation of *Gbx* genes from EHG homeobox in bivalve molluscs. *Mol. Phylogenet. Evol.* **2012**, *63*, 213–217. [CrossRef] [PubMed]

42. Peréz-Parallé, M.L.; Carpintero, P.; Pazos, A.J.; Abad, M.; Sánchez, J.L. The *Hox* cluster in the bivalve mollusc *Mytilus galloprovincialis*. *Biochem. Genet.* **2005**, *43*, 417–424. [CrossRef] [PubMed]

43. Canapa, A.; Biscotti, M.A.; Olmo, E.; Barucca, M. Isolation of *Hox* and *ParaHox* genes in the bivalve *Pecten maximus*. *Gene* **2005**, *348*, 83–88. [CrossRef] [PubMed]

44. Carpintero, P.; Pazos, A.J.; Abad, M.; Sánchez, J.L.; Pérez-Parallé, M.L. Presence *Proboscipedia* and *Caudal* gene homologues in a bivalve mollusc. *J. Biochem. Mol. Biol.* **2004**, *37*, 625–628. [CrossRef] [PubMed]

45. Morino, Y.; Okada, K.; Niikura, M.; Honda, M.; Satoh, N.; Wada, H. A genome-wide survey of genes encoding transcription factors in the Japanese pearl oyster, *Pinctada fucata*: I. Homeobox genes. *Zool. Sci.* **2013**, *30*, 851–857. [CrossRef] [PubMed]

46. Lee, S.E.; Gates, R.D.; Jacobs, D.K. Gene fishing: The use of a simple protocol to isolate multiple homeodomain classes from diverse invertebrate taxa. *J. Mol. Evol.* **2003**, *56*, 509–516. [PubMed]

47. Samadi, L.; Steiner, G. Involvement of *Hox* genes in shell morphogenesis in the encapsulated development of a top shell gastropod (*Gibbula varia* L.). *Dev. Genes Evol.* **2009**, *219*, 523–530. [CrossRef] [PubMed]

48. Samadi, L.; Steiner, G. Expression of *Hox* genes during the larval development of the snail, *Gibbula varia* [L]-further evidence of non-colinearity in molluscs. *Dev. Genes Evol.* **2010**, *220*, 161–172. [CrossRef] [PubMed]

49. Giusti, A.F.; Hinman, V.F.; Degnam, S.M.; Degnan, B.M.; Morse, D.E. Expression of a *Scr/Hox5* gene in the larval central nervous system of the gastropod *Haliotis*, a non-segmented spiralian lophotrochozoan. *Evol. Dev.* **2000**, *2*, 294–302. [CrossRef] [PubMed]

50. Hinman, V.F.; Degnan, B.M. *Mox* homeobox expression in muscle lineage of the gastropod *Haliotis asinina*: Evidence for a conserved role in bilaterian myogenesis. *Dev. Genes Evol.* **2002**, *212*, 141–144. [PubMed]

51. Hinman, V.F.; O'Brien, E.K.; Richards, G.S.; Degnan, B.M. Expression of anterior *Hox* genes during larval development of the gastropod *Haliotis asinina*. *Evol. Dev.* **2003**, *5*, 508–521. [CrossRef] [PubMed]

52. Degnan, B.M.; Morse, D.E. Identification of eight homeobox-containing transcripts expressed during larval development and at metamorphosis in the gastropod mollusc *Haliotis rufescens*. *Mol. Mar. Biol. Biotechnol.* **1993**, *2*, 1–9. [PubMed]

53. Lambert, J.D.; Chan, X.Y.; Spiecker, B.; Sweet, H.C. Characterizing the embryonic transcriptome of the snail *Ilyanassa*. *Integr. Comp. Biol.* **2010**, *50*, 768–777. [CrossRef] [PubMed]

54. Kenny, N.J.; Namigai, E.K.; Marlétaz, F.; Hui, J.H.; Shimeld, S.M. Draft genome assemblies and predicted microRNA complements of the intertidal lophotrochozoans *Patella vulgata* (Mollusca, Patellogastropoda) and *Spirobranchus (Pomatoceros) lamarcki* (Annelida, Serpulida). *Mar. Genom.* **2015**, *24*, 139–146. [CrossRef] [PubMed]
55. Callaerts, P.; Lee, P.N.; Hartmann, B.; Farfan, C.; Choy, D.W.; Ikeo, K.; Fischbach, K.-F.; Gehring, W.J.; Gert de Couet, H. *Hox* genes in the sepiolid squid *Euprymna scolopes*: Implications for the evolution of complex body plans. *Proc. Natl. Acad. Sci. USA* **2002**, *99*, 2088–2093. [CrossRef] [PubMed]
56. Pernice, M.; Deutsch, J.S.; Andouche, A.; Boucher-Rodoni, R.; Bonnaud, L. Unexpected variation of *Hox* genes' homeodomains in cephalopods. *Mol. Phylogenet. Evol.* **2006**, *40*, 872–879. [CrossRef] [PubMed]
57. Fritsch, M.; Wollesen, T.; de Oliveira, A.L.; Wanninger, A. Unexpected co-linearity of *Hox* gene expression in an aculiferan mollusk. *BMC Evol. Biol.* **2015**, *15*. [CrossRef] [PubMed]
58. Biscotti, M.A.; Canapa, A.; Olmo, E.; Barucca, M. *Hox* genes in the Antarctic polyplacophoran *Nuttallochiton mirandus*. *J. Exp. Zool. Part B* **2007**, *308*, 507–513. [CrossRef] [PubMed]
59. Dick, M.H.; Buss, L.W. A PCR-based Survey of Homeobox Genes in *Ctenodrilus serratus* (Annelida: Polychaeta). *Mol. Phylogenet. Evol.* **1994**, *3*, 146–158. [CrossRef] [PubMed]
60. Cho, P.Y.; Cho, S.J.; Lee, M.S.; Lee, J.A.; Tak, E.S.; Shin, C.; Choo, J.K.; Park, S.C.; Lee, K.-S.; Park, H.-Y.; *et al.* Note: A PCR-Based Analysis of *Hox* Genes in an Earthworm, *Eisenia andrei* (Annelida: Oligochaeta). *Biochem. Genet.* **2004**, *42*, 209–216. [CrossRef] [PubMed]
61. Zwarycz, A.S.; Nossa, C.W.; Putnam, N.H.; Ryan, J. Timing and scope of genomic expansion within Annelida: Evidence from homeoboxes in the genome of the earthworm *Eisenia fetida*. *Genome Biol. Evol.* **2015**. [CrossRef]
62. Kourakis, M.J.; Master, V.A.; Lokhorst, D.K.; Nardelli-Haefliger, D.; Wedeen, C.J.; Martindale, M.Q.; Shankland, M. Conserved anterior boundaries of *Hox* gene expression in the central nervous system of the leech *Helobdella*. *Dev. Biol.* **1997**, *190*, 284–300. [CrossRef] [PubMed]
63. Aisemberg, G.O.; Macagno, E.R. *Lox1*, an Antennapedia-class homeobox gene, is expressed during leech gangliogenesis in both transient and stable central neurons. *Dev. Biol.* **1994**, *161*, 455–465. [CrossRef] [PubMed]
64. Wong, V.Y.; Aisemberg, G.O.; Gan, W.B.; Macagno, E.R. The leech homeobox gene *Lox4* may determine segmental differentiation of identified neurons. *J. Neurosci.* **1995**, *15*, 5551–5559. [PubMed]
65. Wong, V.Y.; Macagno, E.R. *Lox6*, a leech *Dfd* ortholog, is expressed in the central nervous system and in peripheral sensory structures. *Dev. Genes Evol.* **1998**, *208*, 51–55. [CrossRef] [PubMed]
66. Wysocka-Diller, J.W.; Aisemberg, G.O.; Baumgarten, M.; Levine, M.; Macagno, E.R. Characterization of a homologue of bithorax-complex genes in the leech *Hirudo medicinalis*. *Nature* **1989**, *341*, 760–763. [CrossRef] [PubMed]
67. Cho, S.J.; Vallès, Y.; Kim, K.M.; Ji, S.C.; Han, S.J.; Park, S.C. Additional duplicated *Hox* genes in the earthworm: *Perionyx excavatus Hox* genes consist of eleven paralog groups. *Gene* **2012**, *493*, 260–266. [CrossRef] [PubMed]
68. Snow, P.; Buss, L.W. HOM/*Hox*-type homeoboxes from *Stylaria lacustris* (Annelida: Oligochaeta). *Mol. Phylogenet. Evol.* **1994**, *3*, 360–364. [CrossRef] [PubMed]
69. Andreeva, T.F.; Cook, C.; Korchagina, N.M.; Akam, M.; Dondua, A.K. Cloning and analysis of structural organization of *Hox* genes in the polychaete *Nereis virens*. *Ontogenez* **2001**, *32*, 225–233. [PubMed]
70. Kulakova, M.; Bakalenko, N.; Novikova, E.; Cook, C.E.; Eliseeva, E.; Steinmetz, P.R.; Kostyuchenko, R.P.; Dondua, A.; Arendt, D.; Akam, M.; *et al.* *Hox* gene expression in larval development of the polychaetes *Nereis virens* and *Platynereis dumerilii* (Annelida, Lophotrochozoa). *Dev. Genes Evol.* **2007**, *217*, 39–54. [CrossRef] [PubMed]
71. Bleidorn, C.; Lanterbecq, D.; Eeckhaut, I.; Tiedemann, R. A PCR survey of *Hox* genes in the myzostomid *Myzostoma cirriferum*. *Dev. Genes Evol.* **2009**, *219*, 211–216. [CrossRef] [PubMed]
72. Cho, S.J.; Lee, D.H.; Kwon, H.J.; Ahn, C.H.; Park, S.C.; Shin, K.S. *Hox* genes in the echiuroid *Urechis unicinctus*. *Dev. Genes Evol.* **2006**, *216*, 347–351. [CrossRef] [PubMed]
73. Irvine, S.Q.; Martindale, M.Q. Expression patterns of anterior *Hox* genes in the polychaete *Chaetopterus*: Correlation with morphological boundaries. *Dev. Biol.* **2000**, *217*, 333–351. [CrossRef] [PubMed]
74. Oliver, G.; Vispo, M.; Mailhos, A.; Martinez, C.; Sosa-Pineda, B.; Fielitz, W.; Ehrlich, R. Homeoboxes in flatworms. *Gene* **1992**, *121*, 337–342. [CrossRef]
75. Olson, P.D. Hox genes and the parasitic flatworms: New opportunities, challenges and lessons from the free-living. *Parasitol. Int.* **2008**, *57*, 8–17. [CrossRef] [PubMed]

76. Kim, K.H.; Lee, Y.S.; Jeon, H.K.; Park, J.K.; Kim, C.B.; Eom, K.S. *Hox* genes from the tapeworm *Taenia asiatica* (Platyhelminthes: Cestoda). *Biochem. Genet.* **2007**, *45*, 335–343. [CrossRef] [PubMed]
77. Bartels, J.L.; Murtha, M.T.; Ruddle, F.H. Multiple Hox/HOM-class homeoboxes in Platyhelminthes. *Mol. Phylogenet. Evol.* **1993**, *2*, 143–151. [CrossRef] [PubMed]
78. Webster, P.J.; Mansour, T.E. Conserved classes of homeodomains in Schistosoma mansoni, an early bilateral metazoan. *Mech. Dev.* **1992**, *38*, 25–32. [CrossRef]
79. Pierce, R.; Wu, W.; Hirai, H.; Ivens, A.; Murphy, L.D.; Noël, C.; Johnston, D.A.; Artiguenave, F.; Adams, M.; Cornette, J.; *et al.* Evidence for a dispersed *Hox* gene cluster in the platyhelminth parasite *Schistosoma mansoni*. *Mol. Biol. Evol.* **2005**, *22*, 2491–2503. [CrossRef] [PubMed]
80. Badets, M.; Verneau, O. *Hox* genes from Polystomatidae (Platyhelmintes, Monogenea). *Int. J. Parassitol.* **2009**, *39*, 1517–1523. [CrossRef] [PubMed]
81. Balavoine, G.; Telford, M.J. Identification of planarian homeobox sequences indicates the antiquity of most *Hox*/homeotic gene subclasses. *Proc. Natl. Acad. Sci. USA.* **1995**, *92*, 7227–7231. [CrossRef] [PubMed]
82. Orii, H.; Kato, K.; Umesono, Y.; Sakurai, T.; Agata, K.; Watanabe, K. The planarian HOM/*HOX* homeobox genes (*Plox*) expressed along the anteroposterior axis. *Dev. Biol.* **1999**, *210*, 456–468. [CrossRef] [PubMed]
83. Nogi, T.; Watanabe, K. Position-specific and non-colinear expression of the planarian posterior (Abdominal-B-like) gene. *Dev. Growth Differ.* **2001**, *43*, 177–184. [CrossRef] [PubMed]
84. Saló, E.; Tauler, J.; Jmenéz, E.; Bayascas, J.R.; González-Linares, J.; Garcia-Fernàndez, J.; Baguñà, J. *Hox* and *ParaHox* Genes in Flatworms. Characterization and Expression. *Am. Zool.* **2001**, *41*, 652–663.
85. Tarabykin, V.S.; Lukyanov, K.A.; Potapov, V.K.; Lukyanov, S.A. Detection of planarian Antennapedialike, homeobox genes expressed during regeneration. *Gene* **1995**, *158*, 197–202. [CrossRef]
86. Bayascas, J.R.; Castillo, E.; Muñoz-Mármol, A.M.; Saló, E. Planarian *Hox* genes: Novel patterns of expression during re generation. *Development* **1997**, *124*, 141–148. [PubMed]
87. Bayascas, J.R.; Castillo, E.; Saló, A.M. Platyhelminthes have a *Hox* code differentially activated during regeneration, with genes closely related to those of spiralian protostomes. *Dev. Genes Evol.* **1998**, *208*, 467–473. [CrossRef] [PubMed]
88. Balavoine, G. Identification of members of several homeobox genes in a planarian using a ligation-mediated polymerase chain reaction technique. *Nucleic Acids. Res.* **1996**, *8*, 1547–1553. [CrossRef]
89. Iglesias, M.; Gomez-Skarmeta, J.L.; Saló, E.; Adell, T. Silencing of *Smed-catenin1* generates radial-like hypercephalized planarians. *Development* **2008**, *135*, 1215–1221. [CrossRef] [PubMed]
90. Martín-Durán, J.M.; Amaya, E.; Romero, R. Germ layer specification and axial patterning in the embryonic development of the freshwater planarian *Schmidtea polychroa*. *Dev. Biol.* **2010**, *340*, 145–158. [CrossRef] [PubMed]
91. Welch, M.J.L.; Welch, M.D.B.; Center for Comparative Molecular Biology and Evolution, Woods Hole, MA, USA. Unpublished work. 2008.
92. Smith, S.A.; Wilson, N.G.; Goetz, F.E.; Feehery, C.; Andrade, S.C.; Rouse, G.W.; Giribet, G.; Dunn, C.W. Resolving the evolutionary relationships of molluscs with phylogenomic tools. *Nature* **2011**, *480*, 364–367. [CrossRef] [PubMed]
93. Smith, S.A.; Wilson, N.G.; Goetz, F.E.; Feehery, C.; Andrade, S.C.; Rouse, G.W.; Giribet, G.; Dunn, C.W. Corrigendum: Resolving the evolutionary relationships of molluscs with phylogenomic tools. *Nature* **2013**, *493*, 708. [CrossRef]
94. Biscotti, M.A.; Canapa, A.; Forconi, M.; Barucca, M. *Hox* and *ParaHox* genes: A review on molluscs. *Genesis* **2014**, *52*, 935–945. [CrossRef] [PubMed]
95. Struck, T.H.; Paul, C.; Hill, N.; Hartmann, S.; Hösel, C.; Kube, M.; Lieb, B.; Meyer, A.; Tiedemann, R.; Purschke, G.; *et al.* Phylogenomic analyses unravel annelid evolution. *Nature* **2011**, *471*, 95–98. [CrossRef] [PubMed]
96. Kristof, A.; Wollesen, T.; Wanninger, A. Segmental mode of neural patterning in sipuncula. *Curr. Biol.* **2008**, *18*, 1129–1132. [CrossRef] [PubMed]
97. Hessling, R. Metameric organisation of the nervous system in developmental stages of *Urechis caupo* (Echiura) and its phylogenetic implications. *Zoomorphology* **2002**, *121*, 221–234. [CrossRef]
98. Struck, T.H.; Wey-Fabrizius, A.R.; Golombek, A.; Hering, L.; Weigert, A.; Bleidorn, C.; Klebow, S.; Iakovenko, N.; Hausdorf, B.; Petersen, M.; *et al.* Platyzoan paraphyly based on phylogenomic data supports a noncoelomate ancestry of spiralia. *Mol. Biol. Evol.* **2014**, *31*, 1833–1849. [CrossRef] [PubMed]

99. Sperling, E.A.; Pisani, D.; Peterson, K.J. Molecular paleobiological insights into the origin of the Brachiopoda. *Evol. Dev.* **2011**, *13*, 290–303. [CrossRef] [PubMed]
100. Riutort, M.; Alvarez-Presas, M.; Lazaro, E.; Solà, E.; Paps, J. Evolutionary history of the Tricladida and the Platyhelminthes: An up-to-date phylogenetic and systematic account. *Int. J. Dev. Biol.* **2012**, *56*, 5–17. [CrossRef] [PubMed]
101. Ferrier, D.E.K.; Holland, P.W.H. *Ciona intestinalis* ParaHox genes: Evolution of *Hox/ParaHox* cluster integrity, developmental mode, and temporal colinearity. *Mol. Phylogenet. Evol.* **2002**, *24*, 412–417. [CrossRef]
102. Focareta, L.; Sesso, S.; Cole, A.G. Characterization of homeobox genes reveals sophisticated regionalization of the central nervous system in the European cuttlefish *Sepia officinalis. PLoS ONE* **2014**, *9*, e109627. [CrossRef] [PubMed]
103. Steinmetz, P.R.H.; Kostyuchenko, R.P.; Fischer, A.; Arendt, D. The segmental pattern of *otx*, *gbx*, and *Hox* genes in the annelid *Platynereis dumerilii. Evol. Dev.* **2011**, *13*, 72–79. [CrossRef] [PubMed]
104. Badets, M.; Mitta, G.; Galinier, R.; Verneau, O. Expression patterns of *Abd-A/Lox4* in a monogenean parasite with alternative developmental paths. *Mol. Biochem. Parasit.* **2010**, *173*, 154–157. [CrossRef] [PubMed]
105. Bayascas, J.R.; Castillo, E.; Muñoz-Mármol, A.M.; Baguñà, J.; Saló, E. Synchronous and early activation of planarian *Hox* genes and re-specification of body axes during regeneration. *Hidrobiologia* **1998**, *383*, 125–130. [CrossRef]
106. Baguñà, J.; Saló, E.; Romero, R.; Garcia-Fernàndez, J.; Bueno, D.; Muñoz-Mármol, A.M.; Bayascas-Ramírez, J.R.; Casali, A. Regeneration and pattern formation in planarians: Cells, molecules and genes . *Zool. Sci.* **1994**, *11*, 781–795.
107. Saló, E.; Baguñà, J. Regeneration in planarians and other worms: New findings, new tools, and new perspectives. *J. Exp. Zool.* **2002**, *292*, 528–539. [CrossRef] [PubMed]
108. Saló, E.; Abril, J.F.; Adell, T.; Cebrià, F.; Eckelt, K.; Fernandez-Taboada, E.; Handberg-Thorsager, M.; Iglesias, M.; Molina, D.; Rodriguez-Esteban, G. Planarian regeneration: Achievements and future directions after 20 years of research. *Int. J. Dev. Biol.* **2009**, *53*, 1317–1327. [CrossRef] [PubMed]

Permissions

All chapters in this book were first published in JDB, by MDPI; hereby published with permission under the Creative Commons Attribution License or equivalent. Every chapter published in this book has been scrutinized by our experts. Their significance has been extensively debated. The topics covered herein carry significant findings which will fuel the growth of the discipline. They may even be implemented as practical applications or may be referred to as a beginning point for another development.

The contributors of this book come from diverse backgrounds, making this book a truly international effort. This book will bring forth new frontiers with its revolutionizing research information and detailed analysis of the nascent developments around the world.

We would like to thank all the contributing authors for lending their expertise to make the book truly unique. They have played a crucial role in the development of this book. Without their invaluable contributions this book wouldn't have been possible. They have made vital efforts to compile up to date information on the varied aspects of this subject to make this book a valuable addition to the collection of many professionals and students.

This book was conceptualized with the vision of imparting up-to-date information and advanced data in this field. To ensure the same, a matchless editorial board was set up. Every individual on the board went through rigorous rounds of assessment to prove their worth. After which they invested a large part of their time researching and compiling the most relevant data for our readers.

The editorial board has been involved in producing this book since its inception. They have spent rigorous hours researching and exploring the diverse topics which have resulted in the successful publishing of this book. They have passed on their knowledge of decades through this book. To expedite this challenging task, the publisher supported the team at every step. A small team of assistant editors was also appointed to further simplify the editing procedure and attain best results for the readers.

Apart from the editorial board, the designing team has also invested a significant amount of their time in understanding the subject and creating the most relevant covers. They scrutinized every image to scout for the most suitable representation of the subject and create an appropriate cover for the book.

The publishing team has been an ardent support to the editorial, designing and production team. Their endless efforts to recruit the best for this project, has resulted in the accomplishment of this book. They are a veteran in the field of academics and their pool of knowledge is as vast as their experience in printing. Their expertise and guidance has proved useful at every step. Their uncompromising quality standards have made this book an exceptional effort. Their encouragement from time to time has been an inspiration for everyone.

The publisher and the editorial board hope that this book will prove to be a valuable piece of knowledge for researchers, students, practitioners and scholars across the globe.

List of Contributors

Tristan Aguero, Yi Zhou and Mary Lou King
Department of Cell Biology, University of Miami School of Medicine, Miami, FL 33136, USA

Malgorzata Kloc
Houston Methodist Research Institute, The Houston Methodist Hospital, Bertner Avenue, Houston, TX 77030, USA

Patrick Chang and Evelyn Houliston
Laboratoire de Biologie du Développement de Villefranche-sur-mer (LBDV), Sorbonne Universités, UPMC Univ Paris 06, CNRS, Villefranche-sur-mer 06230, France

Yoshinori Abe and Nobuyuki Tanaka
Department of Molecular Oncology, Institute for Advanced Medical Sciences, Nippon Medical School 1-396 Kosugi-machi, Nakahara-ku, Kawasaki 211-8533, Japan

Rajeev Singh and Matthias Lauth
Philipps University Marburg, Institute of Molecular Biology and Tumor Research (IMT), Center for Tumor and Immune Biology (ZTI), Hans-Meerwein-Str. 3, 35043 Marburg, Germany

Héctor Herranz and Stephen M. Cohen
Department of Cellular and Molecular Medicine, University of Copenhagen, Blegdamsvej 3, 2200 N Copenhagen, Denmark

Chu Hsien Lim
Yale NUS College, 16 College Avenue West, Singapore 138610, Singapore

Ajay S. Mathuru
Yale NUS College, 16 College Avenue West, Singapore 138610, Singapore
Institute of Molecular and Cell Biology, 61 Biopolis Drive, Singapore 137673, Singapore

Florencia Cavodeassi
Institute of Medical and Biomedical Education, St George's, University of London, London SW17 0RE, UK

Riley Kellermeyer, Leah M. Heydman, Grant S. Mastick and Thomas Kidd
Department of Biology/MS 314, University of Nevada, 1664 N. Virginia St., Reno, NV 89557, USA

Alla Yalonetskaya and Johnny Elguero
Cell and Molecular Biology Program, Department of Biology, 5 Cummington Mall, Boston University, Boston, MA 02215, USA

Kimberly McCall
Cell and Molecular Biology Program, Department of Biology, 5 Cummington Mall, Boston University, Boston, MA 02215, USA
Molecular Biology, Cell Biology, and Biochemistry Program, 5 Cummington Mall, Boston University, Boston, MA 02215, USA

Albert A. Mondragon
Molecular Biology, Cell Biology, and Biochemistry Program, 5 Cummington Mall, Boston University, Boston, MA 02215, USA

Robert A. Kozol
Department of Biology, University of Miami, Coral Gables, FL 33315, USA

Cândida Lucas, Célia Ferreira, Giulia Cazzanelli, Ricardo Franco-Duarte and Joana Tulha
CBMA—Centre of Molecular and Environmental Biology, University of Minho, Campus de Gualtar, 4710-054 Braga, Portugal
School of Biology, Faculty of Biological Sciences, University of Leeds, Leeds LS2 9JT, UK

Rachel Brunner and Zane Deliu
Department of Biological Sciences, University of Illinois at Chicago, Chicago, IL 60607, USA
PTM Biolabs Inc., Chicago, IL 60612, USA

Hsiao-Lei Lai
Department of Biological Sciences, University of Illinois at Chicago, Chicago, IL 60607, USA
PTM Biolabs Inc., Chicago, IL 60612, USA

Elan Melman
Department of Biological Sciences, University of Illinois at Chicago, Chicago, IL 60607, USA
The School of Molecular and Cellular Biology, University of Illinois Urbana-Champaign, Champaign, IL 61801, USA

Q. Tian Wang
Department of Biological Sciences, University of Illinois at Chicago, Chicago, IL 60607, USA

David L. Geenen
Department of Physiology and Biophysics, University of Illinois at Chicago, Chicago, IL 60612, USA
Physician Assistant Studies, Grand Valley State University, Grand Rapids, MI 49503, USA

Elena Sena and Béatrice C. Durand
Institut Curie, Université Paris Sud, INSERM U1021, CNRS UMR3347, Centre Universitaire, Bâtiment 110, F-91405 Orsay Cedex, France

Kerstin Feistel
Institute of Zoology, University of Hohenheim, Garbenstr. 30, 70593 Stuttgart, Germany

Marco Barucca, Adriana Canapa and Maria Assunta Biscotti
Dipartimento di Scienze della Vita e dell'Ambiente, Università Politecnica delle Marche, via Brecce Bianche, 60131 Ancona, Italy

Index

N

O

P

R

S

T

W

X

www.ingramcontent.com/pod-product-compliance
Lightning Source LLC
LaVergne TN
LVHW060942220926
848485LV00015B/1369